Environmental Engineering

Series Editors: U. Förstner, R. J. Murphy, W. H. Rulkens

Springer

Berlin
Heidelberg
New York
Barcelona
Budapest
Hong Kong
London
Milan
Paris
Santa Clara
Singapore
Tokyo

Horst Klingenberg

Automobile Exhaust Emission Testing

Measurement of Regulated and Unregulated Exhaust Gas Components, Exhaust Emission Tests

With 301 Figures

Springer

Series Editors

Prof. Dr. U. Förstner
Arbeitsbereich Umweltschutztechnik
Technische Universität Hamburg-Harburg
Eißendorfer Straße 40
D-21073 Hamburg, Germany

Prof. Robert J. Murphy
Dept. of Civil Engineering and Mechanics
College of Engineering
University of South Florida
4202 East Fowler Avenue, ENG 118
Tampa, FL 33620-5350, USA

Prof. Dr. ir. W. H. Rulkens
Wageningen Agricultural University
Dept. of Environmental Technology
Bomenweg 2, P.O. Box 8129
NL-6700 EV Wageningen, The Netherlands

Author

Prof. Dr. rer. nat. H. Klingenberg
Teichtal 31
D-39165 Lehre, Germany

Translator

Christina Grubinger-Rhodes
Epplestraße 9
D-70597 Stuttgart, Germany

Translated from the German Edition "Automobilmeßtechnik, Bd. C: Abgasmeßtechnik"

ISBN-13:978-3-642-80245-4

Library of Congress Cataloging-in-Publication Data
Klingenberg, Horst. Automobile exhaust emission testing : measurement of regulated and unregulated exhaust gas components, eshaust emission tests / Horst Klingenberg.
(Environmental Engineering)
Includes bibliographical references.
ISBN-13:978-3-642-80245-4 e-ISBN-13:978-3-642-80243-0
DOI: 10.1007/978-3-642-80243-0
1. Automogiles -- Motors -- Exhaust gas -- Testing.
I. Title II. Series: Environmental engineering (Berlin, Germany)
TD886.5.K583 1996 96-28587

Typesetting: Camera-ready by author
SPIN:10127147 61/3020-5 4 3 2 1 0 - Printed on acid -free paper

Preface

Differing legislation of the countries or unions of countries involved in pollution reduction has turned exhaust gas measuring technology into such an extremely extensive and complex field that only a few specialists in the environmental agencies and automobile industry have a grasp of it.

The book at hand is intended as an overview of the basics of exhaust gas measuring technology with the relationships between emission, ambient concentration and effect as a background. It aims at providing experts and students alike with an understanding of the interrelationships and details of this field. This work is basically a translation of Volume C: Exhaust Gas Measuring Techniques in the Automobile Measuring Technique series from the publishers.

The results presented are based on the experience gathered by the author during his work spanning more than two decades in the automobile industry.

In particular, this book is based on the findings of two reports compiled under the author's guidance which have been published by VW and which have become known the world over. To my knowledge, it was the first time ever in the first report that fundamental theories were described for the entire field of exhaust gas test methods. The second report contains the results of extensive measurements of unregulated exhaust gas components.

In the production of this book my former staff of VW as well as the staff of my present working environment, the Otto-von-Guericke University, Magdeburg, Germany, were of help. My special thanks are due to them. I would like to express my thanks to Ms. Grubinger-Rhodes for her translation into English; also I would like to thank Springer Publishers, particularly Dr. Riedesel, for their kind cooperation.

Lehre, Summer 1996

Horst Klingenberg

Contents

List of Symbols

Symbol	Explanation
Indices	
A	Assembly Diluting air
C	Gasoline engine with three-way catalytic converter
C100	100 km/h constant speed
d	Diluted air
D	Diesel engine
DB	Diurnal breathing loss
eur	Extra-urban roads
EDC	European driving cycle
EG	Exhaust gas
Eng.	Engine
EUDC	New European driving cycle including high speed cycle (Extra Urban Driving Cycle)
fhw	Federal highways
FTP	US driving cycle
G	Gasoline engine without catalytic converter
HS	Hot soak loss
i	Exhaust-gas component *i* Initial value Reference year
iur	Intra-urban roads
j	Driving cycle phase

Symbol	Explanation
k	Corrected Assembly year
L	Log-normal density function
LFE	Laminar Flow Element
m	measured molar
MAN	Manufacturer
Meas.	Measurement
M2	TÜV driving cycle for traffic on arterial roads
M3	TÜV driving cycle for flowing through-traffic
M4	TÜV driving cycle for flowing in-town traffic
n	Number of vehicles
N	Normal density function
r	Rotational
r, rel	Relative
RY	Reference year
TEG	Total exhaust gas
TF	Test fuel
W	Weibull density function
x, y, z	Coordinates
ν	Vehicle

Latin letters

a	Parameter
A	Absorption Beam of light Calibration constant Cross-sectional area of vehicle Instrument constant
$A(\tilde{\nu})$	Correction factor

Symbol	Explanation
A_2	Area of throttle cross-section
b	Parameter
B	Calibration constant Magnetic induction, magnetic flux density Proportion of acceleration time
c	Concentration
c_{CO2j}	Concentration of CO_2 in phase j of US-75 driving cycle
c_{ijk}	Concentration of component i in phase j of US-75 driving cycle after correction
c_{ijd}	Concentration of component i in phase j of US-75 driving cycle in diluted air
c_0	Speed of light
c_w	Drag coefficient
C	Pollutant concentration in ambient air Slip flow correction factor
Cp	Combustion process
CWF	Carbon weight fraction
d	Distance Layer thickness of gas Length of absorption path Length of measuring cell
D	Transmission
DF_j	Dilution factor for phase j of US-75 driving cycle
DP	Displacement
D_p	Particle diameter
e	Electronic charge Elementary charge Partial emission factor Threshold concentration
e_{ij}	Exhaust emission factor
e^*_{kj}	Partial exhaust emission factor

Symbol	Explanation
E	Engineering goal Emission Emittance Energy Extinction
$E(\tilde{\nu})$	Extinction
E_e	Energy of valence electrons
E_i	Discrete atomic or molar energy state
E_{kin}	Kinetic energy
E_{pot}	Potential energy
E_r	Rotational energy
E_v	Vibrational energy
f	Frequency Spring constant
fd	Frequency distribution of vehicles over years
f_i	Calibration factor
F	Fluorescence Force Systematic Error
FC	Fuel consumption (Europe)
FE	Fuel economy value (USA)
$F(t)_R$	Tractive force acting on the driving wheels of a vehicle on a chassis dynamometer
F_{Ro}	Rolling-resistance coefficient
g	Acceleration due to gravity
h	Planck's constant ($\hbar = h / 2\pi$)
$h\nu$	Photon
$h(x)$	Density function of a distribution
h_{ik}	Density function of age of the vehicles
H	Building height, roof height Magnetic field intensity Absolute humidityof the ambient air
H/C	Hydrogen to carbon ratio

Symbol	Explanation
i	Intensity of electron current
I	Electric current Inertial moment of a molecule Intencity
$I(\tilde{\nu})$	Light intensity after passing through the medium
$I_0(\tilde{\nu})$	Initial light intensity
$I_0^*(\tilde{\nu})$	Initial light intensity with correction factor
I^+	Intensity of ion current
I_{IR}	Infrared intensity
I_K	Nuclear spin quantum number
J	Rotational quantum number
k	Transmission between roller axis and flywheel mass axis Statistical factor Density factor for the evaluation of fuel evaporation emission
K	Scaling factor Time proportion of constant speed
K_i	Deterioration factor
K_{2j}	Volume correction when removing CO_2 in phase j of US-75 driving cycle
K_{Hj}	Volume correction when removing moisture in phase j of US-75 driving cycle
K_{NOxj}	Humidity correction for NO_x in phase j of US-75 driving cycle
K_p	Pressure correction
K_T	Temperature correction
l	Length of ionisation space Distance Number of systematic error proportions
l_o	Turbulence correction
L	Lagrangian function
m	Mass Ion mass Vehicle mass Number of tests being run by the manufacturer

Symbol	Explanation
m^*	Effective vehicle mass
md	Annual distance driven
m_{CO}	Carbon monoxide mass emission measured
m_{CO_2}	Carbon dioxide mass emission measured
m_{HC}	Hydrocarbon mass emission measured
m_i	Mass of component *i*
$m_{i\,CT}$	Mass of component *i* emitted in cold-start phase of US-75 driving cycle
$m_{i\,HT}$	Mass of component *i* emitted in warm-start phase of US-75 driving cycle
$m_{i\,S}$	Mass of component *i* emitted in stabilized phase of US-75 driving cycle
m_{ij}	Mass of component *i* in phase *j* of US-75 driving cycle
M	Average number of acceleration-deceleration changes Gas molecule Torque
$M(t)$	Torque function
$\overline{M}$	Mean torque
$M(t)_{Br}$	Moment of resistance of chassis dynamometer brakes
$M(t)_{Dyno}$	Moment of resistance driving wheels - dynamometer
$M(t)_{Road}$	Moment of resistance driving wheels - road
M_{mi}	Molar mass of component *i*
n	Negative Number of atoms in a molecule Number of measurements Substance amount
n_i	Principal quantum number of the valence electron shell of an atom
n_j	Total revolutions of blower during phase *j* of US-75 driving cycle
N	Number of sample gas molecules per cm^3 Number of chassis dynamometers or test facilities
NHV	Net heating value

Symbol	Explanation
Or	Operation on each road type
p	Positive Pressure
p_{Pj}	absolute pressure before pump during phase j of US-75 driving cycle
p_a	Atmospheric pressure
p_s	Saturated water vapor pressure
P	Probability
q	Multiple of elementary charge Normal coordinate
q_i	Generalized coordinate
Q	Effective ionization cross-section
r	Distance Radius Dynamic rolling radius
r_e	Radius of orbit in the electrical field
r_m	Radius of orbit in the magnetic field
r_{Dyno}	Radius of the chassis dynamometer rolls
R	Electrical resistance Mass resolving power
R_{aj}	Relative air humidity during phase j of US-75 driving cycle
R_A	Aerodynamic drag
R_G	Gradient resistance
R_I	Inertial resistance of rotating masses
R_m	Molar gas constant
R_{Ro}	Rolling resistance road/wheels
RY	Reference year
RegY	Registration year
s	Path Standard deviation
$2s$	Variability

Symbol	Explanation
$s_{i\,CT}$	Measured distance driven of cold start phase of US-75 driving cycle
$s_{i\,HT}$	Measured distance driven of hot start phase of US-75 driving cycle
$s_{i\,S}$	Measured distance driven of stabilized phase of US-75 driving cycle
s_r	Relative standard deviation
S	Emission source strength Line or oscillator strength Safety margin Time proportion of idle period
S_i	Vibration level
S_{total}	Total safety margin
SG	Specific gravity Test fuel density
t	Time Statististical factor of Student distribution
T	Gas temperature Kinetic energy Term Total number of vehicles Transmittance
T_{Pj}	Gas temperature ahead of pump during phase *j* of US-75 driving cycle
T_a	Ambient temperature
T_{90}	Response time (90% time)
u	Measurement uncertainty of mean value
U	Above-roof wind velocity Acceleration voltage Direct voltage Electrostatic constant field Potential energy
U_g	Gas velocity
U_x	Component of wind vector in x-direction

Symbol	Explanation
v	Speed Velocity
v_{act}	Actual speed
v_{nom}	Nominal speed
$\bar{v}_1$	Average speed of a complete driving cycle
$\bar{v}_2$	Average speed during the driving phases of a driving cycle
$\bar{\dot{v}}_+$	Average acceleration during the acceleration phase of a driving cycle
$\bar{\dot{v}}_-$	Average deceleration during the deceleration phase of a driving cycle
V	High frequency voltage Proportion of deceleration time Voltage Volume
V_{total}	Total volume
$\dot{V} = \frac{dV}{dt}$	Volumetric flow rate
V_i	Volume proportion of component i
V_{jk}	Total volume (exhaust-air mixture) during phase j of US-75 driving cycle, corrected to standard conditions
V_m	Molar volume
V_n	Net volume of the SHED
V_0	Volumetric flow rate of the probe sampling unit
W	Effect Width of street canyon
W_g	Aerosol stream diameter
W_p	Particle concentration
WV	Wind velocity above-roof
x, y, z	Coordinates Distances Emission values
$\bar{x}$	Mean value
$\bar{x}_r$	Relative mean value

Symbol	Explanation
$\bar{\bar{x}}$	Total mean value
x_G, y_G, z_G	Exhaust emission standards
$y(2s)$	Step function
$\bar{y}$	Ambient air concentration mean value
z	Number of charges of an ion
z_i	Random number
z_0	Height of line source above ground

Greek letters

α	Angle Flow coefficient
β	Angle Shape parameter
γ	Minimum response parameter
Γ	Gamma function
Δ	Difference
Δ p	Differential pressure
ε	Coefficient of expansion
$\varepsilon_i(\tilde{\nu})$	Molar decadic extinction coefficient of the component *i*
ε_{kjv}	Vehicle emission
η	Conversion rate
η_g	Gas viscosity
ϑ	Scale factor
Θ	Moment of inertia Total moment of inertia of the dynamometer
Θ_{Dyno}	Moment of inertia of the dynamometer rolls
Θ_{sim}	Additional, simulated moment of inertia of the dynamometer rolls
κ	Magnetic susceptibility

Symbol	Explanation
λ	Air number, air ratio Parameter for accounting for the rotating masses Wavelength
$\bar{\lambda}$	Mean air number
μ	(True) mean value Molecular dipole moment
μ_o	Magnetic field constant (permeability of the vacuum)
μ_r	Permeability number (substance-specific) Relative mean value (true)
ν	Frequency
$\tilde{\nu}$	Wave number
ξ_i	Nuclear charge
ρ	Gas density Air density during standard conditions Mass concentration
ρ_i	Mass concentration of the component *i* Density of the component *i*
ρ_p	Particle density
ρ_{TF}	Density of the test fuel
ρ_{xy}, ρ_{xz}, ρ_{yz},	Correlation coefficients
σ	Standard deviation
σ^2	Variance
σ_i	Volume concentration of the component *i*
σ_z	Dispersion parameter
τ	Mean duration of a driving phase
Φ_p	Inertial parameter
Φ_e	Spectral intensity of radiation
χ^2	Statistical parameter of the χ^2-test
ω	Circular frequency
$\frac{d\omega(t)}{dt}$	Angular acceleration

Symbol	Explanation
Abbreviations	
ADC	Analogue-to-digital converter
AU	Exhaust emission check (Germany)
ASTM	American Society for Testing and Materials
ASU	Exhaust emission special check (Germany)
BAM	Federal Institute for Materials Research and Testing (Germany)
BAST	Federal Institute for Road Research (Germany)
BMFT	Federal Ministry for Research and Technology (Germany)
CARB	California Air Resources Board
CCMC	Committee of Common Market Automobile Constructors
CFC	Chlorofluorocarbons
CFR	Code of Federal Regulations
CFV	Critical Flow Venturi
CI	Chemical ionisation
CLA	Chemiluminescence analyser
CLD	Chemiluminescence detector
COHb	Carboxyhemoglobin
CRC	Crankcase
CV	Commercial vehicle
CVS	Constant Volume Sampling
CWF	Carbon weight fraction
DB	Diurnal breathing loss test
DC	Direct current
DI	Direct-injection diesel engine
DMS	Dimethyl sulfide
DNPH	Dinitrophenylhydrazine
EC	European Community
ECD	Electron capture detector
ECE	Economic Commission for Europe

Symbol	Explanation
EDC	European driving cycle
EEC	European Economic Community
EI	Electron impact
EPA	Environmental Protection Agency
EU	European Union
FAT	Research Association Automotive Technology (Germany)
FFT	Fast-Fourier transformation
FhG-IPM	Fraunhofer Gesellschaft, Institute for Physical Measuring Technology
FhG-ITA	Fraunhofer Gesellschaft, Institute for Toxicology and Aerosol Research
FID	Flame ionisation detector
FPD	Flame photometer detector
FR	Federal Regulation
FRG	Federal Republic of Germany
FTIR	Fourier-Transform infrared spectroscopy
FTP	Federal Test Procedure (US-75 Test)
FVV	Research Association of Combustion Engines Manufacturers (Germany)
GASP	Gas analysis by sampling plasmas
GC	Gas chromatography
GC/MS	Coupling of gas chromatography/mass spectrometry
GLC	Gas-liquid chromatography
GP	Greenhouse potential
GSC	Gas-solid chromatography
Hb	Hemoglobin
HC	Hydrocarbons
H/C	Hydrogen to carbon ratio
HDC	Highway driving cycle
HEI	Health Effects Institute
HO_2	Hydroperoxy radical

Symbol	Explanation
HPLC	High-Pressure Liquid Chromatography High-Performance Liquid Chromatography
HS	Hot soak loss test
ICR	Ion cyclotron resonance
IDI	Indirect injection (prechamber or swirl-chamber) diesel engine
IR	Infrared
JRP	Joint research program
KBA	German Federal Office for Motor Traffic
LDT	Light-Duty Truck
LEV	Low-Emission Vehicle
LIS	Landesanstalt für Immissionsschutz (State Institute for Ambient Air Protection in North Rhine-Westphalia, Germany)
LFE	Laminar-Flow Element
LLC	Liquid-liquid chromatography
LSC	Liquid-solid chromatography
MAK	Maximum Workplace Concentration
MBTH	3-methyl-2-benzothiazolinonhydrazon-hydrochloride
MIK	Maximum pollutant concentration in ambient air
MS	Mass spectrometry, mass spectrogram
MTC	Mercury tellurium cadmium
MVEG	Motor-Vehicle Emission Group
MV	Mean value
MW	Microwaves
MY	Model year
NAAQS	National Ambient Air Quality Standard
NBS	National Bureau of Standards
NDIR	Non-dispersive infrared measuring method
NDUV	Non-dispersive ultraviolet analyzer
NFID	Nitrogen specific FID
NHV	Net heating value

Symbol	Explanation
NMHC	Non-methane HC
NMOG	Non-methane organic gases
NO_x	Nitrogen oxides
NRW	North Rhineland - Westphalia
OBD	On-board diagnosis
OECD	Organisation for Economic Cooperation and Development
OH	Hydroxyl radical
O_2Hb	Oxyhemoglobin
PAH	Polycyclic aromatic hydrocarbons
PAN	Peroxyacetylnitrate
PC	Passenger car
PDP	Positive-Displacement Pump
pH	Potentia hydrogenii (negative logarithm of hydrogen-ion concentration
PTB	Federal Institute for Metrology, Braunschweig-Berlin
PTFE	Polytetrafluoroethylene
PV	Passenger vehicle
R	Hydrocarbon fraction Alkyl or aryl radical
RO	Alkoxy radical
RO_2	Peroxy radical
RW-TÜV	Technical Control Board of Rhineland-Westphalia, Germany
SDL	Semiconductor diode laser
SEA	Selective-Enforcement Auditing
SESAM	System for Emission Sampling and Measurement
SET	Sulfate emission test
SHED	Sealed housing for evaporative determinations
SRI	Stanford Research Institute
StVZO	Automobile Safety Act (Germany)
TA	Technical Directive of the VDI

Symbol	Explanation
TEM	Transmission electron microscope
TFR	Total fertility rate
THC	Total HC
TID	Thermo ionization detector
TLC	Thin-layer chromatography
TLEV	Transmission low-emission vehicle
TLV	Threshold limit value
TOF	Time of flight
TRK	Technical reference concentration
TÜV	Technical Control Board (Germany)
UBA	German Federal Environmental Protection Agency
ULEV	Ultra low emission vehicle
UN	United Nations
UV	Ultraviolet
VDA	Association of German Automotive Manufacturers
VDI	Association of German Engineers
VIS	Visible (visible light)
VW	Volkswagen Inc.
YAG	Yttrium-aluminum garnet
ZEV	Zero-emission vehicle

1 Overview

1.1 Introduction

Technical progress has made living conditions for man more comfortable the world over, not only in the highly industrialized countries, and has made it possible for him to control or at least contain the damage from natural catastrophes.

Automobiles represent a major contribution to this. No one today wants to do without the mobility they provide - they have almost become a basic necessity.

One of the environmental requirements automobiles must meet is the reduction of the emission of certain exhaust-gas components.

In the field of automobile exhaust gases the maximum values or standards set by legislation for some of these components will become increasingly stringent in the future. In addition, the number of those exhaust-gas components whose emissions or resulting concentrations in ambient air are to be limited by law will rise with continuing research on their effect potential on the environment. Of course, the setting of standards only makes sense if adherence to them can be controlled. However, measuring these values is becoming increasingly difficult due to the growing number of limited exhaust-gas components and the setting of constantly lower standards. Thus, the requirements to be met by measuring technology have grown disproportionately. The reason for this disproportion is that traditional measuring methods are no longer adequate for recording increasingly minute concentrations or that this can be done only at a much higher expense.

There are different stages between the formation of exhaust gases and their effects on the environment. One differentiates between

- *emission*, which is the exhaust gas on the point of discharge from the, e.g. automobile exhaust pipe or a chimney or stack,
- *transmission*, the dilution of exhaust gas after being discharged from, e.g. an automobile exhaust pipe or a stack and its transport into the atmosphere, chemical reactions included,
- *air quality,* the final concentration of pollutants in ambient air developing after transmission through the air, which is responsible for the effect,
- and *effects*, which are the influence of this air quality on the environment.

The chain of cause and effect between emission - transmission - air quality - effects is shown in Fig. 1.1.

Apart from the amount of exhaust output, i.e. the emission, transmission has also a significance, that is the way in which pollutants reach our atmosphere and disperse there. Fig. 1.2 shows an example. Only water vapor and particles are

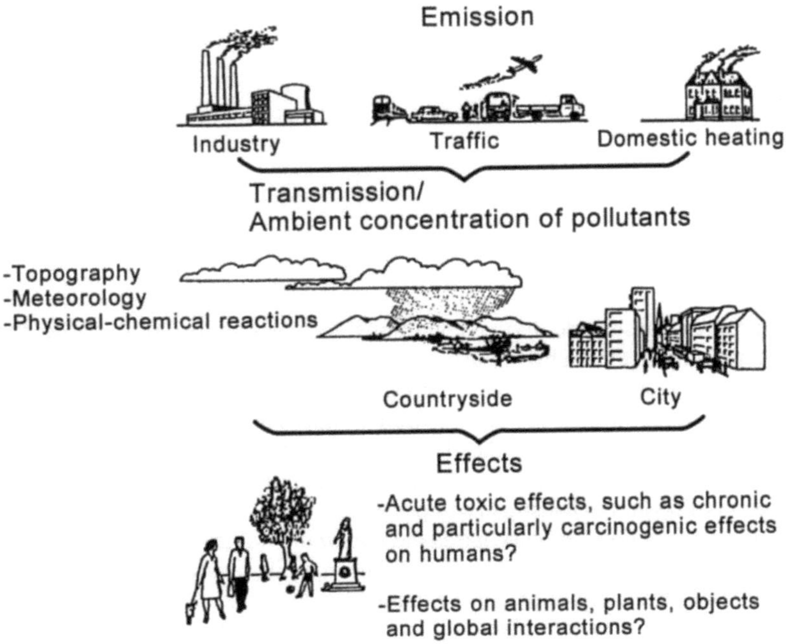

Fig. 1.1: Chain of cause and effect of environmental interrelationships

visible, gases are invisible. This is also true of exhaust-gas clouds leaving the exhaust pipe of an automobile.

As can be seen in Fig. 1.3, the exhaust from different sources disperses differently.

Automobiles emit their exhaust gases close to the ground. The chimneys of houses usually have a height of less than 30 meters. Stacks of power plants and factories emit their pollutants at heights of up to 300 meters or more. Accordingly, pollutants are transported over a larger area the higher the source, with dispersion also depending on wind speed.

Continued dispersion of the exhaust gases is carried out via vertical and horizontal transport prevailing in the troposphere, the lowest layer of the atmosphere. Weather events also take place at this level.

In many countries the deterioration observed in air quality has already led to the introduction of regulations limiting both the pollutant amount emitted (emissions) and the build-up of pollutants in the atmosphere (pollution) to a maximum admissible level. Setting the appropriate threshold values for certain substances and substance groups - locally, regionally and globally - is to prevent harmful damage to human beings, animals, plants, inanimate objects and the climate.

Emissions can be recorded with measuring technology, but for many components this requires great expense and complicated laboratory methods.

Fig. 1.2: Long-range distribution of an exhaust gas plume made visible by water vapor and particles

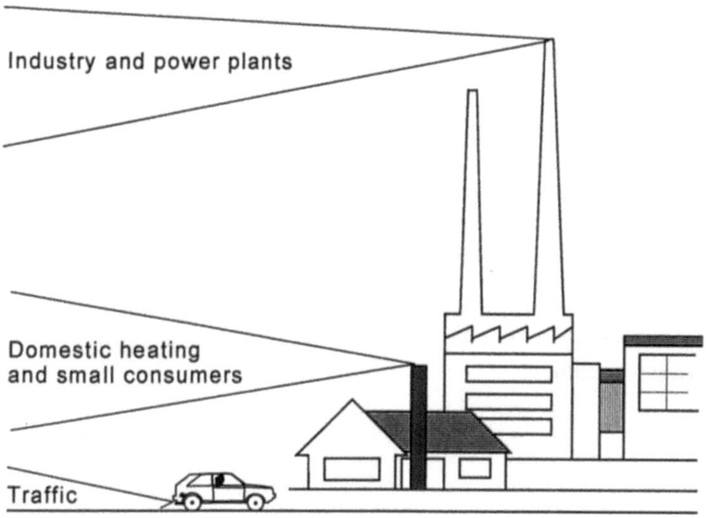

Fig. 1.3: Dispersal of pollutants in the environment

Pollutant concentrations in ambient air can also be measured. Due to the effort required for this, they can only be recorded as random measurements of spatial and temporal distribution. Pollutant concentrations in street canyons are the highest and therefore significant in relation to human health if enough time is spent there. Data for this type of situation is insufficient. E.g., changing the height of a house in a street canyon can change concentration values by a factor of 2, whereby climatic influences are to be taken into consideration. At present, only stochastically determined values exist, these being unsuited as reference values.

A "standard" street canyon would have to be defined as one with pollutant concentrations representative of the mean value of the street canyons of all cities. With this, it would be possible to establish the connection to the cause, the mean emission. The Technical Directive on Air Pollution Control of Germany (TA-Luft, VDI Commission Clean Air) is of no help in this case, as pollutant concentrations are averaged only via grids of 1 x 1 km^2.

Little is known about transmission. Efforts have been made to calculate dispersion with the help of mathematical models, but the chemical reactions cannot be taken adequately into consideration as their complex interreactions remain as yet unexplained.

Even less is known about the effects on humans, animals, plants, inanimate objects, global occurrences, as implied in Fig. 1.1.

Actually, the correct course to take would be to start from the effects and, with the help of transmission models, calculate the pollutant concentrations causing appreciable effects. Only then does it make sense to quantify and limit the emissions concerned.

As a result of inadequate knowledge on the effects in the past the emissions have been the starting point and legislation has limited them as a preventive measure.

1.2 This Volume's Contents

Automobile exhaust gas emissions, including relevant legislation, are dealt with in Chap. 2. Chap. 3 deals with natural and anthropogenic emissions, Chap. 4 with pollutant concentrations in ambient air and Chap. 5 with their effects. Chap. 6 describes measuring techniques and measuring instruments and their theoretical bases. Chap. 7 describes the measurement of non-limited (unregulated) exhaust gas components and diesel exhaust gas particles, including results. Chap. 8 deals with exhaust gas test procedures prescribed by law. Chap. 9 includes exhaust gas tests with an overview and comments. Finally, Chap. 10 takes a look at the future of exhaust gas measuring technology.

2 Automobile Exhaust Gas Emissions

2.1 Origin of Automobile Exhaust Gas Emissions

Automobile exhaust gas emitted through the exhaust pipe take their origin in the combustion within the engine during which process the chemical energy stored in various hydrocarbons of the fuel is set free as oxidation heat. The basic physical processes taking place during the conversion of combustion heat into mechanical work have been thoroughly experimentally researched and can be well described through suitable operating cycles. In comparison to this, the chemical processes in the combustion engine are relatively unknown. From present measuring results (mainly from some exhaust gas components) as well as from idealized model experiments and calculations, with laminar hydrocarbon flames for example, one can try to draw conclusions about the chemical processes during engine combustion. The following comments are limited just to passenger vehicles.

2.1.1 Chemistry of Engine Combustion

The decisive chemical process in the engine's energy conversion is the oxidation of fuel hydrocarbons with the oxygen of the admitted ambient air.

The ideal transition is described, for example, by

$$C_n H_{2n+2} \quad + \quad \frac{3n+1}{2} O_2 \tag{2.1}$$

(Hydrocarbons + Oxygen)

$$\Downarrow$$

$$n\, CO_2 \quad + \quad (n+1)\, H_2O \quad + \quad \text{Heat.} \tag{2.2}$$

(Carbon dioxide + Water + Heat) .

From the known data about the composition of combustion air (Table 2.1) and fuel (Table 2.2) (relating just to the C and H atoms of the molecules of the fuel because of their complex chemical composition), one can calculate how many kg of air are needed on an average for the complete combustion of 1 kg of gasoline or diesel fuel.

Table 2.1: Main components of air

Component	Formula	Dry		Wet (22 °C, 50 % rel. humidity)	
		Vol.-%	Weight-%	Vol.-%	Weight-%
Nitrogen	N_2	78.08	75.46	77.06	74.88
Oxygen	O_2	20.95	23.19	20.68	22.97
Noble gases	-	0.94	1.30	0.93	1.29
Carbon dioxide	CO_2	0.03	0.05	0.03	0.05
Water(-vapor)	H_2O	-	-	1.30	0.81

Table 2.2: Carbon and hydrogen content of commercial fuels

Fuel	Density kg/l	C Weight-%	H Weight-%
Regular	0.745	85.5	14.5
Premium	0.76	85.8	14.2
Diesel	0.83	86.4	13.1

The ratio of the actual air supply and the amount theoretically required is described as the air ratio λ:

$$\lambda = \frac{\text{amount of admitted air}}{\text{theoretical air demand}} \tag{2.3}$$

Stoichiometrical relationships call for $\lambda = 1$. With lower amounts of air ($\lambda < 1$) the term "rich" mixture is used, and with air excess ($\lambda > 1$) the term "lean" mixture.

The mass balance of the complete, that is ideal, fuel combustion in gasoline or diesel engines ($\bar{\lambda}$ is the mean air ratio) is as follows:

Gasoline Engine (Assumption $\bar{\lambda} = 1$):

$$\begin{aligned} &1\text{ kg Fuel} + 14.9\text{ kg Air } (\hat{=}\ 3.4\text{ kg } O_2 + 11.5\text{ kg } N_2) \\ &\Longrightarrow 3.1\text{ kg } CO_2 + 1.3\text{ kg } H_2O + 11.5\text{ kg } N_2 \end{aligned} \tag{2.4}$$

Carbon dioxide and hydrogen are produced as oxidation products, the proportion of nitrogen remains unchanged.

Diesel Engine (Assumption $\bar{\lambda} \approx 3$):

$$\begin{aligned} &1\text{ kg Fuel} + 43.7\text{ kg Air } (\hat{=}\ 10.0\text{ kg } O_2 + 33.7\text{ kg } N_2) \\ &\Longrightarrow 3.1\text{ kg } CO_2 + 1.3\text{ kg } H_2O + 6.6\text{ kg } O_2 + 33.7\text{ kg } N_2 \end{aligned} \tag{2.5}$$

The proportion of nitrogen in the admitted combustion air passes through the engine unchanged. The diesel engine works as an auto-igniter with a relatively lean mixture, so that besides the oxidation products carbon dioxide and water, corresponding amounts of unused oxygen as well as an increased nitrogen content are obtained in the exhaust.

2.1.2 Typical Main Components of Automobile Exhaust Gas

Complete fuel combustion producing just carbon dioxide and water is not practicable even with a very lean mixture. This is due to the fact that in the combustion phase of the engine's operating cycle, the chemical reactions reach no equilibrium conditions and non-homogeneous gas mixtures appear, making secondary chemical reactions possible (incomplete combustion).

In Table 2.3, measurement values for exhaust emission components of a typical gasoline engine vehicle without catalytic converter are listed. The table could be continued for hundreds of exhaust gas components with progressively lower concentrations, refer to Chap. 7.

Table 2.3: Typical composition of gasoline engine exhaust gas (vehicle without catalytic converter)

Component	Formula	kg/kg Fuel	kg/l Fuel	Weight-%	Vol.-%
Carbon dioxide	CO_2	2.710	2.019	17.0	10.9
Water vapor	H_2O	1.330	0.990	8.3	13.1
Oxygen	O_2	0.175	0.130	1.1	1.0
Nitrogen	N_2	11.500	8.568	72.0	72.8
Hydrogen	H_2	$5.6 \cdot 10^{-3}$	$4.2 \cdot 10^{-3}$	$3.5 \cdot 10^{-2}$	0.5
Sum				**98.4**	**97.8**
Carbon monoxide	CO	0.224	0.167	1.4	1.4
Hydrocarbons	HC	$2.0 \cdot 10^{-2}$	$1.5 \cdot 10^{-2}$	0.13	0.27
Nitrogen oxide	NO_x	$1.7 \cdot 10^{-2}$	$1.3 \cdot 10^{-2}$	0.11	0.1
Sum				**1.64**	**1.77**
Sulfur dioxide	SO_2	$3.3 \cdot 10^{-4}$	$2.4 \cdot 10^{-4}$	$2.0 \cdot 10^{-3}$	$9.0 \cdot 10^{-4}$
Sulfates	SO_4^{2-}	$2.3 \cdot 10^{-5}$	$1.7 \cdot 10^{-5}$	$1.5 \cdot 10^{-4}$	$4.0 \cdot 10^{-5}$
Aldehydes	RCHO	$3.4 \cdot 10^{-4}$	$2.5 \cdot 10^{-4}$	$2.0 \cdot 10^{-3}$	$2.0 \cdot 10^{-3}$
Ammonia	NH_3	$1.5 \cdot 10^{-5}$	$1.1 \cdot 10^{-5}$	$1.0 \cdot 10^{-4}$	$1.5 \cdot 10^{-4}$
Lead compounds		$1.0 \cdot 10^{-4}$	$7.5 \cdot 10^{-5}$	$6.0 \cdot 10^{-5}$	-

Over 98 % by weight of the exhaust gas (5th column) is made up of carbon dioxide, water, oxygen, nitrogen and hydrogen.

These are followed by limited exhaust components as characteristic products of incomplete combustion with a total of about 1.6 % by weight. These are carbon monoxide - an intermediate stage of carbon dioxide formation - , then total hydrocarbons - the unburned and cracked fuel components along with their newly formed chemical compounds - and finally nitrogen oxides (NO_x) - oxidation products of the intake air nitrogen - mainly nitrogen monoxide (NO) and nitrogen dioxide (NO_2).

With a λ-sensor-controlled three-way catalytic converter the values for CO, HC and NO_x are lowered by about a factor of 20 meaning that for the sum of CO, HC and NO_x the values of 0.082 % by weight and 0.078 % by volume are obtained.

The comparatively very small remaining amount of exhaust components, comprising less that 0.05 % by weight of the exhaust emissions in the case of an engine without a catalytic converter, are the unlimited (unregulated) exhaust components (refer to Chap. 7). Their main representatives are hydrogen (pyrolysis product of hydrocarbons), the sulfur compounds (oxidation products of the fuel's sulfur content), the aldehydes (partly oxidized hydrocarbons) and ammonia (reduction product of nitrogen oxide). Their concentrations (% by weight) are around 5 powers of ten less than the limited substances and are trace substances. In addition, with leaded gasoline lead compounds, mainly lead halogens are also emitted.

With the exception of the lead compounds, all the substances mentioned are also to be found in diesel engine exhaust gas, however, because of the even leaner air-fuel mixture with much lower concentrations (Table 2.4). A continuation of this table is also possible, refer to Chap. 7.

Table 2.4: Typical composition of diesel engine exhaust gas (without oxidation catalytic converter)

Component	Formula	kg/kg Fuel	kg/l Fuel	Weight-%	Vol.-%
Carbon dioxide	CO_2	3.147	2.612	7.1	4.6
Water vapor	H_2O	1.170	0.971	2.6	4.2
Oxygen	O_2	6.680	5.554	15.0	13.5
Nitrogen	N_2	33.540	27.838	75.20	77.6
Hydrogen	H_2	$9 \cdot 10^{-4}$	$7 \cdot 10^{-4}$	$2 \cdot 10^{-3}$	$3 \cdot 10^{-2}$
Sum				**99.9**	**99.9**
Carbon monoxide	CO	$1.3 \cdot 10^{-2}$	$1.1 \cdot 10^{-2}$	$3 \cdot 10^{-2}$	$3 \cdot 10^{-2}$
Hydrocarbons	HC	$3.1 \cdot 10^{-3}$	$2.5 \cdot 10^{-3}$	$7 \cdot 10^{-3}$	$1.4 \cdot 10^{-2}$
Nitrogen oxides	NO_x	$1.3 \cdot 10^{-2}$	$1.1 \cdot 10^{-2}$	$3 \cdot 10^{-2}$	$3 \cdot 10^{-2}$
Sum				**0.067**	**0.074**
Sulfur dioxide	SO_2	$4.4 \cdot 10^{-3}$	$3.7 \cdot 10^{-3}$	$1 \cdot 10^{-2}$	$5 \cdot 10^{-3}$
Sulfate	SO_4^{2-}	$7.2 \cdot 10^{-5}$	$6.0 \cdot 10^{-5}$	$1.6 \cdot 10^{-4}$	$5 \cdot 10^{-5}$
Aldehyde	RCHO	$6.3 \cdot 10^{-4}$	$5.2 \cdot 10^{-4}$	$1.4 \cdot 10^{-3}$	$1.4 \cdot 10^{-3}$
Ammonia	NH_3	$2.4 \cdot 10^{-5}$	$2.0 \cdot 10^{-5}$	$5 \cdot 10^{-5}$	$9 \cdot 10^{-5}$
Particulate		$2.5 \cdot 10^{-3}$	$2.1 \cdot 10^{-3}$	$6 \cdot 10^{-3}$	-

These concentrations of the limited exhaust components are as a whole comparable to those from the gasoline engine with a λ-sensor-controlled three-way catalytic converter. The proportion of nitrogen oxides is higher, however.

Due to the higher sulfur content of diesel fuel more sulfur compounds are emitted; for new fuels - from October 1996 on - the sulfur content is limited to 0.05 Weight-%. In addition to the limited components of gasoline engine exhaust, particles in diesel engine exhaust gas are also of relevance. Standards have been set for them in the USA and Europe.

2.1.3 Calculation of Mass Concentration in Volume Concentrations

The emission concentration of the component i is normally given either as mass concentration

$$\rho_i = \frac{m_i}{V_{total}} \tag{2.6}$$

in mg/m^3 or in $\mu g/m^3$, or as volume concentration

$$\sigma_i = \frac{V_i}{V_{total}} \tag{2.7}$$

in Vol.-% or in parts per million (1 ppm $\hat{=}$ proportion of 10^{-6} $\hat{=}$ 10^{-4} Vol-% related to 100 Vol-%) or in parts per billion (1 ppb $\hat{=}$ proportion of 10^{-9} $\hat{=}$ 10^{-7} Vol-%), where the following ratio is valid for gases:

$$\frac{\sigma_i}{\rho_i} = \frac{V_i}{m_i} = \frac{V_m}{M_{mi}} \quad , \tag{2.8}$$

m_i	mass of the component i in mg,
V_{total}	total volume of the gas probe in m^3,
V_i	volume proportion of the component i in ml.
σ_i	volume concentration in ml/m^3 ($\hat{=}$ ppm)
ρ_i	mass concentration in mg/m^3
V_m	molar volume in $m^3/kmol$
M_{mi}	molar mass of the component i in kg/kmol ($\hat{=} 10^6 \text{ mg / kmol}$)

The following equations result for the calculation of mass concentrations (mg/m^3) in volume concentrations (ppm) and vice versa:

$$\sigma_i \text{ in ppm} \mathrel{\hat{=}} \frac{V_m}{M_{mi}} \cdot \rho_i \quad , \qquad \left(\rho_i \text{ in mg / m}^3\right) \tag{2.9}$$

and

$$\rho_i \text{ in mg / m}^3 \mathrel{\hat{=}} \frac{M_{mi}}{V_m} \sigma_i \qquad \left(\sigma_i \text{ in ppm}\right). \tag{2.10}$$

Both sides of this equation are nondimensional. The factor 10^{-6} is implicitly included by the calculation of mg to kg.

The molar volume for ideal gases in the ambient conditions of $p_a = 1.013 \cdot 10^5$ Pa and $T_a = 293$ K ($\hat{=}$ 20 °C) is calculated from the ideal gas equation to:

$$V_m = \frac{R_m \, T_a}{p_a} = 24.36 \ m^3 / kmol \quad , \tag{2.11}$$

R_m molar gas constant (8314.5 Pa m^3/(kmol K)),
T_a ambient temperature in K,
p_a atmospheric pressure in Pa .

For the determination of molar mass, the atomic masses from the periodic table can be used if accuracy permits, for example:

$$M_{mCO} = 12 + 16 = 28 \text{ kg / kmol} \tag{2.12}$$

The molar masses and calculation factors for a few important gases are given in Table 2.5.

Table 2.5: Molar masses and calculation factors of selected gases in an ambient condition of 20 °C and $1.013 \cdot 10^5$ Pa

Component	Formula	M_{mi} in kg/kmol	(V_m / M_{mi}) in ppm / mg m^{-3}	(M_{mi} / V_m) in mg m^{-3} / ppm
Carbon monoxide	CO	28	0.870	1.149
Hydrocarbons	HC [1)]	-	-	-
Carbon dioxide	CO_2	44	0.554	1.806
Nitrogen monoxide	NO	30	0.812	1.231
Nitrogen dioxide	NO_2	46	0.530	1.888
Ozone	O_3	48	0.508	1.970

1) Conversion ppm into mg/m³ should be done acc. to VDI guideline no. 3481 pg. 1 "Messen der Kohlenwasserstoff-Konzentration, Flammen-Ionisations-Detektor (FID)" (Measure-ment of Hydrocarbon Concentrations, Flame-Ionization Detector) (refer to Chap. 6.2).

2.1.4 Formation of Soot Particles

Reactions with radicals are mainly responsible for the formation of higher, straight-chain, branched and cyclic hydrocarbons and also ultimately for the formation of elementary carbon (soot) during combustion. Acetylene (ethine) plays a central role here, originating by a complex route from the paraffin hydrocarbons in the fuel with ethylene formed as an intermediate stage. Figure 2.1 shows a hypothetical example of such a process schematically for propane.

Without going into all of the details concerning this reaction scheme, it must be mentioned that hydrocarbon, hydrogen, hydroxyl, and oxygen radicals appear again and again in the individual steps (crack, dehydration, and polymerization processes) as reactants or as reaction products.

Furthermore, cyclic and polycyclic aromatic hydrocarbons can be formed from the acetylene by polymerization and ring closure. After dimerization of the acetylene butadiyne is formed first, with larger structures following due to successive attachment of more acetylene molecules (Fig. 2.2).

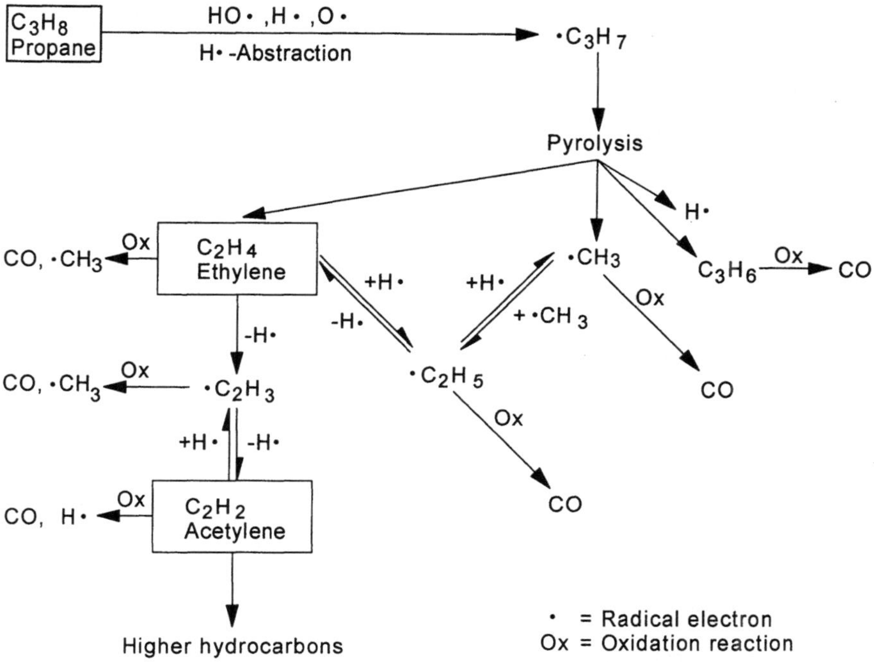

Fig. 2.1: Formation of acetylene by radical crack, dehydration and polymerization processes with the example of propane

Polycyclic aromatics, partly with lateral chains, result as an intermediate product. These are relatively stable thermally and are consequently detectable in exhaust emissions (for example naphthalene, anthracene, fluoranthene, pyrene, chrysene and the corresponding derivatives). If the acetylene addition is continued, then the carbon content in the molecules continues to rise at the cost of the hydrogen content, until finally a graphite-like soot particle is formed.

As Figure 2.3 schematically shows, these "nuclei" (about 0.001 to 0.001 μm in diameter) are enlarged in a second phase, the coagulation, to spherical "primary soot particles" (about 0.01 to 0.05 μm in diameter). The primary particles coagulate in a third development phase, the agglomeration, into loose chains of three-dimensional branches consisting of thousands of individual particles.

Figures 2.4 and 2.5 give an impression of what such agglomerates look like. These figures are photos of various magnifications from a transmission electron microscope (TEM). The projection of the agglomerate is then seen.

The agglomerates have an extraordinary adsorption ability because of their sponge-like structure and the very large surface area (50 to 200 m^2/g soot). Up to 75 % of the total particle mass can consist of adsorbed organic and inorganic compounds, depending on the ambient temperature and pressure (refer to Fig. 7.32, Chap. 7.6.5).

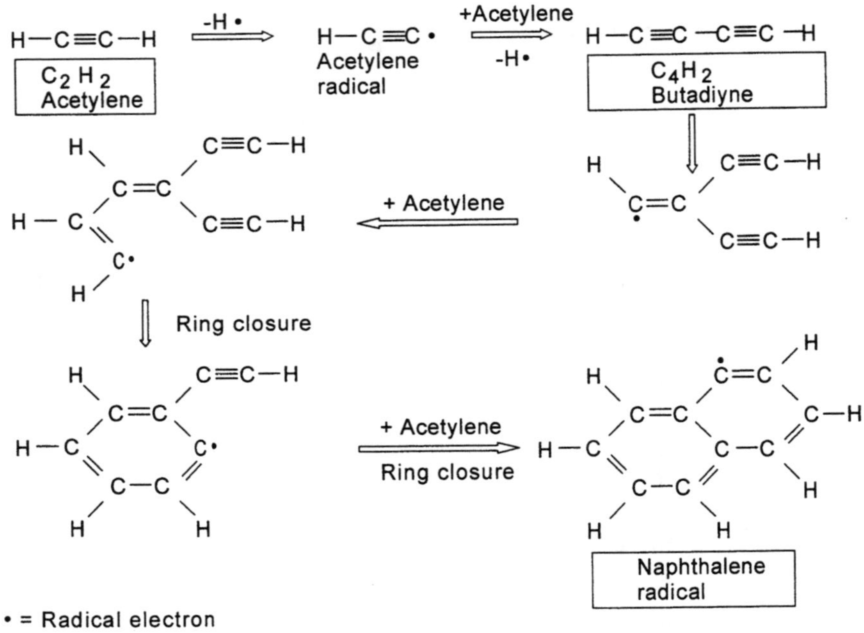

Fig. 2.2: Formation of polycyclic aromatic hydrocarbons by the polymerization of acetylene

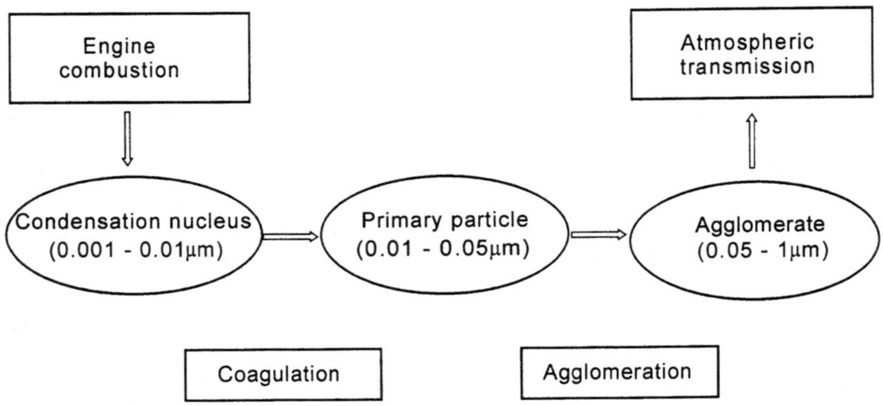

Fig. 2.3: Physical process of diesel particle formation

Figure 2.6 shows a typical density function, calculated from a bar graph of values measured, with a mean aerodynamic diameter of 0.261 µm ±0.005 µm of particles from a diesel engine (refer to Chap. 7.6)

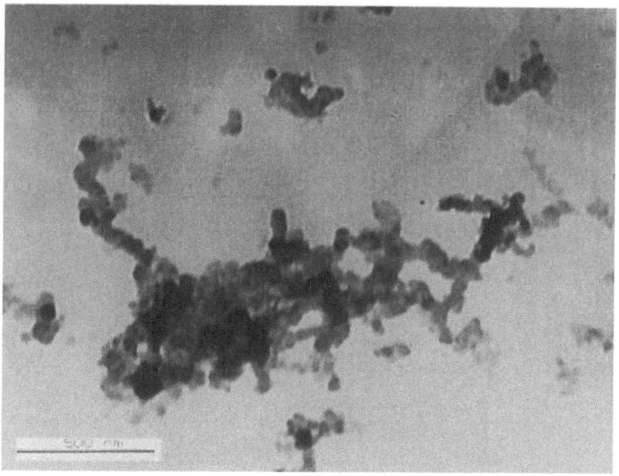

Fig. 2.4: TEM photo
500 nm

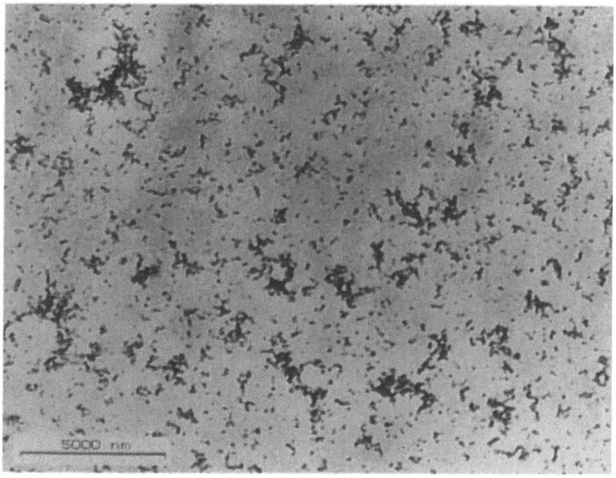

Fig. 2.5: TEM photo
5000 nm

2.2 Evaporation and Fueling Emissions

2.2.1 Hydrocarbon Emission Sources of a Motor Vehicle

When hearing the words "Emission sources of a motor vehicle" most think merely of the exhaust emission from the exhaust pipe. Frequently, the hydrocarbon (HC) evaporation emissions are forgotten. These emissions are in part limited by legislation. An overview of the HC emission sources of a motor vehicle and the emission proportions is shown in Fig. 2.7.

Evaporation and fueling emissions form part of the hydrocarbon pollution of the air by motor vehicles.

If all gasoline engine-powered vehicles were equipped with λ-sensor-controlled three-way catalytic converters, HC emissions from the exhaust pipe would be 3000

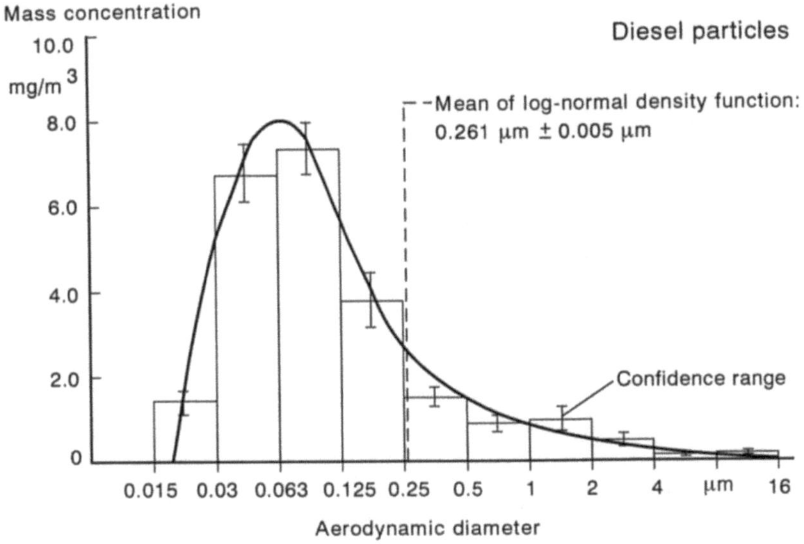

Fig. 2.6: Histogram (bar graph) of values measured and calculated density function of the aerodynamic diameter of diesel engine particles, measured after sampling from the exhaust pipe behind the oxidation catalytic converter

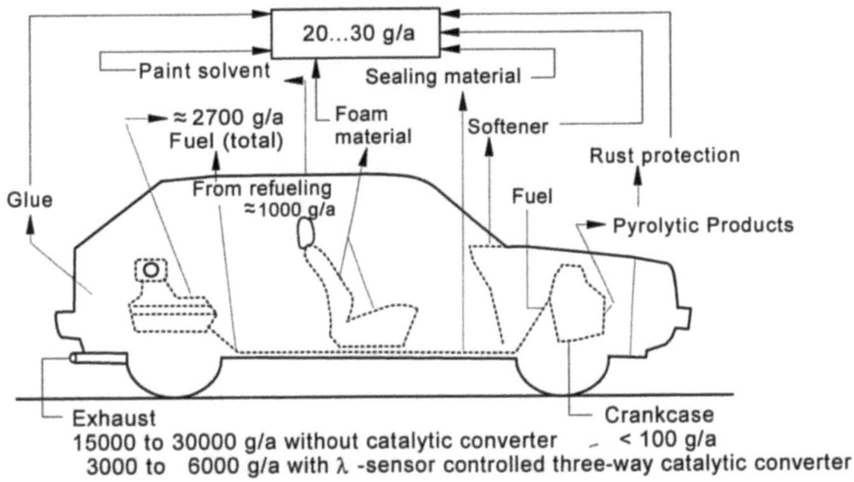

Fig. 2.7: Annual HC emissions of a gasoline engine vehicle with and without three-way catalytic converter

to 6000 g annually. It must also be taken into account that the vapors primarily consist of harmless components. With this in mind, the values for harmful components are probably around 100 to 300 g per year and vehicle, less than 15 % compared to the total HC emission from the exhaust pipe of a vehicle equipped with a three-way catalytic converter.

2.2.2 Evaporation Emissions

Evaporation is defined as the release of gaseous and vaporous hydrocarbons into the environment by evaporation from various parts of the vehicle. The HC emissions from the crankcase and the engine are not included here.

The evaporation emissions also include all volatile hydrocarbon compounds from the parts and materials utilized in the manufacture or the use of the vehicle. These are, for example, solvents and thinners for paint, glue and sealing compounds, foam material, underbody coating, preservation compounds and their pyrolysis products.

Furthermore, the fuel which evaporates from the technically required openings of the fuel system and diffuses through the walls of the container and pipes into the outside air are also included.

These evaporation losses are characterized by a HC vapor release from various parts of a vehicle distributed over a long period of time. A maximum of 0.4 g/min vaporizes from the fuel system. The amount emitted is responsible for a yearly average of 7.4 g HC per day per vehicle (Fig. 2.8). For this, the emission behavior corresponding to the temperature is monitored throughout the year. Values under 3 g and over 20 g per day can appear, depending on the fuel and parking conditions. The mean annual value of one vehicle is approx. 2700 g of fuel vapor (Fig. 2.7).

The emissions from vehicle bodies without engine and fuel system can amount to well over 100 g of hydrocarbons per day (approx. 6 g/Test), directly after leaving the assembly-line. The emissions subside quickly, however, and after a few days reach values of under 1 g per test (Fig. 2.9).

The sum total of an assumed operating life of 15 years is 300 to 450 g (Fig. 2.7).

In the USA evaporation emissions are limited to 2 g per test. Sweden, Switzerland, and Austria have followed this example. The Federal Republic of

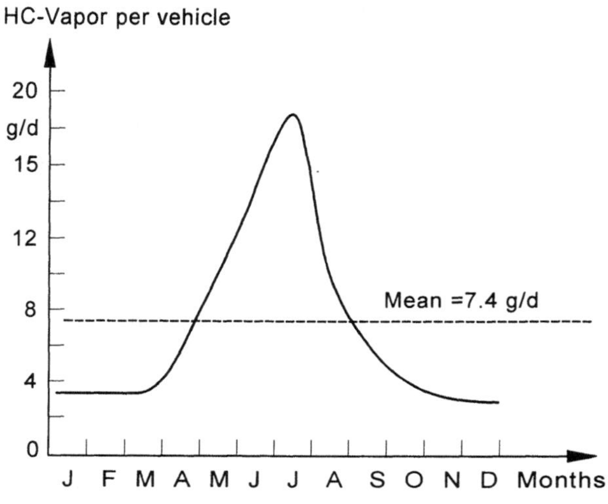

Fig. 2.8: Vapor emissions from the fuel system per vehicle over one year

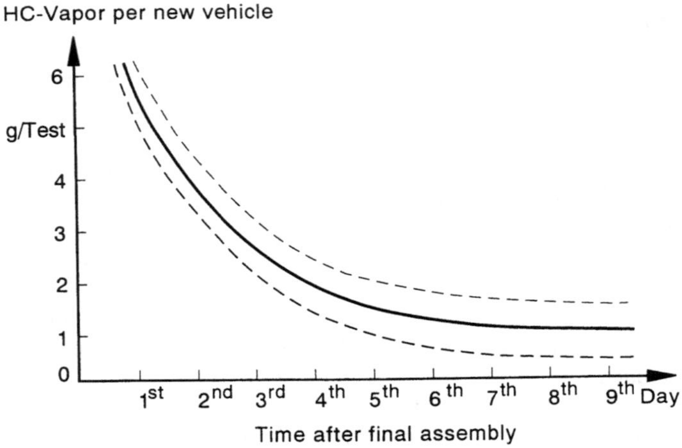

Fig. 2.9: Hydrocarbon emission of a new vehicle without engine or fuel system (so-called background emission)

Germany also demands compliance with this requirement for vehicles in accordance with annex XXIII of Germany's Automobile Safety Act. In Chap. 8.2.10 the test procedure is described.

The experience of the US authorities shows that in a large number of vehicles, which easily comply with standards in tests, very high hydrocarbon emissions are measured, that is emissions that are far above the set standards values. This is the case when these vehicles are checked under conditions different from the set test conditions - for example, when normal fuel from a fueling station is used instead of specified testing fuel, or when temperature, humidity or vehicle handling are different than in the laboratory test.

Test conditions are fulfilled when a canister with active charcoal is installed in the vehicle (Fig. 2.10). The fuel vapors are adsorbed by the active charcoal and then gradually desorbed if dry air flows through the canister, the vapors thus being fed via the intake manifold into the engine during operation of the vehicle.

This active charcoal system is capable of intermediate retention of 40 to 50 g of hydrocarbon vapor. Under test conditions (refer to Chap. 8.2.10) this retention capacity still offers a sufficient safety reserve.

The working capacity of the canister is insufficient under the following conditions:

- when fuel with a higher vapor pressure - with more butane - is used,
- when fuel with oxygen containing components such as methanol or ethanol is used,
- when damp air is sucked in for desorption,
- when the temperatures during parking or operation are higher than those in the test,
- when the ratio of immobilization time to operation time of the vehicle is too unfavorable.

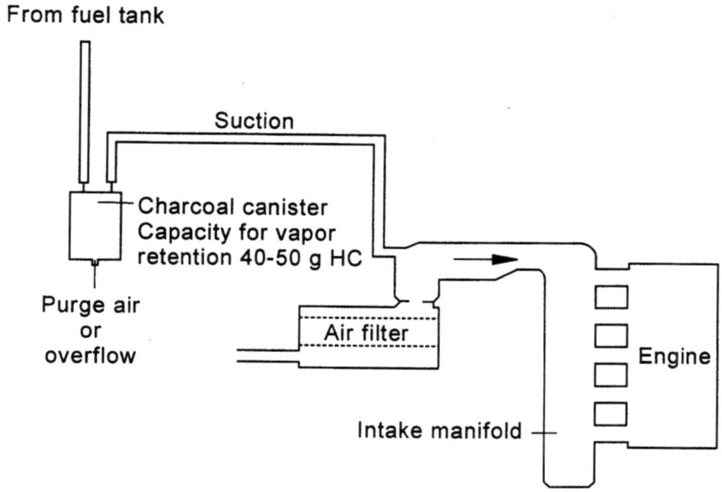

Fig. 2.10: Fuel vapor retention system

Under these circumstances the vapors escape into the ambient air.

In summer, when the fuel temperature climbs to about 40° C, about 850 g of vapor can develop from a residue volume of 20 l in the tank. For this, 8.5 kg of active charcoal would be needed for adsorption. In order to regenerate the capacity of the canister without a deterioration of vehicle handling or the exhaust emissions, after every parking phase with such conditions the vehicle would have to be driven at high speeds for at least 200 km through an area with low humidity. Otherwise, the canister would not be fully cleared.

2.2.3 Fueling Emission

Fueling emissions are defined as gaseous and vaporous hydrocarbons escaping from the fuel tank during fueling due to the displacement of the gas (fuel vapor and air) in the tank. It is characterized by an emission of HC vapors concentrated over a period of a few minutes from a limited source, the filler neck. A maximum of 80 g/min is emitted in this way.

Liquid fuel allowed into the open by overfilling or spilling out of the fuel nozzle should actually be preventable through suitable measures at the vehicle and the filling station, and so they are not taken into account here. On this premise, fueling emissions appear only from vehicles with gasoline fueled engines because of the low vapor pressure of diesel fuel.

When filling the fuel tank, the air laden with fuel vapors to varying degrees is displaced from the tank into the open (Fig. 2.11). The hydrocarbon amounts fluctuate depending on ambient conditions, fuel quality, filling rate, internal tank design and the fueling nozzle as well as the way in which the nozzle is introduced into the tank.

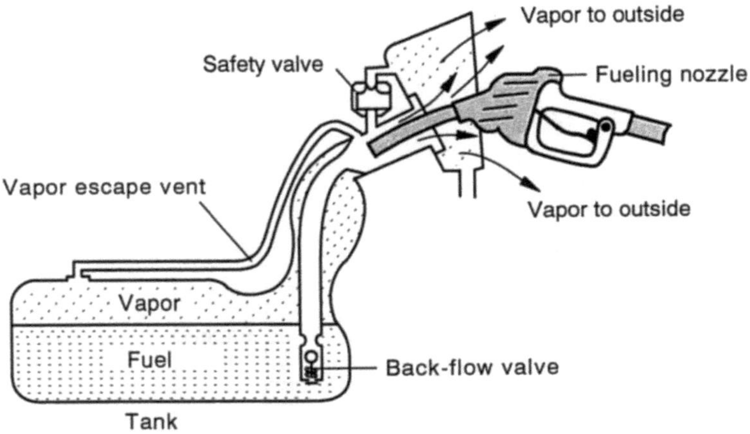

Fig. 2.11: Fueling emissions

The vapor contains approx. 1 g of hydrocarbons per liter of fuel under normal fuel vapor pressure (of $6 \cdot 10^4$ Pa acc. to Reid), at a filling rate of 30 l/min and a fuel and tank temperature of below 15 °C. On the average 50 g are emitted per fueling, annually around 1000 g per vehicle with 10,000 km driven per year on an average.

The amount of hydrocarbons clearly rises with the filling rate and the temperature (Fig. 2.12).

This increase is traced back to the expulsion of the gases - above all butane - dissolved in the gasoline. This effect is visible by filling a completely empty plastic bag, sealed from the outside, with gasoline (Fig. 2.13). About a third of the

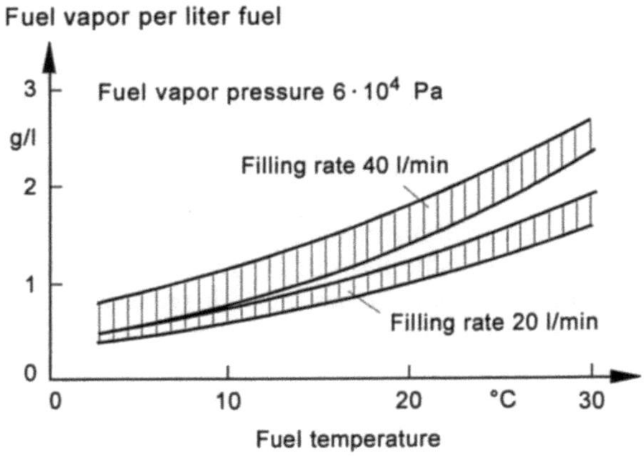

Fig. 2.12: HC vapor formation (g per liter fuel) during fueling as a function of fuel temperature, including variation ranges with filling rate as parameter

Fig. 2.13: Gas expulsion from gasoline after filling a plastic bag. The upper third is filled with gas

bag is filled with gas at the final point when the nozzle is shut off. (The process can be compared to the varying emissions of carbon dioxide (CO_2) when carbonated beverages are being poured from one container to another.) Butane is added to fuel by the mineral oil industry. Because of its low combustion value, butane contributes insignificantly to energy production of the engine.

The fueling emissions are at present only limited by legislation in a few states of the USA, where the return flow of the vapors to the main tank of the refinery is required. This process is known as return shuttling (Figure 2.14). This measure is being introduced step by step in Germany as well.

A special fueling nozzle is required for return shuttling (Fig. 2.15). During fueling the filler neck is sealed from the outside by a sprung rubber ring. The vapors cannot escape past the filling pipe of the nozzle, they flow through the gaiter seal and the special nozzle into the underground tank via a second hose; they are "shuttled back". In this fashion, most of the total fueling emissions of a fueling

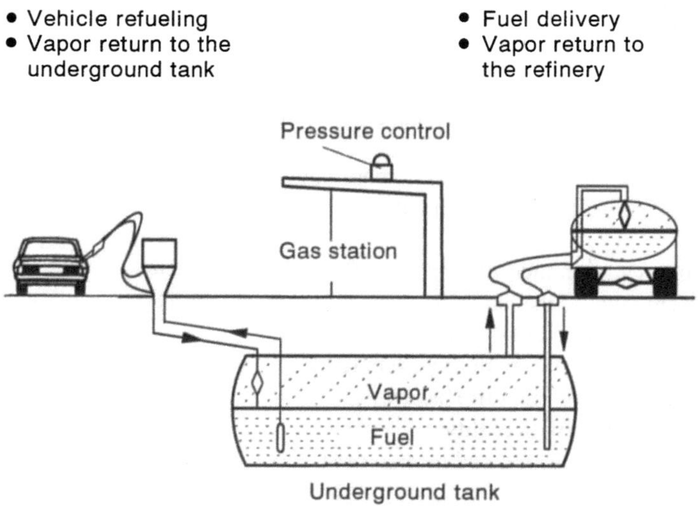

Fig. 2.14: Fueling with the gas return shuttle system

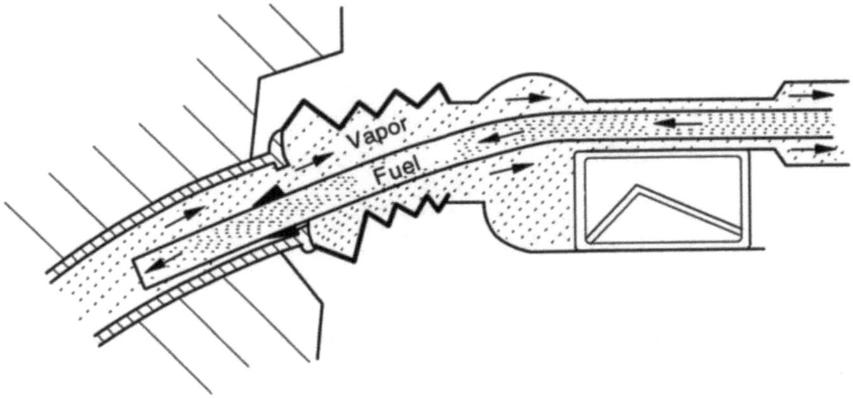

Fig. 2.15: Filling nozzle for gas return shuttling

station thus equipped are not allowed to escape into the atmosphere, because emissions from nearly all vehicles, even older vehicles, can be captured. Furthermore, the gas concentration in the underground tank is adjusted so that the vapors are led to the return hose.

The US exhaust emission authority, the EPA, wants to further change the relevant regulations. Lower vapor pressure of the station's fuel is to be standardized. The fueling emissions are to be limited by a combined active charcoal retention system installed into the vehicle, that is also to be effective for evaporation emissions.

The requirements to be met by an active charcoal system in the vehicle are as follows:

- prevention of fuel vapor emissions while fueling;
- complete adsorption and desorption of the vapors in the canister;
- safety during refueling;
- fuel filling rate of 6 to 50 l/min;
- maximum back pressure of about 147 Pa;
- no adverse effects on driving characteristics, emission behavior and vehicle safety.

After much experience with small active charcoal canisters and with experiment models of larger active charcoal canisters the following unexplained questions have emerged:

- optimization of the retention medium for the active charcoal system;
- design of the system (size of the canister, space requirements in the vehicle for the entire system, installation possibilities);
- installation and function of the condensate separator;
- safety in a crash test (preventing a fire hazard)
- test regulations: correlation with reality;
- standardization of fuel (for testing and normal use), especially narrower tolerances for vapor pressure;
- serial production and everyday suitability.

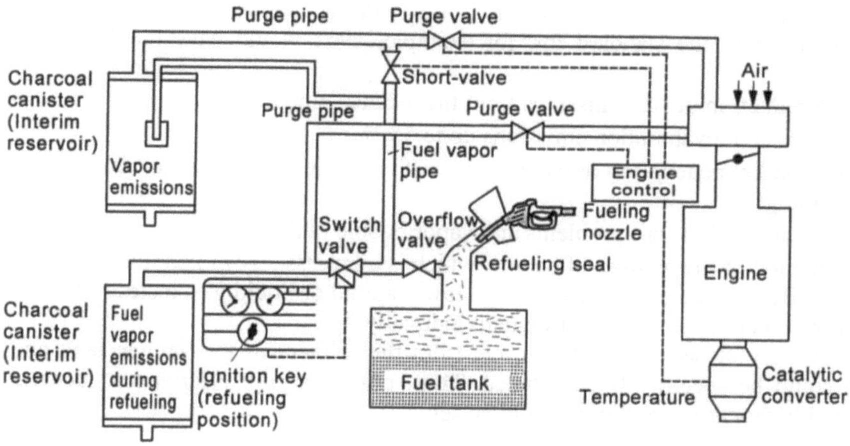

Fig. 2.16: Fuel vapor retention system, function schematic

The function schematic in Figure 2.16 shows how complicated such an arrangement can be.

Using the example of a standard-size automobile Fig. 2.17 shows how an additional charcoal canister can be installed using 1/4 of the trunk volume.

All other systems suggested so far for the reduction of fueling emissions are in no way close to production. Other possibilities of minimizing fuel emissions from inside the car (such as internal tank bladders or pressure tanks) combined with return shuttling are not ready for production either. There are no retrofit solutions in sight.

Faster and more efficient measures which would include vehicles already on the road concern fuel, particularly vapor pressure, and the distribution chain.

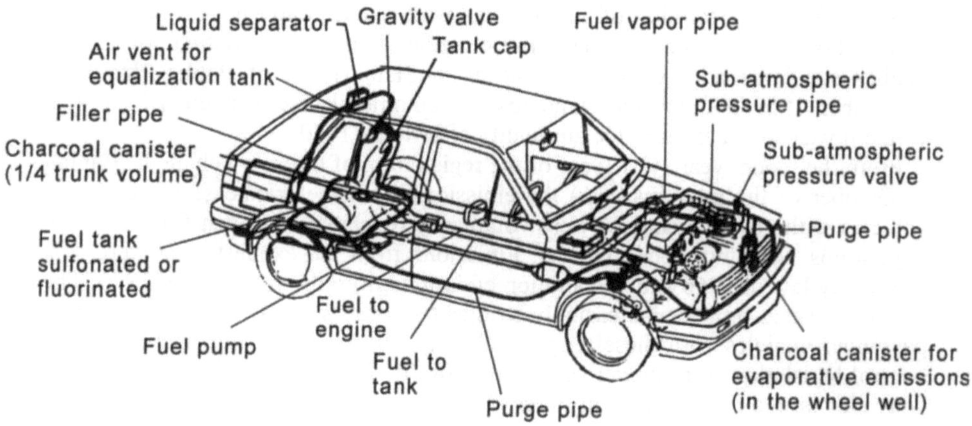

Fig. 2.17: Fuel vapor retention system (installation example)

The limitation of vapor pressure to $5 \cdot 10^5$ Pa in summer and $7 \cdot 10^5$ Pa in winter, that was the standard for, for example, 1962 would have the following effects:

- a 5 % reduction of HC emissions from the exhaust pipe,
- a reduction of evaporation emissions up to 55 %,
- quick market penetration of the measure with considerable efficiency from the beginning,
- a reduction of hot start problems and vapor bubble buildup,
- a reduction of the wasting of butane, which is unnecessarily mixed into the fuel today. Butane could be employed as a raw material or liquid gas, instead of simply escaping into the environment indirectly.

A completely closed distribution cycle from the refinery to the vehicle tank and back could be specifically introduced. Thus, priorities with reference to general regional hydrocarbon pollution could be set.

2.3 Legislation

2.3.1 Historical Development

The emission control legislation valid in various countries for passenger vehicles and passenger vehicle engines originated in California.

Very early on, population growth and traffic volume in this American state as well as special climatic conditions led to the irritation or damage caused by air pollution later to be found in other US states and other countries of the world. In 1943 there were the first reports of serious air pollution in Los Angeles, causing plant damage, throat and eye irritation to people as well as low visibility.

From 1948 the authorities introduced measures to combat these problems due to public pressure by reduction of emissions from stationary sources. The milestone research investigations of 1952 by J.A. Haagen-Smit from the California Institute of Technology led to the realization that fuel evaporation and combustion products and therefore the operation of automobiles were substantially responsible for the typical Los Angeles smog. These investigations laid the foundation for the air quality standards adopted by the California State Board of Public Health in December 1959 and for the standards for automobile exhaust emissions with which it was hoped to reach the air quality of 1940 by 1970.

From the model year 1961 (year of the registration of the new vehicle beginning in October of the previous year) the emissions from the crankcase of passenger vehicles with gasoline engines were voluntarily regulated, and then from 1964 on regulations were introduced by law. Emissions from the exhaust pipe were first limited by law in 1966 and evaporation emissions from the fuel tank in 1970. The following years brought increasingly stringent standards for these emissions, with passenger vehicles with diesel engines also included in the Californian emission control legislation from model year 1980 onward. Test and measuring regulations were also issued for certification (refer to Chap. 8).

US federal legislation against general air pollution was triggered by an incident in Donara, Pennsylvania, in 1948 in which industrial exhaust and smoke were prevented from escaping by inversion weather conditions, and according to doctors, thousands of cases of illness and a few fatalities were reported. The first investigations into the composition and biological effects of automobile exhaust gas began in 1959 through the Public Health Service of the Department of Health, Education and Welfare. Special emphasis was given to the exhaust emissions from automobile engines only in the period after 1960, with the investigations of the federal agency being mainly built up on the foundation laid in California. The federal government also adopted the 1966 California test procedure without working out an emission baseline of its own.

In the following years the federal environmental authority, the Environmental Protection Agency (EPA) was clearly in control of developing emission control legislation in the USA along with conditions for its implementation and also the improvement of test regulations. Legislation was passed by Congress in the form of the Clean Air Act, which has been modified repeatedly. Japanese legislation developed in a similar fashion.

Meanwhile, many countries, also those of the European Union (EU), have issued exhaust regulations. Fig. 2.18 shows the application date of exhaust legislation for the countries of Europe since 1970.

Test regulations belong to legislative regulations and are described in Chap. 8. Unfortunately, test regulations differ from country to country throughout the world.

Fig. 2.18: Application dates of exhaust legislation in Europe in the seventies

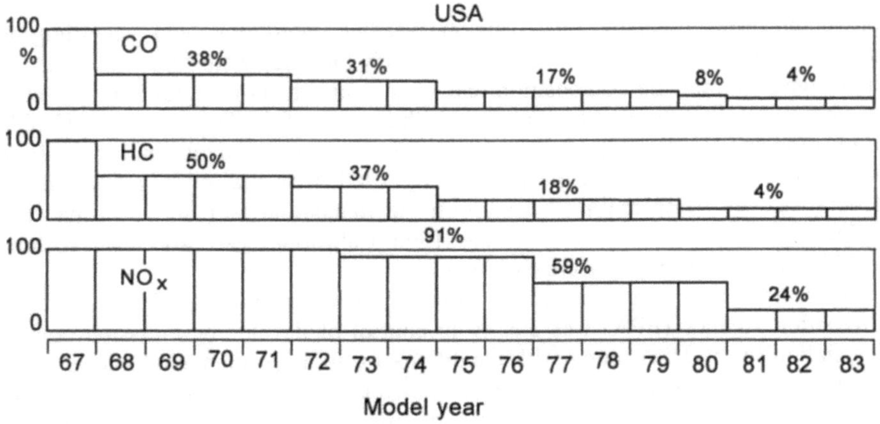

Fig. 2.19: Lowering of standards in percent in the USA relative to values before the introduction of legislative regulations until model year 1983

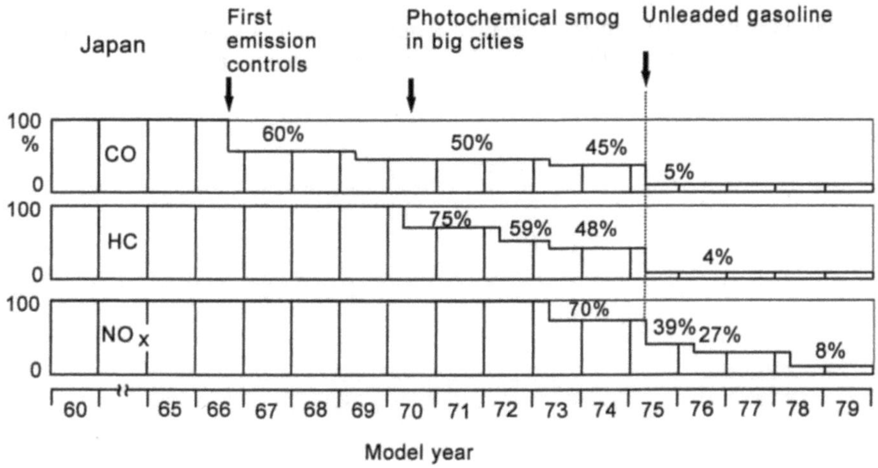

Fig. 2.20: Lowering of standards in percent in Japan relative to before the introduction of legislative regulations until model year 1979

Fig. 2.19 shows the tendency of the relative lowering of the emission standards in the USA, Figure 2.20 shows those of Japan, for successive model years (model year is the year of registration, that is the year preceding the calendar year).

Because of the differing test regulations, absolute values are not comparable. Furthermore, minor reductions were stipulated in following years.

2.3.2 Legislation from the Present Viewpoint

In California the coming years will bring drastically more stringent emission standards for motor vehicles, which will inevitably require innovative technology. At the California environmental authority, the California Air Resources Board (CARB), this goal is grouped under the following headings:
– Low Emission Vehicles (LEV)
– Transmission Low Emission Vehicles (TLEV)
– Ultra Low Emission Vehicles (ULEV)
– Zero Emission Vehicles (ZEV)
(refer to Table 2.8 and Fig. 2.21).

Table 2.8: Schedule for tightening the California exhaust emission limits, MY = Model Year

MY	Measure
1991	- on-board diagnosis 1st stage (OBD I) (indication in vehicle)
1993	- more stringent standards for HC and CO with 100,000 mile (mi) durability - new evaporation test and evaporation emission standards for high ambient temperatures (40.5 °C ≙ 105 °F) in the SHED (refer to Chap. 8.2.10) - formaldehyde standard
1994	- on-board diagnosis 2nd stage (OBD II, refer to Chap. 9.11.5) for passenger cars (PC) and light duty trucks (LDT) - stricter NO_x standards
1997 - 2003	phase in of - Low Emission Vehicles (LEV) - Ultra Low Emission Vehicles (ULEV) - Zero-Emission Vehicles (ZEV, starting in 2003)
1998 - 2003	for LEV: - cold start at lower temperatures (10 °C), refer to Chap. 8 - standards for NMOG (non-methane organic gases), stricter standards for HC, NO_x and particles

The emission standards for future regulatory measures are summarized in Fig. 2.21.

Table 2.9 shows the Californian emission standards for passenger cars which could be issued in other states of the US.

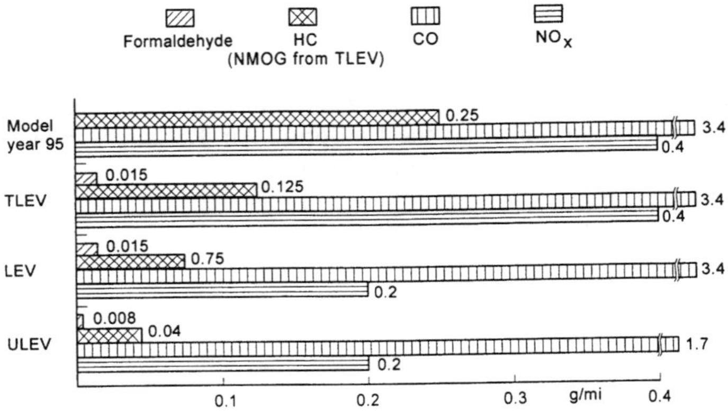

Fig. 2.21: Emission standards for future regulatory measures in California

Table 2.9: Standards for California, PC (PC = passenger cars) ≤ 12 occupants (for certification testing of new vehicles and testing of production vehicles) in g/mi, MY = Model Year, CRC = Crankcase

MY	Durability in 10^3 mi	Σ HC	CO	NO_x	HDC- NO_x[e]	Evaporation[d]	CRC - Emission	Particles
PC with gasoline engines								
1989	50	0.41 (0.39)[b]	7.0	0.7 (0.4)[c]	$NO_x \cdot 1.33$	2.0	0	-
	100	-	-	-	-	-		-
1993[a]	50	0.39 (0.25)	7.0 (3.4)	0.7 (0.4)[c]	$NO_x \cdot 1.33$	2.0	0	-
	100	(0.31)	(4.2)	-	-	-		-
1995	50	0.25	3.4	0.4	$NO_x \cdot 1.33$	2.0	0	-
	100	0.31	4.2	-	-	-	-	-
PC with diesel engines								
1989	100	0.46	8.3	1.0	$NO_x \cdot 1.33$	-	-	0.08
1993[a]	100	0.46 (0.31)	8.3 (4.2)	1.0	$NO_x \cdot 1.33$	-	-	0.08
1995	100	0.31	4.2	1.0	$NO_x \cdot 1.33$	-	-	0.08

[a] Phase in of the standards in parenthesis: MY '93: 40 %, MY '94: 80 %, MY'95: 100 % of the planned sales figures of a manufacturer

[b] Standards in parenthesis for methane-free HC measurements. From MY '93 just methane-free HC standards

[c] Phase in of the standards in parenthesis: MY '89: 50 %, MY '90: 90 %, MY '94: 100 % of the planned sales figures of a manufacturer

[d] in g/test
Durability: 50,000 mi

[e] HDC-NO_x: NO_x measured in Highway Driving Cycle

Table 2.10 shows the standards for the next years in the USA and Table 2.11 shows those for the European Union.

Table 2.10: Present and future standards in the USA - 49 states for PC (Passenger cars) ≤ 12 occupants with gasoline or diesel engines[a] (for certification testing of new vehicles and testing of production vehicles) in g/mi (US -75 Test)

MY	Durability in 10^3mi	Σ HC	NMHC	NMOG	HCHO	CO	NO_x Gasoline	NO_x Diesel	Particles	Evaporation in g/Test
1992/ '93	50	0.41	-	-	-	3.4	1.0	1.0	0.2	2.0
1994/ '95	50	0.41[b]	0.25	-	-	3.4[c]	0.4	1.0	0.08[d]	2.0
	100	-	0.31	-	-	4.2	0.6	1.25	0.10	-
	in-use[e]	-	0.32	-	-	3.4	0.4		0.08	2.0
1996/ '97	50	0.41	0.25	-	-	3.4	0.4	1.0	0.08	2.0
	100	-	0.31	-	-	4.2	0.6	1.25	0.10	-
	in-use	-	0.32	-	-	3.4	0.4	-	0.08	2.0
	50[f]	-	-	0.125	0.015	3.4	0.4		-	2.0
	100	-	-	0.156	0.018	4.2	0.6		0.08	-
1998/ '99/ 2000	50	0.41	0.25	-	-	3.4	0.4	1.0	0.08	2.0
	100	-	0.31	-	-	4.2	0.6	1.25	0.10	-
	50[f]	-	-	0.125	0.015	3.4	0.4	1.0	-	2.0
	100	-	-	0.156	0.018	4.2	0.6	1.25	0.08	-
2001/ '02/ '03	50[f]	0.41	0.25	-	-	3.4	0.4		0.08	2.0
	100	-	0.31	-	-	4.2	0.6		0.10	-
	50[f]	-	-	0.075	0.015	3.4	0.2		-	2.0
	100	-	-	0.090	0.018	4.2	0.3		0.08	-

NMHC: Non Methane HC; NMOG: Non Methane Organic Gases;
HCHO: Formaldehyde

[a] Phase in of the standards: MY'94: 40 %, MY'95: 80 %, MY'96: 100 %

[b] HC: 0.41 g/mi and NMHC: 0.25 g/mi, required in combination

[c] From MY'94 CO cold: standard: 10 g/mi at 20°F phase in like [a]

[d] Particle standard for diesel and gasoline vehicles

[e] In-use standard: MY'94: 40 %, MY'95: 80 % of the vehicles in a manufacturer's fleet. All other vehicles must meet the MY'93 standard. From MY'96, 60 %, from MY'97 20 % of the vehicles must meet the in-use standards, while all others have to fulfill certification standards as in-use standards. From MY'98 this applies to all vehicles.

[f] Application of clean fuel standards: In areas with higher ozone pollution and CO pollution > 16 ppm all the new vehicles must satisfy these standards. Clean fuel vehicles in fleets which can be centrally fueled, must be in the following proportions from MY'98: MY'98: 30 %, MY'99: 50 %, MY2000: 70 %

Table 2.11: Present and future standards in the European Union (PC with gasoline or diesel engines, ≤ 6 occupants, ≤ 2500 kg total mass)

Law	Application date NM	Application date AP	Standard for	Test: Type I[a] CO	Type I[a] HC + NO_x	Particles[b]	Type III CRC-Emissions[c]	Type IV Evaporation[c]
				g/km	g/km	g/km	-	g/test
91/441/ EEC	Dec.31, 92	Jan.7, 92	T	2.72	0.97	0.14	0	2.0
			S	3.16	1.13	0.18	0	2.0
			T = S					
94/12/ EC	Jan.1, 96	Jan.1, 97	Gasoline	2.2	0.5	-		
			Diesel DI	1.0	0.9	0.1		
			Diesel IDI		0.7	0.08		

NM New Model certification
AP All Production vehicles
T Type certification standard
S Series vehicle standard

a New European driving cycle (MVEG A)
b Only diesel engines
c Only gasoline engines
DI: Direct injection engines
IDI: Precombustion chamber and swirl-chamber engines (indirect injection engines)

Exception for DI Diesel:
HC+ NO_x- and particle standards are multiplied by the factor 1.4. Valid until Jul.1, 94 for new certification and until Dec.31, 94 for all production vehicles.

Transition Requirements for the EU Directive 91/441/EEC:
Alternatively, the following transition requirements continue to apply:
(new certification: until Jul.1, 93; all production vehicles: until Dec.31, 94)

- Vehicles ≥ 1.4 l (EU directive: 88/436/EEC, annex IIIA)
  HC: 0.25; CO: 2.11; NO_x: 0.62; Particle: 0.124 g/km in the US-75 driving cycle
- Vehicles > 2.0 l (gasoline engine) (EU directive: 88/76/EEC)
  HC + NO_x: 6.5 (NO_x max: 3.5); CO: 25 g/Test in the ECE driving cycle
- Vehicles ≤ 1.4 l (EU directive: 89/458/EEC)
  (HC +NO_x: 5; CO: 19; Particle: 1.1 g/Test in ECE driving cycle)
- Driving cycle MVEG A: For vehicles ≤ 30 kW power and ≤ 130 km/h top speed, the maximum speed for the high speed part of the driving cycle is reduced from 120 km/h to 90 km/h. This exemption is valid until Jul.1, 94.

Tightening of standards for EU directive 94/12/EC
National backing possible from Jul. 94; exception for DI diesel as above until Sep.30, 99

2.4 Present Condition of the Exhaust Pipe Exhaust Emissions of Passenger Vehicles

Figs. 2.22 to 2.26 show data on limited exhaust components for different types of vehicles. The results are mean values of different repeated tests and three different driving cycles (FTP, SET, HDC, refer to Chap. 8) on several vehicles per type. The total mean value $\bar{\bar{x}}$ is included for every vehicle category.

Except for particle emissions, emissions from gasoline engines with three-way catalytic converters and of diesel engines are roughly comparable. The NO_x values, however, are higher with diesel engines.

In diesel engines particles consist mainly of soot particles in a general sense, while in gasoline engines without catalytic converters particles are mainly lead compounds when leaded gasoline is used. All other particles from gasoline engines with catalytic converters consist of different chemical compounds (refer to Chap. 7.6).

Modern diesel concepts with a turbo charger and an oxidation catalytic converter (oxi-cat) show lower particle emissions (Fig. 2.26). A reduction of PAH (polycyclic aromatic hydrocarbons) emissions is effected by the catalytic converter; particle emission reduction is effected mainly by the turbo charger which is used largely to increase the air flow rate and not to increase power.

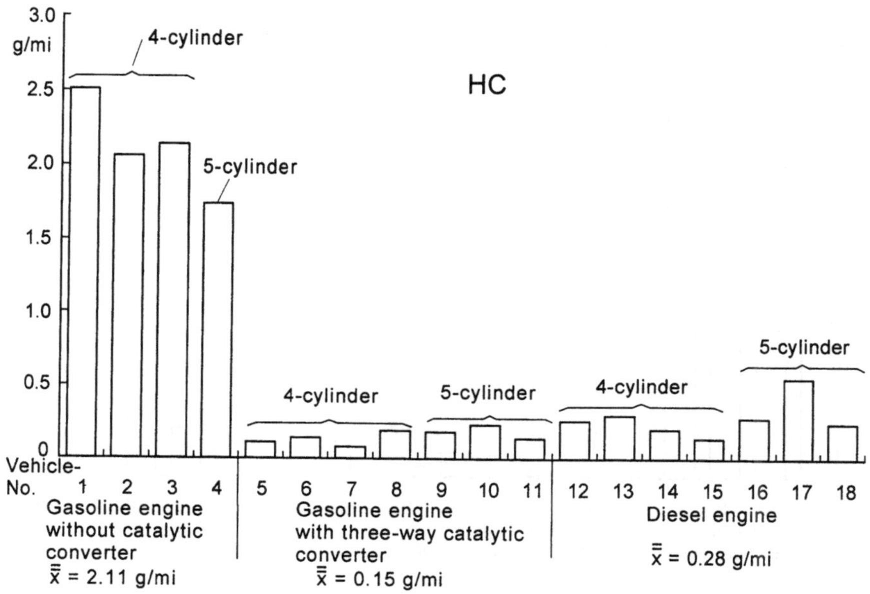

Fig. 2.22: HC mass emissions per distance driven in g/mi; 18 vehicles from different categories. Mean values of different repeated tests and three different driving cycles (FTP, SET, HDC) on several vehicles per type

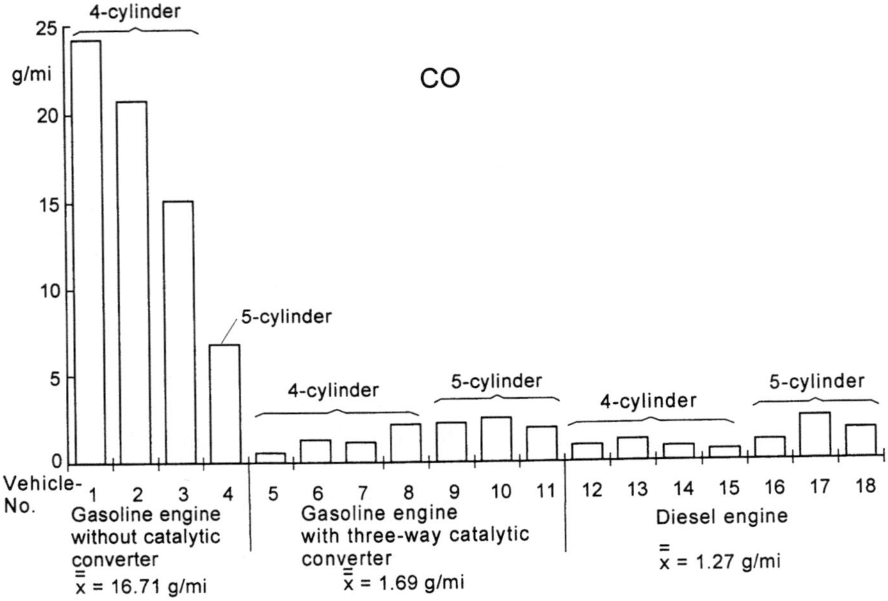

Fig. 2.23: CO mean mass emissions per distance driven in g/mi; 18 vehicles from different categories, otherwise as in Fig. 2.22

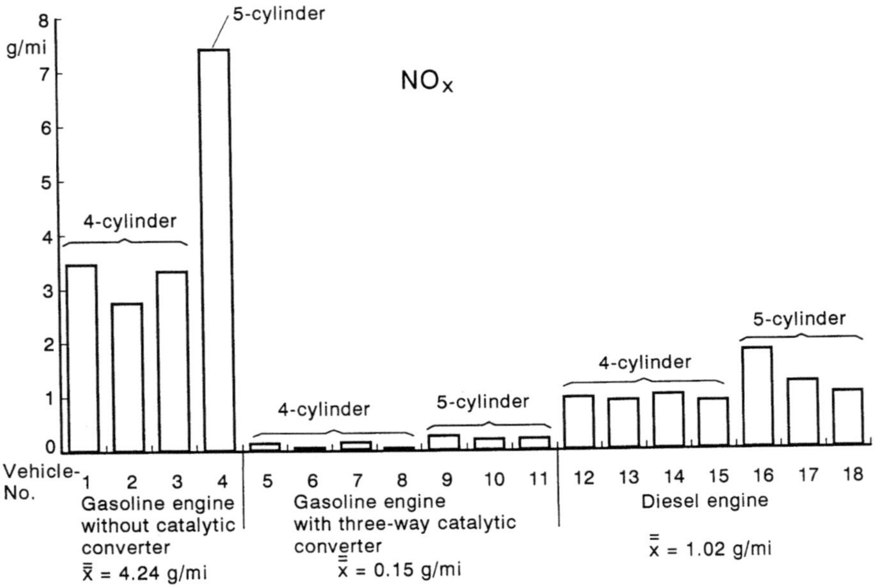

Fig. 2.24: NO_x mean mass emissions per distance driven in g/mi; 18 vehicles from different categories, otherwise as in Fig. 2.22

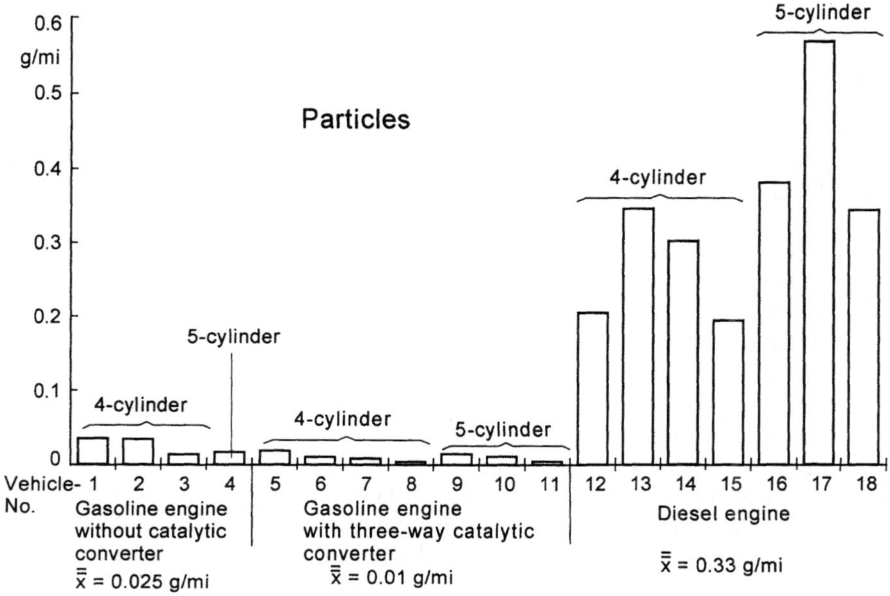

Fig. 2.25: Particle mean emissions per distance driven in g/mi; 18 vehicles from different categories, otherwise as in Fig. 2.22

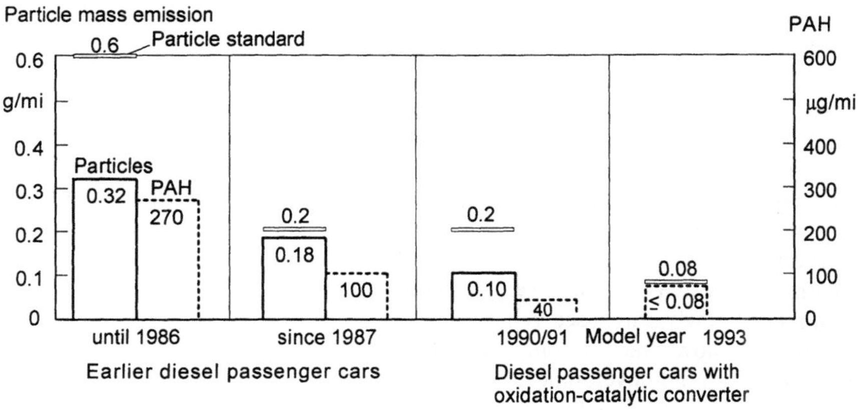

Fig. 2.26: Particle and PAH emissions for vehicles of different model years

2.5 Emission Prognosis for Germany (Calculated in 1990)

2.5.1 Overview

For future regulation it is important to calculate the development of total emissions for the next two decades, e.g. for vehicles in Germany, in spite of the fact that only the pollutant concentrations in ambient air are important with regard to the effects, refer to Chaps. 1 and 5.

For such a prognosis suitable calculation models are used which include the most important influencing factors resulting from the vehicles' specific characteristics, from road traffic situations and vehicle emissions, etc.

The prognosis for total emissions is carried out in 3 sections:

- determination of exhaust gas emission factors
- determination of road-related exhaust gas emission factors, abbreviated to specific emissions
- determination of total emissions.

By mathematically linking traffic data with data containing the mean emissions of vehicles in relation to their driving behavior and operational condition, the data for the prognosis of the specific emissions are developed. Total emission is derived by multiplying these with the annual distance driven (km/year).

Fig. 2.27 shows a simple flow chart for calculating exhaust emission factors.

2.5.2 Parameters

To calculate the emission for each combustion and displacement class, the respective numbers of vehicles in use must be known.

Number of vehicles in use

Until 1989 statistical data could be taken from the German Federal Office for Motor Traffic (Kraftfahrtbundesamt (KBA)) and from prognosis data from Shell, describing the number of vehicles in use and of newly registered vehicles classified according to combustion process and including a partial breakdown into the registration years and displacement classes. The prognosis until 2010 is possible by estimating the decrease of older vehicles in use (s. below) and plausibly estimating the number of newly purchased vehicles.

Figure 2.28 shows the development of the number of vehicles in use in Germany classified acc. to the development of the numbers of different classes of vehicles (with gasoline engines - with and without three-way catalytic converters - , two-stroke engines and diesel engines). Vehicles with gasoline engines without catalytic converters are distinctly on the decrease. Correspondingly, the number of vehicles with gasoline engines and three-way catalytic converters is increasing. Though the number of vehicles with diesel engines is low, it is increasing continuously but slowly. The number of vehicles with two-stroke engines is even lower and is decreasing because of the reunification of Germany. Overall, the maximum number of vehicles in use will be reached in the years 2000 to 2005 due to the saturation of the market and the development of the population structure.

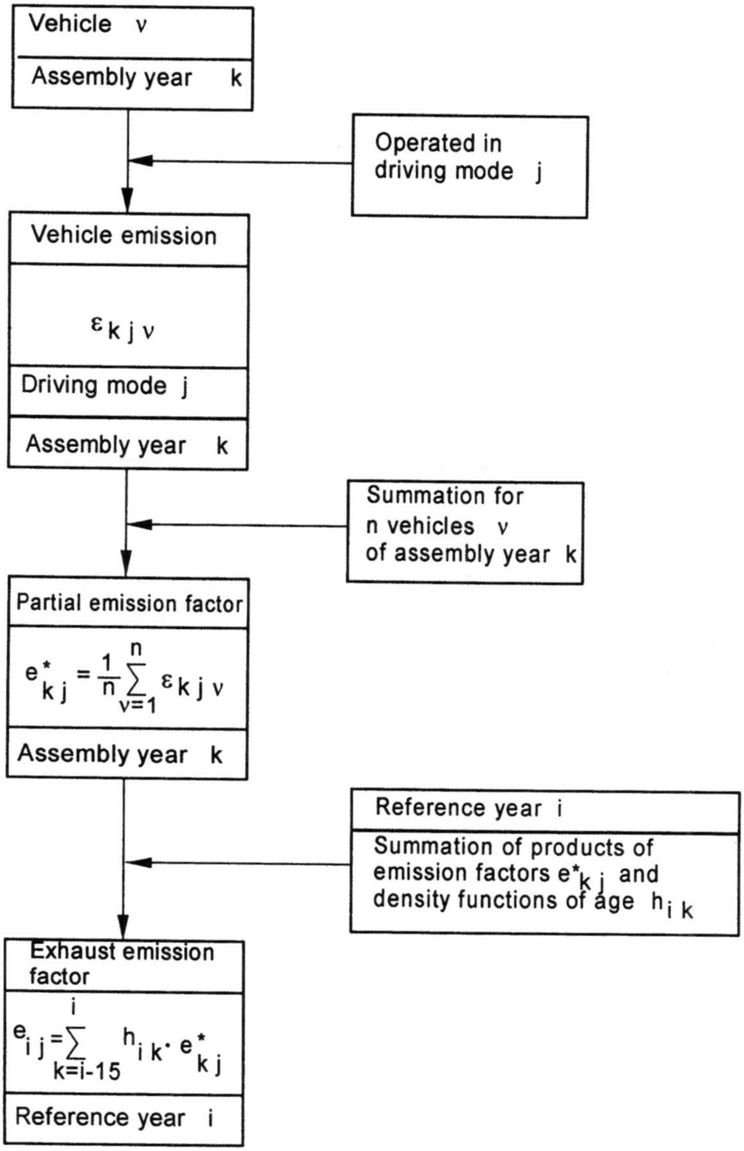

Fig. 2.27: Flow chart for calculation of exhaust emission factors

Frequency distribution and decrease curve

As emissions and kilometers driven are different for every displacement class of each combustion type, frequency distribution per combustion process and per displacement class must be known along with the respective numbers of vehicles in use acc. to registration years, s. Fig. 2.29.

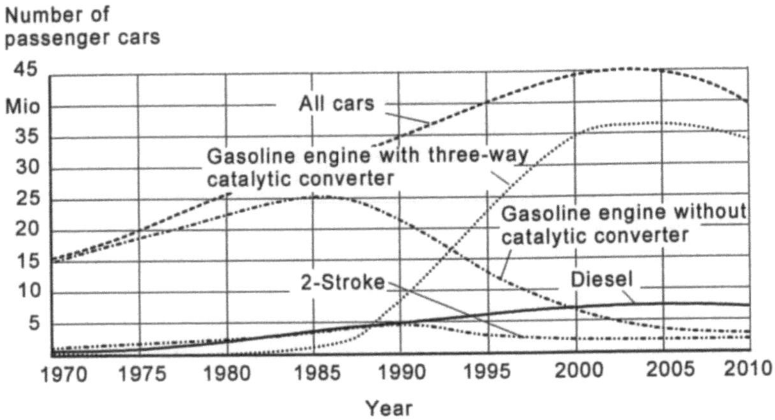

Fig. 2.28: Development of the number of vehicle in use in all federal states of Germany from 1970 to 2010 (Source: Metz)

This frequency distribution may be determined from the number of vehicles in use per registration year. In the year 1985 the number of vehicles in use by July 1 may be used. Fig 2.29 shows the frequency distribution according to registration years of vehicles with gasoline engines of displacement category < 1.4 l for the reference year 1985, ending with the 16th year which includes all even older vehicles (calculating backward, i.e. < 1970). The fourth year shows a break

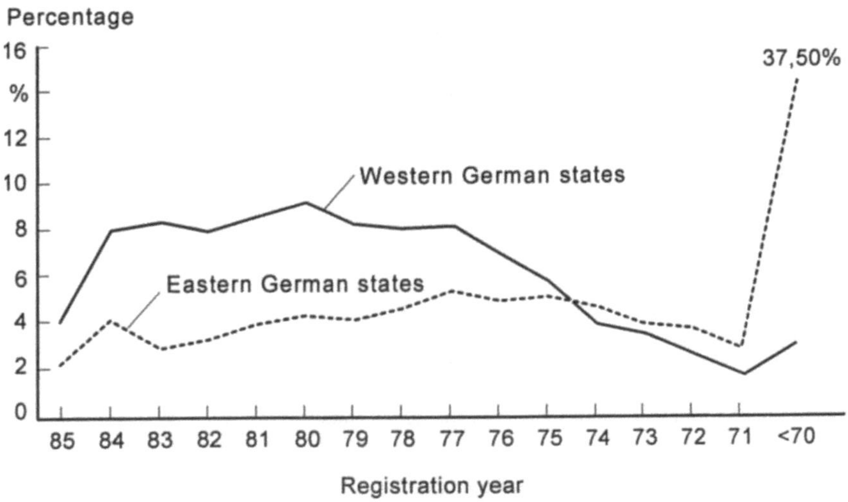

Fig. 2.29: Comparison of frequency distribution according to registration years in the western and eastern states of Germany, vehicles with gasoline engines of displacement category < 1.4 l (Source: Metz)

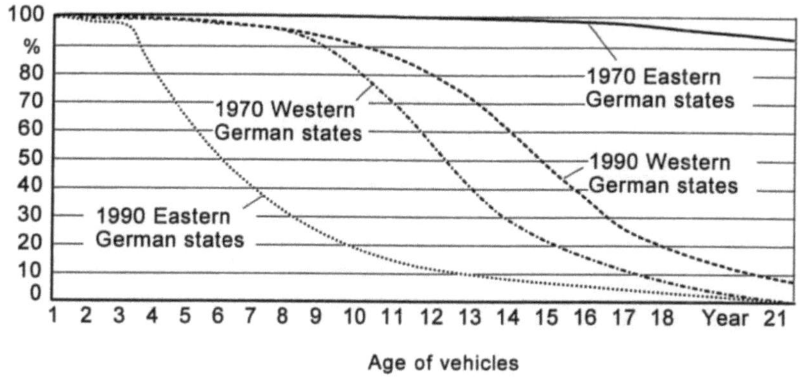

Fig. 2.30: Decrease curve of two registration years (1970 and 1990) according to vehicle age in the western and eastern states of Germany (Source: Metz)

resulting from a disproportionately frequent change of vehicles. The following years show a slight rise with a subsequent steady drop.

In contrast to this, the frequency distribution in the new states of Germany is entirely different. The high proportion of old vehicles, a result of long waiting times for a new vehicle and long usage, is striking.

For improved estimates of future vehicles driven the decrease curve of vehicles of a registration year is of help, s. Fig. 2.30.

This decrease curve is known from statistically recorded data and can therefore be applied to years to come unless other serious interfering factors arise.

Mean annual distance driven per vehicle type

The mean annual distance driven is estimated by different institutes. It, too, depends on the year of calculation, on the type of combustion process and the displacement category. Unfortunately, estimates are highly inaccurate as there are no reliable data.

Fig. 2.31 (upper part) shows the annual distance driven as a function of the vehicles' age using as example the year 1985 (registration year) and the three displacement classes of vehicles with gasoline engines without catalytic converter and, as comparison, of vehicles with diesel engines. The distances driven in 1985 are shown below for the new states of Germany. Although it is lower initially it does not decrease as rapidly with increasing vehicle age.

Development of the annual distance driven

Using the number of total vehicles and the mean annual distance driven per individual vehicle, both the total annual distance driven and for each of the three different types of roads: extra-urban, intra-urban and highways can be determined for each combustion process and displacement class, s. Fig. 2.32.

Despite the slightly decreasing distance driven per individual vehicle due to the growing vehicle density on the largely unchanging road network, the total annual distance driven rises until the year 2000 and then drops off slightly.

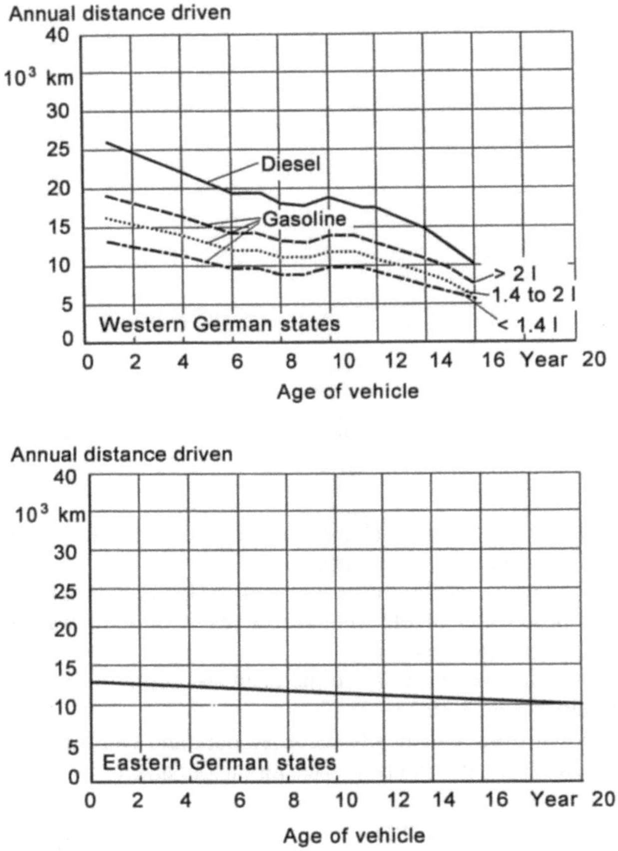

Fig. 2.31: Annual distance driven relative to vehicle age in the western and eastern states of Germany, registration year: 1985 (Source: Metz)

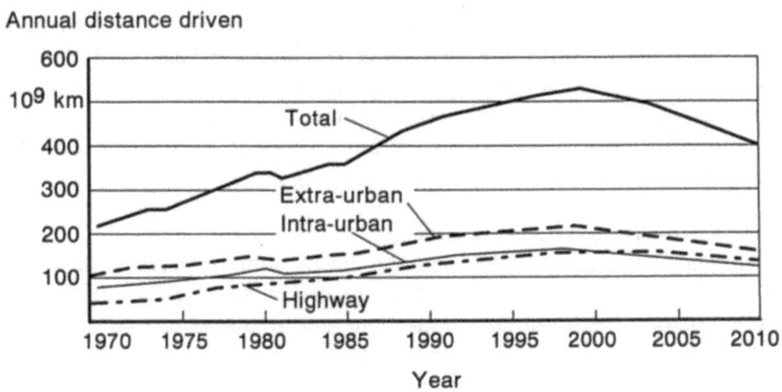

Fig. 2.32: Development of annual distance driven total and relative to road types (Source: Metz)

Percentage distribution for the type of road is as follows:

extra-urban	approx. 40 %
intra-urban	approx. 33 %
highway	approx. 27 %.

2.5.3 Calculation Method

A detailed analysis must take into account legislation from the earlier years.

Total vehicle emissions for one component are calculated from the sum of emissions of vehicles with gasoline engines with and without three-way catalytic converter and vehicles with diesel engines. Total emissions from any combustion process are made up of the total emission of the displacement category $DP < 1.4$ l; $DP = 1.4 \dots 2.0$ l and $DP > 2.0$ l, as different emission factors and data used must be taken into account, a result of former European legislation. The emissions from each displacement class of each combustion type may be determined by summation of 16 registration years by means of multiplying the number of vehicles registered in this class by the percentage frequency of the particular registration year, the mean annual distance driven, the proportion driven on the type of road and the corresponding emission factor.

The emission factor depends on the type of road under consideration. Owing to data basis available and already existing prognoses, a weighted consideration of typical driving conditions in the city can be chosen. The driving curves determined by the Technical Control Board (Technischer Überwachungsverein = TÜV) of the German state of Rhineland based on extensive studies of Cologne traffic for flowing traffic and traffic on arterial roads can be weighted according to proportion in the same way as the driving cycles prescribed by law (refer to Chap. 8.2.3).

For extra-urban road operation TÜV data must be supplemented by more recent data gathered with the high speed phase of the new European driving cycle (refer to Chap. 8.2.3, Fig. 8.6) which can be divided into an extra-urban road and a highway section on the basis of speed. The extra-urban road proportion, e.g., is defined by the first 205 s of the high speed phase (from 800 s to 1005 s) and covers speeds of up to 70 km/h. As the data basis of the TÜV Rhineland contains emission data with a mean speed of 60 km/h and a constant 100 km/h a polygon is formed with these two theoretical points of support, and the emissions of the assumed mean speed of 70 km/h are calculated from this by interpolation for extra-urban roads.

The same procedure can be applied to highway operation. By observing traffic conditions for many years the German BAST (Bundesanstalt für Straßenwesen or Federal Institute for Road Research) has determined a mean speed of 112 km/h on highways. Here, too, data can be supplemented by more recent results of the high speed phase of the new European driving cycle where the highway proportion is defined by emissions from second 1005 to second 1170 with speeds up to 120 km/h.

In the following an emission prognosis is described which is based on conditions of the year 1990 and was carried out for the years 1970 - 2010.

The following system of equations implies the calculation method:

$$E = E_G + E_C + E_D \qquad E_C = E_{C<1.4} + E_{C\,1.4-2} + E_{C>2} \tag{2.13}$$

$$E_{C>2} = \sum_{i=RY}^{i=RY-16} T_{i,\,RY>2} \cdot fd_{i,\,RY>2} \cdot md_{i,\,RY>2} \cdot Or(RY) \cdot e_{i,\,iur,\,RY>2} \tag{2.14}$$

$$e_{iur} = 0.65\left(0.65\,e_{M3} + 0.35\,e_{M4}\right) + 0.35\left(0.65\,e_{FTP} + 0.35\,e_{EDC}\right) \tag{2.15}$$

$$e_{eur} = 0.5\left\{\left[1/40 \cdot \left(e_{C100} - e_{M2}\right) \cdot (70-60) + e_{M2}\right] + e_{EUDC,eur}\right\} \tag{2.16}$$

$$e_{fhw} = 0.5\left\{\left[1/40 \cdot \left(e_{C100} - e_{M2}\right) \cdot (112-60) + e_{M2}\right] + e_{EUDC,fhw}\right\} \tag{2.17}$$

E	Emission in t/a = f (*Cp, DP, RY, RegY, Or*)	*RY*	Reference year
G	Gasoline engine without catalytic converter	*RegY*	Registration year
C	Gasoline engine with three-way catalytic converter	*C100*	100 km/h constant speed
D	Diesel engine	*M2*	Driving curve of traffic on arterial roads provided by the TÜV
T	Total number of cars = f (*Cp, DP, RY, RegY*)	*M3*	Driving curve of moving traffic on connecting roads provided by TÜV
fd	Frequency distribution = f (*Cp, DP, RY*)	*M4*	Driving curve of moving city traffic provided by TÜV
md	Annual distance driven	*FTP*	US-75 driving cycle
Or	Operation on each road type in percent	*EDC*	European driving cycle
e	Partial emission factor in g/km as f (*Cp, D, RegY, RY, Or*)	*EUDC*	New European driving cycle
		iur	Intra-urban roads
Cp	Combustion process	*eur*	Extra-urban roads
DP	Displacement	*fhw*	Federal highways

There are three types of road classification for such a prognosis:

- federal highways road type index *fhw*
- extra-urban roads road type index *eur*
- intra-urban roads road type index *iur*

2.5.4 Results

Figs. 2.33 to 2.42 show results (Source: Metz).

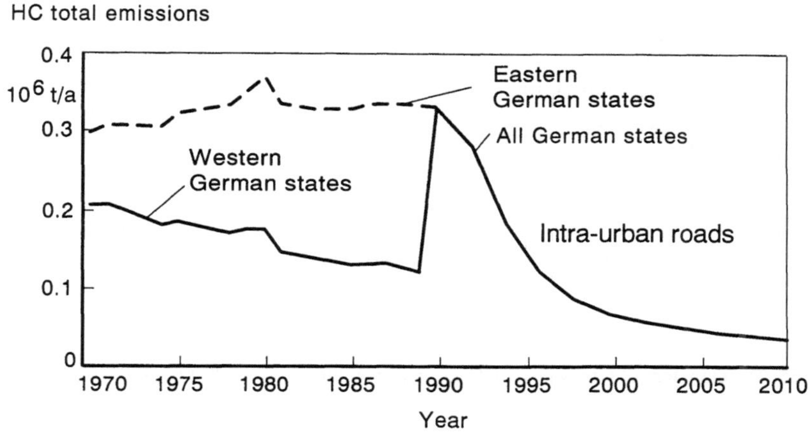

Fig. 2.33: Development of total intra-urban vehicle HC emissions in all states of Germany

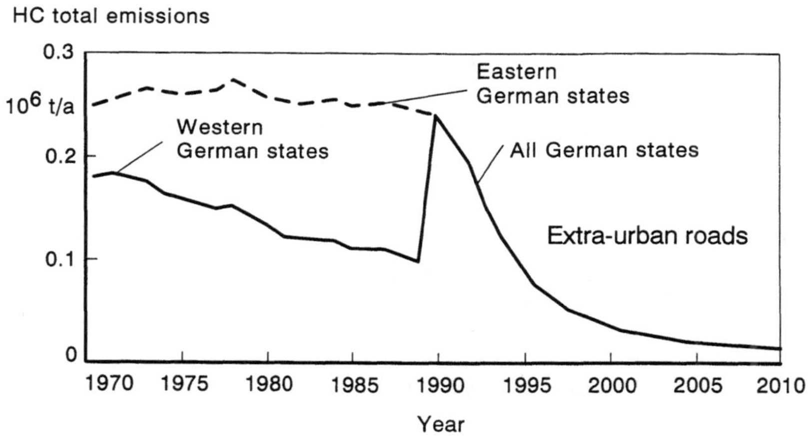

Fig. 2.34: Development of total extra-urban vehicle HC emissions in all states of Germany

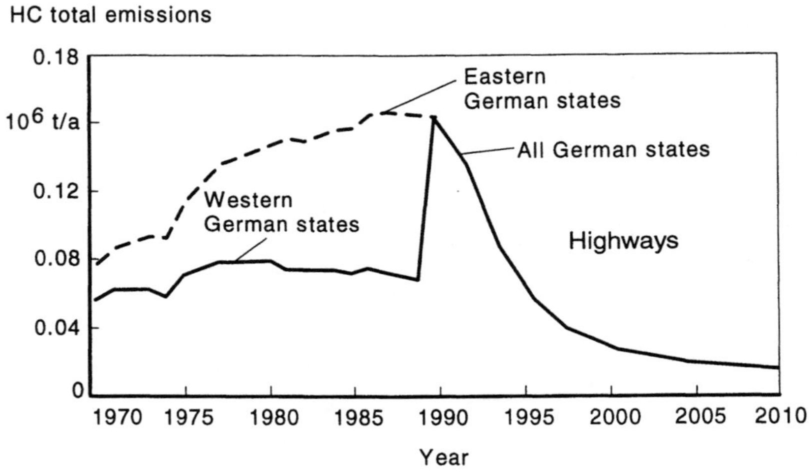

Fig. 2.35: Development of total vehicle HC emissions on highways in all states of Germany

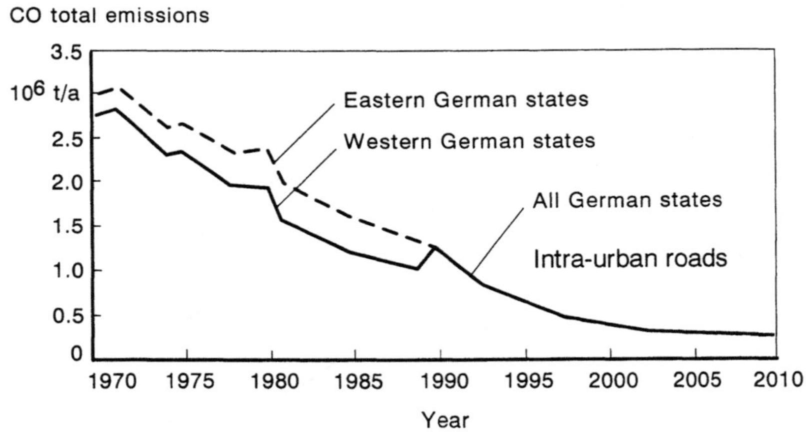

Fig. 2.36: Development of total intra-urban vehicle CO emissions in all states of Germany

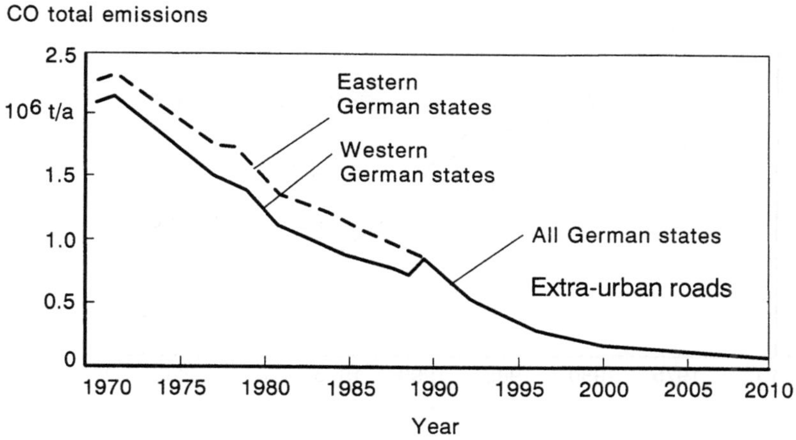

Fig. 2.37: Development of total extra-urban vehicle CO emissions in all states of Germany

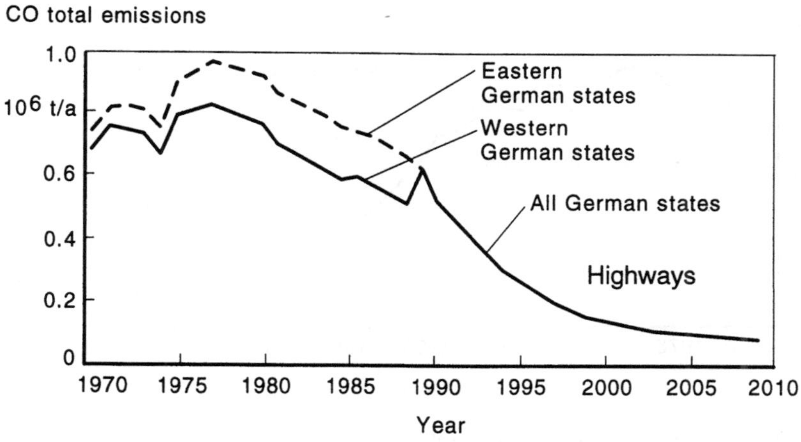

Fig. 2.38: Development of total vehicle CO emissions on highways in all states of Germany

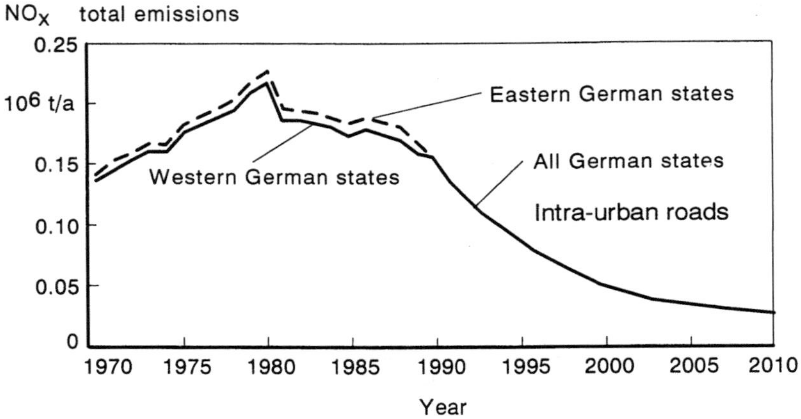

Fig. 2.39: Development of total intra-urban vehicle NO_x emissions in all states of Germany

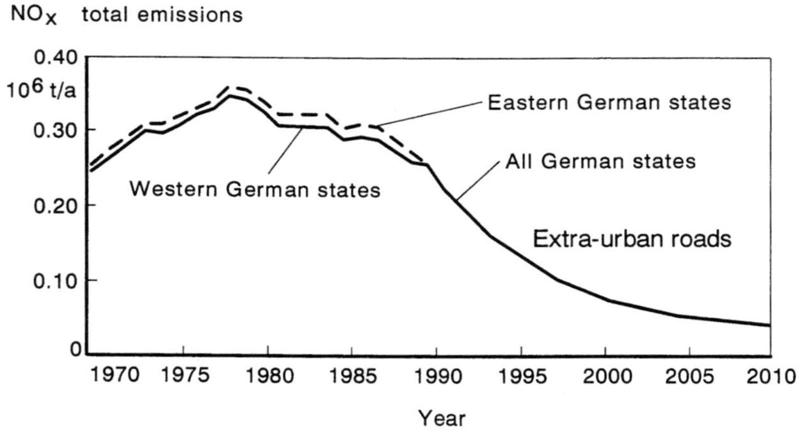

Fig. 2.40: Development of total extra-urban vehicle NO_x emissions in all states of Germany

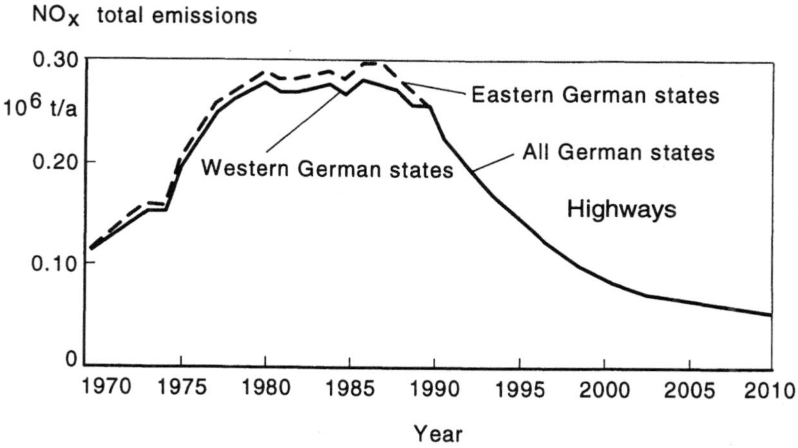

Fig. 2.41: Development of total vehicle NO_x emissions on highways in all states of Germany

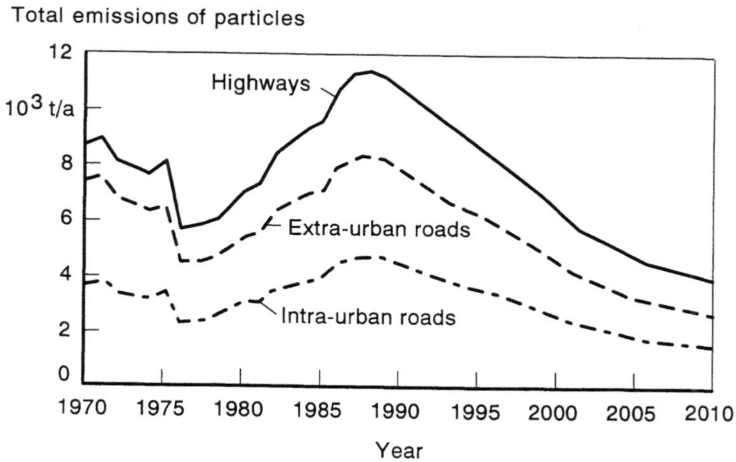

Fig. 2.42: Development of total vehicle particle emission relative to road types

2.5.5 Consequences

The results show the necessity of a differentiated consideration of the total emission rates and the annual distance driven per combustion process and registration year, as emission rates rise with increasing age while at the same time the annual distance driven drops.

The introduction of three-way catalytic converters in gasoline engines and emission improvement in new passenger vehicles with diesel engines will reduce total exhaust gas emissions of all relevant components despite the growing number of passenger vehicles which will continue to rise until the year 2000.

Table 2.12 shows prognosticated reductions in total exhaust-gas emissions of passenger vehicles from 1991 on.

Table 2.12: Total emission prognosis of passenger vehicles

Parameter	Reduction by % until year 2000	Reduction by % until year 2010
CO	65	79
HC	78	89
NO_x	64	79
Particles	49	63
Particles with low-sulfur diesel fuel	59	74
SO_2	-	11
SO_2 with low-sulfur diesel fuel	12	50
SO_2 additionally with low-sulfur gasoline	27	65

For the time being CO_2 emissions will increase until the year 2000 by 7 %, then drop by 8 % below the level of 1991 in the year 2010.

3 Natural and Anthropogenic Emissions on a Global and Country-Related Scale and the Resulting Pollutant Concentrations in the Atmosphere

3.1 Introduction

Apart from various gases which are emissions caused by man, there are also emissions of natural origin, a fact frequently disregarded. To be able to estimate environmental pollution caused by exhaust gases reliable figures for the contribution of the individual sources are required. Along with classifying emissions caused by man (anthropogenic) according to causation groups (emission maps), the contribution of natural sources to the total amount of those trace gases emitted into the atmosphere which also occur in automobile exhaust gas is also of interest. In this matter, consideration must be given to globally calculated emissions, but also to emissions occurring within the boundaries of individual countries, e.g., the Federal Republic of Germany.

The dispersal (transmission) of the emitted gases takes mainly place in the lowest layer of the atmosphere, the so-called troposphere. Weather events, too, take place in the troposphere. Fig. 3.1 gives an overview of the lower atmosphere. Air pressure and temperature relative to altitude are also included. Some cloud formations and mountains have been included for graphic illustration.

Trace gases, whether of anthropogenic or natural origin, are generally emitted on the ground or close to the ground, refer to Chap. 1. Dispersal in the atmosphere occurs by way of horizontal or vertical transport. On a global scale horizontal transport along the parallels of latitude takes place relatively rapidly; transport along the meridians, however, is considerably slower, and vertical transport very slow. Thus, it takes gases emitted by a source an average of 20 days from the ground to the tropopause. In the stratosphere dispersal time into the uppermost layers takes several years due to the positive temperature gradient. In individual cases dispersal time depends on meteorological conditions.

The distance covered depends on wind velocity, turbulent diffusion rates and mean residence time of the particular trace gas in the atmosphere. Mean residence time is the ratio of the amount of gas present in the atmosphere and its production or decomposition rate under stationary conditions.

Mean residence times or life-spans of the individual trace gases in the atmosphere depend very much on local weather conditions and the presence of suitable reaction partners. Only with trace gases with markedly long life-spans - > 1 month - are these influences averaged out on a global scale.

Table 3.1 gives an overview of the atmospheric life-spans of the gases discussed here.

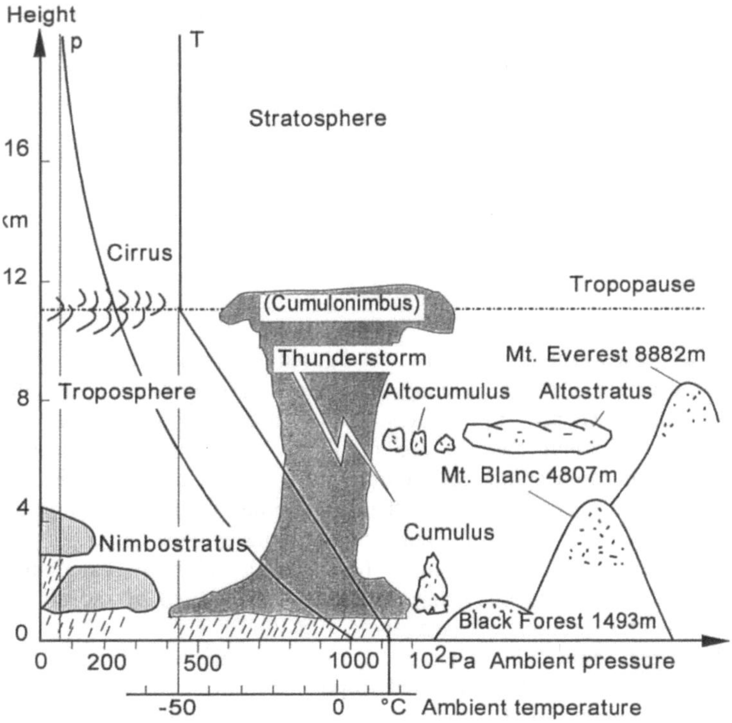

Fig. 3.1: Lower atmosphere with standardized pressure and temperature profile and some cloud formations. As a comparison some mountains were included schematically

Table 3.1: Compilation of the atmospheric life-spans of some important gases

Substance	Formula	Atmospheric life-span
Methane	CH_4	approx. 7 years
Non-methane hydrocarbons	NMHC	some hours to several days
Carbon monoxide	CO	approx. 60 days
Carbon dioxide	CO_2	2-4 years; 200 years in exchange for the part contained in the oceans
Nitrogen oxide	NO	3-30 hours
Nitrogen dioxide	NO_2	1-2 days
Dinitrogen oxide	N_2O	100-200 years
Sulfur dioxide	SO_2	approx. 5 days
Ozone	O_3	35-40 days in clean air, several hours in polluted air

Emissions may be classified into two or three groups, refer to Table 3.2.

Table 3.2: Classification of emissions

Natural emissions	–	natural emissions, not influenced by man
In a more extended sense	–	emissions caused by man interfering with nature, e.g., in agriculture, cattle breeding and clearing forests by burning
Technical-anthropogenic emissions	–	anthropogenic emissions, caused by industrial processes, by the generation of energy and heat and by traffic

As the boundary between the first two groups is fluid, these two groups can be combined as natural emissions and the differentiation is made only between natural emissions in this more extended sense and anthropogenic emissions in a narrower sense (technical-anthropogenic emissions).

When considering and discussing these emissions chemical reactions in the air must also be taken into consideration. This applies mainly when another emitted component is transformed into one of the gases considered here and plays a role as a secondary product.

Natural emissions primarily develop as a result of the formation and decomposition of biomass. Closely related to this is the activity of microorganisms which emit large amounts of gases. Also plants, in particular trees, emit hydrocarbons.

A third important source - and at the same time sink - are air-chemical reactions, involving photochemical reactions as an important factor. A major part of the gases emitted by natural and anthropogenic sources undergoes air-chemical transformations. Frequently, residence time in the atmosphere is limited by air-chemical reactions with the latter depending to high degree on temperature, solar radiation and humidity. During this, the same substances are formed which are also emitted by natural and anthropogenic sources. Thus, the atmosphere itself represents a significant source of trace gases. E.g., if the effect of a reduction of anthropogenic sources on total pollution is to be determined for a certain substance, the secondary products must be taken into consideration. The common factor for all these sources is their dependency on seasons, climatic zones and weather conditions.

Concentrations with higher values than in the surrounding area for which anthropogenic emissions can be excluded as source are indicators of natural emissions. Bodies of water may be considered as sources when the concentration of a dissolved gas in the water is higher than it would be if there were an equilibrium with the concentration of this gas in the air above it.

Source strengths may be determined by calculating the production rates required for maintaining the concentrations observed in known transport and decomposition processes. Another method consists of directly measuring the formation or consumption of gases on plants in closed containers with a defined air through-put (lysimeter). These types of measurements can be carried out in the laboratory and in the natural habitat of the plant. In the latter case plants or parts of them are temporarily enclosed in glass or foil.

All these methods not only require elaborate equipment and a lot of time but they also contain numerous possible sources of faults. Further faults may be added when applying isolated results such as these to larger areas or even on a global scale. This explains the in part extreme differences between data of different authors.

A (as yet rough) control is the observation of global substance cycles, often called mass balances. These balances comprise all known sources and sinks. The sum of all sources must be equal to the sum of all sinks, as long as the concentration of the atmospheric substance considered remains constant.

Determining the natural emission values of a country such as Germany presents a particular problem, as only few data are available. Frequently, all that is available are global estimates which in part only exist as area-related data, e.g., for certain forms of soil utilization or soil composition.

One possibility of proceeding is, as a first step, to find emission values per area for conditions as similar as possible, i.e., humid-temperate climatic conditions (such as in the FRG) and then calculate from global data for the area in question.

If the relevant areas are known , corresponding emissions can be calculated in the second step. Difficulties might arise when different criteria are used for the division of the area according to type of utilization. Frequently, sufficiently verified data even for anthropogenic sources are not available.

3.2 Sources and Source Strengths of Individual Trace Gases

3.2.1 Hydrocarbons

The term "hydrocarbons" comprises a number of individual substances with different properties. Of these, methane (CH_4) occupies a special place due to its frequent occurence and its inertness to air-chemical reactions. In emission and air quality measurements one therefore frequently differentiates between "total hydrocarbons" (THC) and "non-methane hydrocarbons" (NMHC). Depending on the problem at hand individual HC substances or substance groups such as polycyclic aromatic hydrocarbons (PAH) must be examined.

With the exception of methane whose mean residence time in the atmosphere is estimated to be approx. 7 years and which has a global background concentration of a mean of 1.6 ppm, residence times of most of the other hydrocarbons are only few hours to several days, so that globally no uniform concentrations occur. In street canyons total HC concentrations of, e.g., 3 - 5 ppm have been observed (refer to Chap. 4). At present, an increase in the methane content of the atmosphere of ≈ 1.5 % per year (analogous to + 0.02 ppm/a) has been observed.

Tables 3.3 and 3.4 list the most important global sources of hydrocarbon emissions divided into methane and non-methane hydrocarbons.

On a global scale emissions of natural sources outweigh anthropogenic ones by far with approx. 84 % and 94 % for methan and NMHC respectively. It is very likely that other NMHC sources, not listed here, exist in nature.

Even in Germany approx. half to two thirds of hydrocarbon emissions come from natural sources (Table 3.5). This table indicates total HC including methane.

Table 3.3: Global annual methane emission (ranges based on reference literature are in parenthesis). Figures in % related to 10^6 t C/a.

Sources		Emissions in 10^6 t CH_4/a	10^6 t C/a	%
Extended natural sources:	bodies of water	4 (1-7)	3	1
	swamps and wetlands	34 (11-57)	26	10
	rice fields	83 (45-120)	62	24
	ruminants	86 (72-99)	65	25
	termites	4 (2-5)	3	1
	burning of biomass	75 (53-97)	56	21
	other	8 (6-10)	6	2
Σ Extended natural sources		294 (190-395)	221	**84** (80-88)
Anthropogenic sources:	natural gas leaks	24 (19-29)	18	7
	mining	30 (25-35)	23	9
Σ Anthropogenic sources		54 (44-64)	41	**16** (12-20)
Total		348 (234-459)	262	**100**

Table 3.4: Global annual emission of non-methane hydrocarbons (NMHC) (mixtures of different hydrocarbons), ranges are listed in parenthesis. Figures in % relative to 10^6 t C/a

Sources		Emissions in 10^6 t HC/a	10^6 t C/a	%
Natural sources:	trees (terpenes and isopren)	941 (565 - 1317)	830	94
Σ Natural sources		941 (565 - 1317)	830	**94** (91 - 95)
Anthropogenic sources:	vehicle emissions including fuel vapors	40 (32 - 48)	34	4
	solvents	10 (8 - 12)	8	1
	other	14 (11 - 17)	12	1
Σ Anthropogenic sources		64 (51 - 77)	54	**6** (5 - 9)
Total		1005 (600 - 1400)	884	**100**

Fig. 3.2 shows the data calculated from Table 3.5 in the form of a pie chart, hypothetically using passenger vehicle emissions of the year 2010 based on the emission prognosis of Chap. 2.5, refer to Table 2.12, and omitting the reduction of emissions from other anthropogenic sources. In this case contribution of passenger vehicle emissions is negligibly small.

In a conclusion, Fig. 3.3 provides an overview of the natural and anthropogenic emissions of volatile organic compounds in European countries.

Table 3.5: Hydrocarbon emissions including methane in Germany (without the new states). Figures in % relative to 10^6 t C/a; passenger vehicles without catalytic converter

Sources		Emissions in 10^6 t HC/a	10^6 t C/a	%
Extended natural sources:	forests	1.4 (0.6 - 2.2)	1.2	39
	ruminants	0.7 (0.4 - 1.0)	0.5	16
	swamps	< 0.1	< 0.1	1
Σ Extended natural sources		2.1 (1.0 - 3.2)	1.7	**56** (44 - 65)
Anthropogenic sources:	power plants, piped heat	0.01 (-)	0.009	< 1
	industry	0.45 (0.35 - 0.55)	0.39	12
	households, small consumers	0.52 (0.35 - 0.70)	0.46	15
	transportation (passenger and commercial vehicles)	0.58 (0.40 - 0.75)	0.49	16
	other transportation	0.06 (0.04 - 0.08)	0.05	1
Σ Anthropogenic sources		1.6 (1.2 - 2.0)	1.4	**44** (35 - 56)
Total		3.7 (2.2 - 5.2)	3.1	**100**

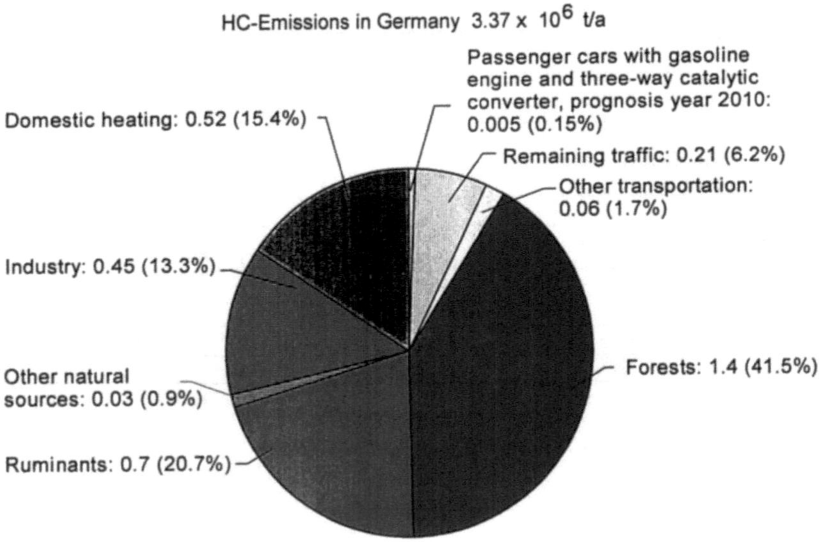

Fig. 3.2: Hydrocarbon emissions in Germany, vehicles with gasoline engines and three-way catalytic converters, calculated with data of the emission prognosis of Chap. 2.5 (Table 2.12) and based on the hypothetical assumption that emissions of other anthropogenic sources are not reduced

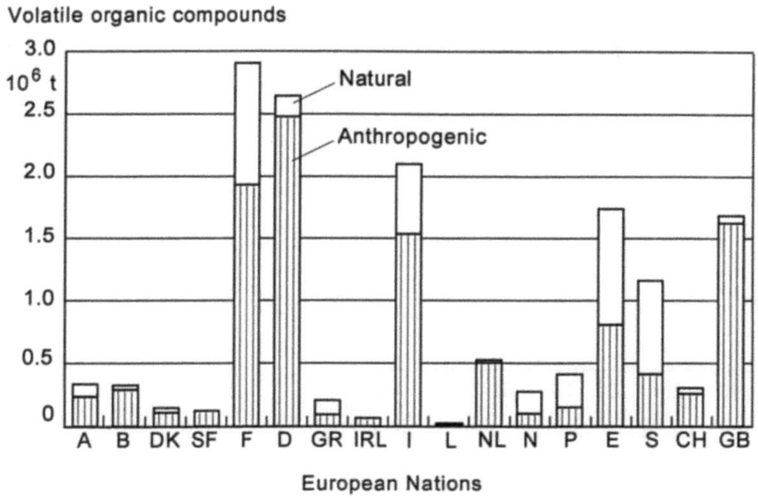

Fig. 3.3: Natural and anthropogenic emissions of volatile organic compounds of European OECD countries for the year 1980, source: OECD report, Paris 1990

3.2.2 Carbon Monoxide

In the troposphere carbon monoxide has a mean life-span of two months. This is not sufficient for the formation of a globally uniform concentration. Nevertheless, almost identical CO concentrations are found in zones of the same geographic latitude.

With 0.1 to 0.2 ppm the northern hemisphere has distinctly higher concentrations than the southern hemisphere (50 to 80 ppb). Fig. 3.4 shows how the concentration of carbon monoxide is dependent on latitude. This difference is due to the fact that CO is formed mainly on continents or by air-chemical reactions of hydrocarbons emitted on the mainland. Two-thirds of the earth's land mass is in the northern hemisphere, this is also where the greatest proportion of anthropogenic CO is produced.

In comparison, the sea, where carbon monoxide is also formed, plays only a subordinate role. CO concentrations vary seasonally with maxima in winter and spring, minima in summer and fall. As a result, differences between the two hemispheres reach maximum values in the northern winter to spring, whereas in the in the northern summer concentration differences are almost equalized.

Table 3.6 gives an overview of known CO sources, the extended natural sources being predominant with a proportion of 79 % of total CO generation. Uncertainties resulting from the ranges must be taken into consideration.

The air-chemical transformation of hydrocarbons is also included here as source.

Carbon monoxide is mainly degraded by predominantly air-chemical reactions mostly with the OH radical and by microbiological processes in the uppermost layers of the soil.

Fig. 3.5 gives an overview of the mass balance of the global CO.

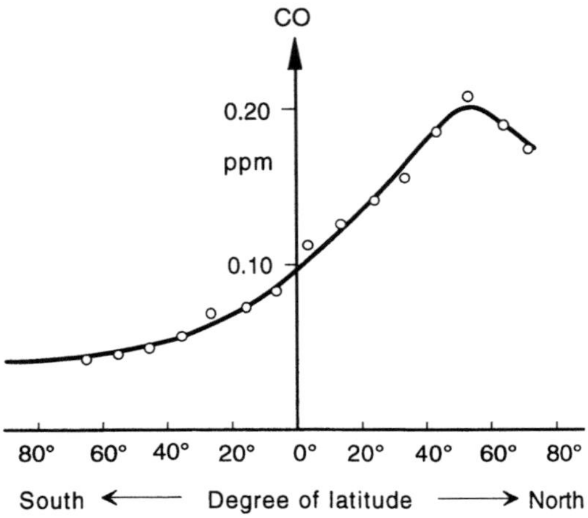

Fig. 3.4: Mean CO concentration in the troposphere relative to geographic latitude (smoothed)

Table 3.6: Global annual carbon monoxide emissions

Sources		Emissions in 10^6 t CO/a		10^6 t C/a	%	
Extended natural sources:	oceans	95	(10 - 180)	41	1	
	plants	50	(30 - 70)	21	3	
	oxidation of methane	650	(370 - 930)	279	20	
	oxidation of higher HC	850	(400 - 1300)	364	25	
	burning of biomass	1000	(400 - 1600)	429	30	
Σ Extended natural sources		2645	(1200 - 4100)	1134	**79**	(69 - 84)
Anthropogenic sources:	combustion of fire-wood	60	(45 - 75)	26	2	
	and fossil fuels	640	(480 - 800)	274	19	
Σ Anthropogenic sources		700	(525 - 875)	300	**21**	(16 - 31)
Total		3345	(1700 - 5000)	1424	**100**	

Table 3.7 gives an overview of the CO emissions in the Federal Republic of Germany. Here, anthropogenic sources predominate over natural ones with a direct share of approx. 66 %, of which approx. 9 % are caused by oxidation of anthropogenic hydrocarbons.

When determining CO it is assumed that the same proportions of HC emissions are transformed into CO as on a global scale. For NMHC the amounts emitted in Germany are used as a basis. For methane, global emissions relative to the surface areas must be used because of methane's long life-span. Again, uncertainty factors must be taken into consideration. As far as the burning of biomass is concerned, only the burning of agricultural waste relative to the areas agriculturally utilized is considered. In addition, it is assumed that only half of the global proportion of agricultural waste is burned, as in Germany it is put to different use.

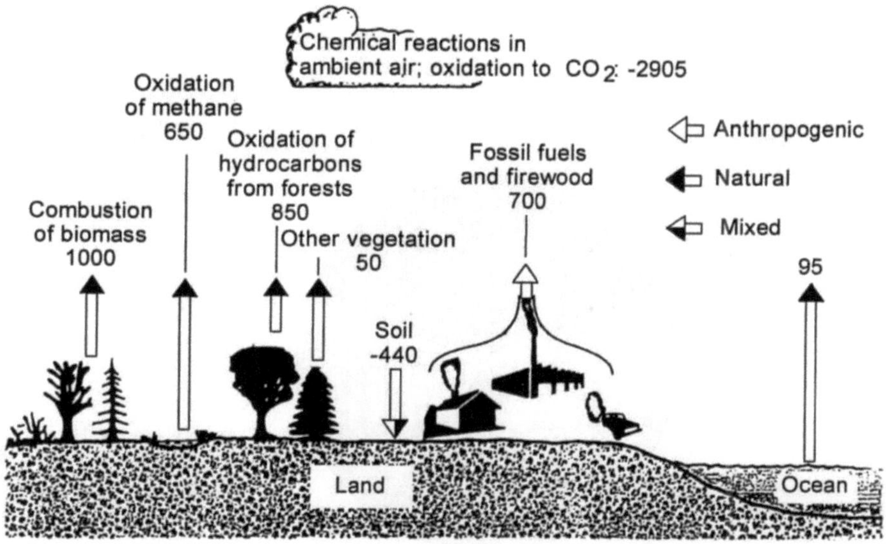

Fig. 3.5: Global carbon monoxide mass balance in million tons per year with sources and sinks

Table 3.7: Carbon monoxide emissions in Germany (without the new states, gasoline engines without catalytic converter)

Sources		Emissions in 10^6 t CO/a	10^6 t C/a	%
Extended natural sources:	plants	0.3 (0.2 - 0.4)	0.14	2
	oxidation of methane	0.2 (0.1 - 0.3)	0.09	1
	oxidation of HC from forests	1.2 (0.6 - 1.8)	0.53	10
	burning of biomass	1.4 (0.6 - 2.2)	0.6	11
Σ Extended natural sources		3.1 (1.5 - 4.7)	1.36	**24** (16 - 33)
Anthropogenic sources:	oxidation of anthropogenic HC	1.2 (0.6 - 1.8)	0.53	10 (5 - 14)
	power plants, long-distance heating	0.03 (0.02 - 0.04)	0.01	0.2
	industry	1.12 (0.8 - 1.5)	0.48	9
	households / small consumers	1.72 (1.2 - 2.2)	0.74	14
	road transport (passenger and commercial vehicles)	5.15 (3.6 - 6.7)	2.20	41
	other transportation	0.25 (0.2 - 0.3)	0,10	2
Σ Anthropogenic sources (direct emission)		8.2 (5.8 - 10.7)	3.53	**76** (57 - 84)
Total		12.5 (8.7 - 19.2)	5.4	**100**

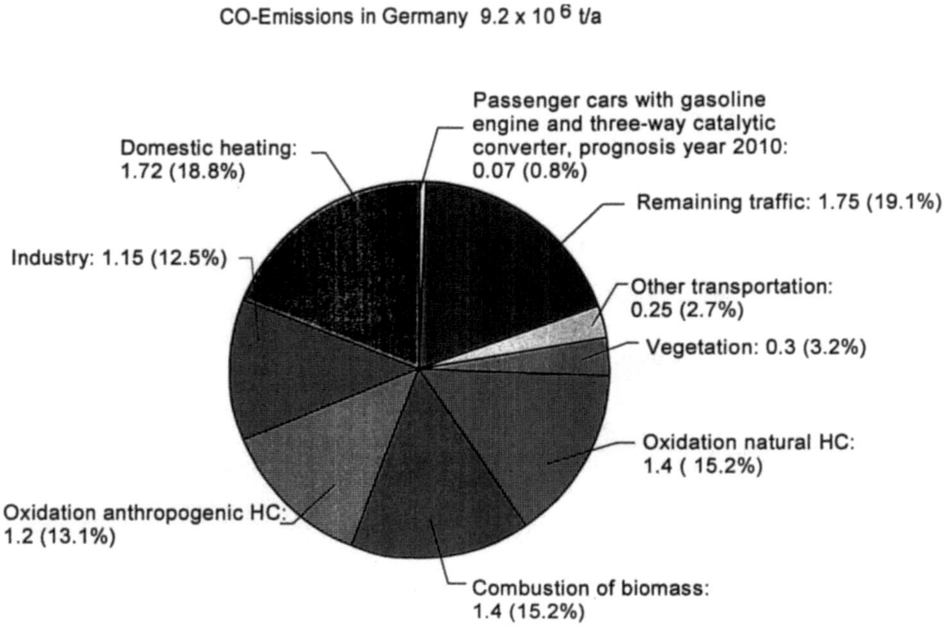

Fig. 3.6: Carbon monoxide emissions in Germany (gasoline engines with three-way catalytic converter acc. to the prognosis for the year 2010, Table 2.12), conditions as in Fig. 3.2

Clearing woods by burning, forest fires, bush fires etc. are virtually insignificant and have not been taken into consideration. Unlike global conditions, oxidation of anthropogenic HC plays a role. It is assumed that one third of the anthropogenic hydrocarbons listed in Table 3.5 are oxidized to CO.

Fig. 3.6 shows data calculated for a case comparable to Fig. 3.2 in a pie chart, with the emission prognosis of Chap. 2.5 (Table 2.12) used for the year 2010 for the passenger vehicle contribution, which then becomes practically zero.

3.2.3 Carbon Dioxide

After nitrogen, oxygen and argon the fourth most frequent gas in the atmosphere is carbon dioxide (0.03 vol.-%), refer to Table 2.1. Its concentration varies slightly depending on the seasons, overlapping with a continuous rise which has been observed over a longer period. Fig. 3.7 shows the annual CO_2 increase observed by the Mauna Loa Observatory on Hawaii. The seasonal variations caused by the vegetation period can be clearly seen, as also the continuous long-term rise (refer to Fig. 5.18 for comparison).

Large amounts of carbon dioxide are released on land by the plants' metabolism and the decomposition of biomass and in the sea by degasification processes (Table 3.8).

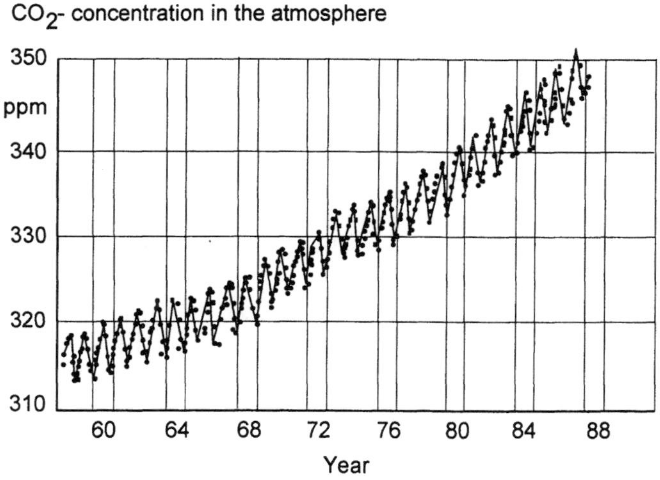

Fig. 3.7: Annual increase of carbon dioxide concentrations in the atmosphere with overlapping seasonal variations, (refer to Fig. 5.18)

Table 3.8: Global annual carbon dioxide emissions

Sources		Emissions in 10^9 t CO_2/a	10^9 t C/a	%
Extended natural sources:	ocean	385 (311 - 458)	105	45
	vegetation	227 (183 - 272)	62	26
	soil	227 (183 - 272)	62	26
	burning of biomass	9 (7 - 12)	2.5	1
Σ Extended natural sources		848 (684 - 1014)	231.5	**98**
Anthropogenic sources:	combustion of fire wood	2 (1.5 - 2.5)	0.5	< 1
	and fossil fuels	19 (17 - 21)	5.2	2
Σ Anthropogenic sources		21 (18.5 - 23.5)	5.7	**2**
Total		869 (702 - 1038)	237.2	**100**

In the atmosphere carbon exists mainly in the form of carbon dioxide. The mass balance is not equalized. CO_2 accumulates in the atmosphere, refer to Fig. 3.7 and Fig. 5.18. Large amounts of it are stored in the depths of the oceans.

The mean life-span of CO_2 in the atmosphere is approx. 3 years. Anthropogenic CO_2 emissions are merely ≈ 2% of global CO_2 emissions, refer to Chap. 5, nonetheless.

On the other hand, there is an equally large consumption of CO_2 for the build-up of biomass, so that the natural carbon situation is mostly balanced, s. Fig. 3.8.

With 50 %, the anthropogenic proportion is considerably higher in the Federal Republic of Germany (FRG) (Table 3.9). Errors for anthropogenic source data are ≤ 10 % without exception, therefore merely a few ranges are indicated. Figures have been calculated from consumption or production data.

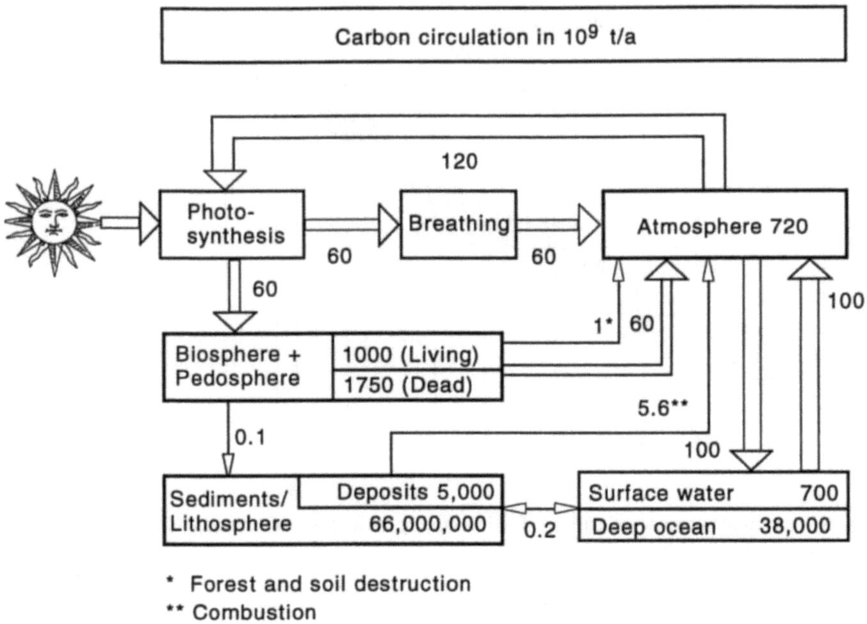

Fig. 3.8: Global carbon mass balance in billion tons of carbon per year, sources and sinks

Table 3.9: Carbon dioxide emissions in Germany without the new states

Sources		Emissions in 10^6 t CO_2/a	10^6 t C/a	%
Natural	soils	457 (365 - 550)	125	25
sources:	vegetation	457 (365 - 550)	125	25
Σ Natural sources		914 (730 - 1100)	250	**50** (45 - 55)
Anthropoge-	combustion of:			
nic sources:	gasoline	70	19	4
	diesel fuel	42	11	2
	fuel oil	164	45	9
	mineral coal	282	77	16
	lignite	220	60	12
	natural gas	102	28	6
	cement production	16	4	1
Σ Anthropogenic sources		896	244	**50** (45 - 55)
Total		1810 (1626 - 1996)	494	**100**

The carbon balance is upset by the combustion of fossil fuels and the cutting of tropical forests. Approx. half of the CO_2 annually released in this manner accumulates in the atmosphere. If CO_2 concentrations continue to rise, considerable effects on the global climate are prognosticated if they increase above ≈ 400 ppm, as model calculations reveal (refer to Chap. 5). Recently, some researchers dispute this.

3.2.4 Nitrogen Oxides

During combustion processes nitrogen oxides are predominently emitted as NO, which is, however, transformed relatively rapidly (in only a few hours) into NO_2, mainly. Oxidation by ozone predominates here. For emission data NO and NO_2 are usually summarized by the term NO_x. This does not include dinitrogen monoxide (N_2O). NO_x has a mean life-time of 1 to 2 days. This should therefore entail a discussion of further secondary products. In contrast to NO and NO_2 dinitrogen monoxide - N_2O - whose atmospheric life-time is estimated to be 100 to 200 years, is very slow to react.

Table 3.10 gives an overview of global NO_x sources. Extended natural sources predominate with a proportion of approx. 68 %, with uncertainty factors needing to be taken into consideration again.

Table 3.10: Global annual NO_x emissions (NO_x expressed as NO_2)

Sources		Emissions in 10^6 t NO_2/a		10^6 t N/a	%	
Extended natural sources:	burning of biomass	39.4	(13 - 79)	12	22	
	lightning	26.3	(7 - 66)	8	15	
	soils	26.3	(13 - 53)	8	15	
	mineral fertilizer	6.5	(4 - 9)	2	4	
	oxidation of NH_3	18.1	(3 - 33)	5.5	10	
	oceans	1.6	(0.7 - 2.6)	0.5	1	
	from the stratosphere	1.6	(0.7 - 2.6)	0.5	1	
Σ Extended natural sources		119.8	(39 - 240)	35.8	**68**	(50 - 80)
Anthropogenic sources:	fossil fuel burning	57	(36 - 76)	17.3	**32**	(20 -50)
Total		176.8	(76 - 315)	53.1	**100**	

NO_2 concentration is predominantly reduced by means of moist and dry deposition after air-chemical reactions with OH radicals, including transformation into nitric acid and nitrate aerosol. Direct absorption by soils and plants replaces fertilizing. Apart from this, direct deposition, particularly near emission sources, also plays a role. Fig. 3.9 depicts global NO_x mass balance.

Nitrogen oxides play an important part in the formation of photooxidizing agents.

Nitrogen oxides from near-ground sources such as motor vehicles are partially removed from the atmosphere before they can contribute to the formation of photooxidizing agents. Accordingly, the significance of an NO_x emission source for the formation of photooxidizing agents may not be judged only according to the annually emitted amount of NO_x.

Table 3.11 shows the estimated emissions of nitrogen oxides (NO_x) in the FRG. According to it, approx. 9 % stem from extended natural sources and approx. 91 % from anthropogenic sources. The contribution of vehicles with gasoline engines without catalytic converter amounts to approx. 20 %, with catalytic converter to approx. 1 %.

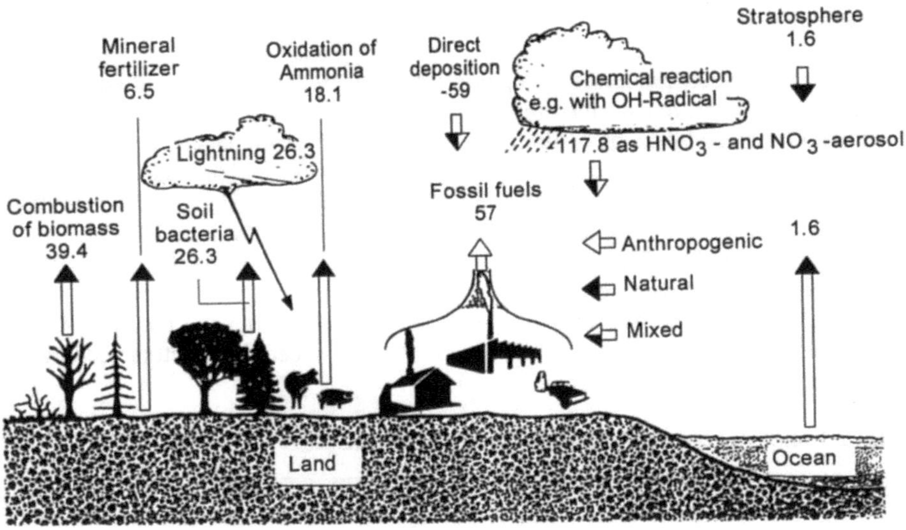

Fig. 3.9: Global mass balance of the nitrogen oxides NO and NO_2 expressed as NO_2 in million tons per year, sources and sinks

Table 3.11: Emissions of nitrogen oxides (NO_x) in the FRG (without the new states); gasoline vehicles without catalytic converter (NO_x expressed as NO_2)

Sources		Emissions in 10^3 t NO_2/a	10^3 t N/a	%
Extended natural sources:	soils	76 (39 - 150)	23	2
	lightning	26 (7 - 66)	8	1
	burning of biomass	56 (2 - 177)	17	2
	mineral fertilizer	135 (8 - 19)	41	4
	forest fires	3 (2 - 5)	1	< 1
Σ Extended natural sources		300 (150 - 590)	90	**9** (6 - 14)
Anthropogenic sources	power plants, long-distance heating	860 (570 - 1150)	260	25
	industry	430 (290 - 570)	130	13
	households / small consumers	110 (70 - 150)	40	3
	road transport (passenger and commercial vehicles)	1470 (960 - 1980)	440	43
	other transportation	220 (150 - 300)	70	7
Σ Anthropogenic sources (direct emission)		3100 (2100 - 4100)	940	**91** (86 - 94)
Total		3400 (2200 - 4800)	1030	**100**

As in Germany the NO_x concentration in ambient air consistently amounts to approx. 60 ppt (≙ 0,06 ppb) in Germany, oxidation of ammonia as a source of NO_x can probably be ruled out.

Otherwise a total source strength of approx. 200 · 10^3 t NO_x/a, in addition to the sources listed in Table 3.1, would result when transferring global conditions.

Fig. 3.10 shows calculated data as a pie chart, with the passenger vehicle proportion based on the year 2010 according to the emission prognosis of Chap. 2.5, Table 2.12 (refer to Fig. 3.2). Its contribution then becomes negligibly small.

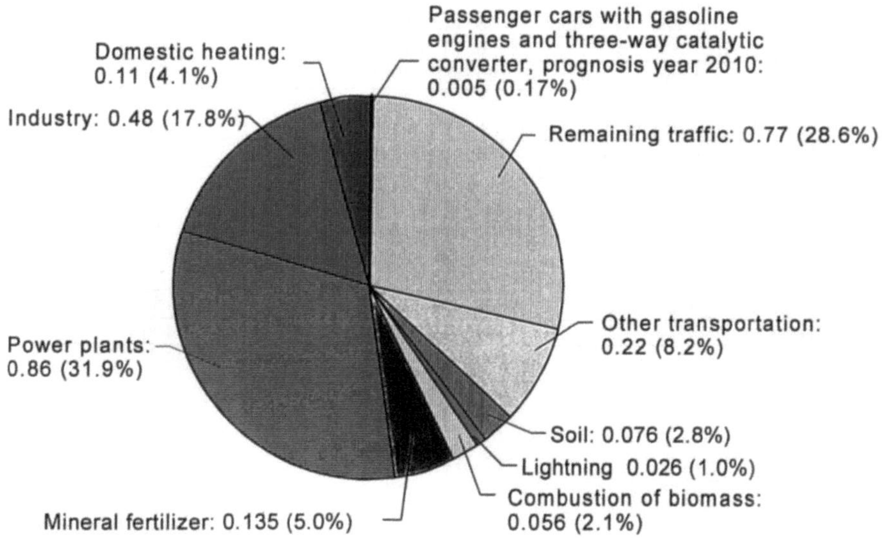

Fig. 3.10: Nitrogen oxide emissions in Germany, vehicles with gasoline engines with three-way catalytic converter, calculated on the basis of the emission prognosis of Chap. 2.5, Table 2.12, conditions as in Fig. 3.2

3.2.5 Sulfur Dioxide

Sulfur and sulfur compounds occur in small concentrations (a maximum of a few percent) in biomass and in fossil fuels. This is the reason why sulfur dioxide is formed during almost all combustion processes. In the biological decomposition of biomass other gaseous sulfur compounds are initially formed, e.g. hydrogen sulfide H_2S or dimethyl sulfide (DMS = $(CH_3)_2S$), which are partially oxidized to sulfur dioxide in the atmosphere. It is assumed that approx. $^3/_4$ of the sulfur volumes indicated in literature for these compounds are transformed into SO_2. Table 3.12. gives an overview of global sulfur sources.

Approx. half of the SO_2 is removed from the atmosphere directly by washing out or dry deposition, the other half is previously transformed into sulfurous acid, sulfuric acid and sulfates; its mean life-span is approx. 5 days. Fig. 3.11 depicts global sulfur dioxide mass balance.

Table 3.12: Global annual sulfur dioxide emissions

Sources		Emissions in 10^6 t SO_2/a	10^6 t S/a	%
Extended natural sources:	volcanoes	40 (20 - 60)	20	10
	oxidation of H_2S and $(CH_3)_2S$ from			
	oceans	56 (48 - 64)	28	14
	coastal areas	15 (10 - 22)	8	4
	tropical forests	25 (15 - 35)	12	6
	swamps and rice fields	38 (22 - 54)	19	10
	fields	6 (4 - 8)	3	1
Σ From oxidation		140 (80 - 200)	70	35
	burning of biomass	6 (4 - 10)	3	2
Σ Extended natural sources		186 (104 - 270)	93	**47** (39 - 53)
Anthropogenic sources:	combustion of coal	128 (116 - 140)	64	32
	combustion of oil	52 (46 - 58)	26	13
	ore beneficiation	22 (20 - 24)	11	6
	others	6 (4 - 8)	3	2
Σ Anthropogenic sources		208 (186 - 230)	104	**53** (47 - 61)
Total		394 (290 - 500)	197	**100**

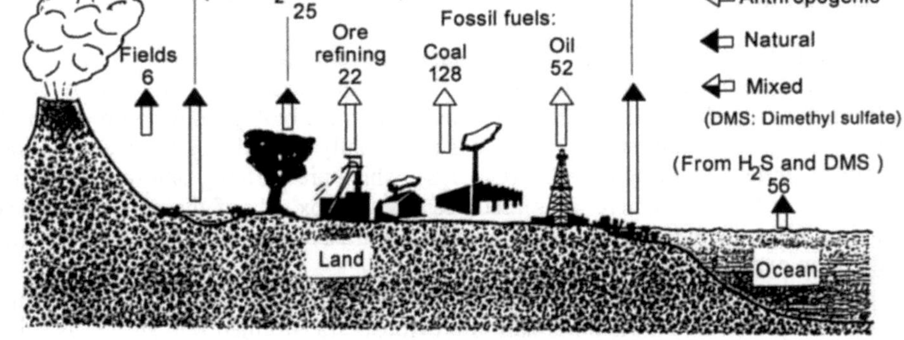

Fig. 3.11: Global sulfur dioxide mass balance in million tons per year

Even on a global scale anthropogenic SO_2 sources outweigh natural ones. In densely populated industrial areas natural emissions can be virtually ignored, as can be seen in the compilation of SO_2 emissions of the FRG (Table 3.13). Road transportation is only a minor factor.

Table 3.13: Sulfur dioxide emisisons in Germany (excluding the new states)

Sources		Emissions in 10^6 t SO_2/a	10^6 t S/a	%
Natural sources:		0.02 (0.01 - 0.03)	0.01	**1**
Anthropogenic sources:	power plants/long-distance heating	1.86 (1.68 - 2.04)	0.93	62
	industry	0.76 (0.68 - 0.84)	0.38	25
	households / small consumers	0.28 (0.24 - 0.32)	0.14	9
	transportation	0.10 (0.08 - 0.12)	0.05	3
Σ Anthropogenic sources		3.0 (2.6 - 3.4)	1.5	**99**
Total		3.02 (2.6 - 3.4)	1.51	**100**

3.2.6 Ozone

Ozone (O_3) is not emitted directly, but is formed in the atmosphere by photochemical reactions with the influence of solar radiation. As ozone formation depends on many, not yet sufficiently researched parameters, a quantitative allocation to individual sources is not yet possible at present. Some of the reactions related to ozone formation are listed below.

a) Photochemical ozone formation from NO_2

$$NO_2 + h\nu \Rightarrow NO + O \tag{3.1}$$

$$O + O_2 + M \Rightarrow O_3 + M \tag{3.2}$$

b) Oxidation of NO to NO_2 with the help of hydrocarbons

$$O + HC \Rightarrow R + OH \tag{3.3}$$

$$OH + HC \Rightarrow H_2O + R \tag{3.4}$$

$$R + O_2 \Rightarrow RO_2 \tag{3.5}$$

$$NO + RO_2 \Rightarrow NO_2 + RO \tag{3.6}$$

c) Decomposition of ozone by oxidation of NO

$$O_3 + NO \Rightarrow NO_2 + O_2 \tag{3.7}$$

d) Oxidation of CO, formation of NO_2

$$CO + OH \Rightarrow CO_2 + H \tag{3.8}$$

$$H + O_2 + M \Rightarrow HO_2 + M \tag{3.9}$$

$$HO_2 + NO \Rightarrow OH + NO_2 \tag{3.10}$$

with:	$h\nu$	solar radiation energy	OH	hydroxyl radical;
	M	gas molecules as impact partners	RO_2	peroxy radical;
	R	alkyl or aryl radical	HO_2	hydroperoxy radical
	RO	alkoxy radical.		

Reactions b) and d) provide NO_2 for the basic reaction a).

Ozone is formed in large amounts in the stratosphere, from where is reaches the troposphere in part by diffusion and by breaking through the tropopause. In addition, ozone can be formed in the troposphere.

Extended periods of intense solar radiation in the short wave range of light (ultraviolet (UV) radiation) and the presence of suitable precursor substances such as nitrogen oxides and hydrocarbons are preconditions for the formation of high O_3 concentrations. These conditions are as a rule fulfilled during stable high pressure weather situations even in our latitudes (approx. 50° north). The mean tropospheric ozone concentration shows a distinct dependency on latitude, i.e., maxima in high latitudes and a wide minimum in the southern tropics at approx. 20° south, s. Fig. 3.12.

The geographic latitude distribution of O_3 concentrations can be explained with the intrusion of high-ozone air from the stratosphere, as observations in the polar areas have shown ("sudden warmings"). The mean life-span of ozone in clean air is estimated to be 35 to 40 days. In polluted air the ozone life-span can drop to a few hours. Globally, $^2/_3$ of the ozone is degraded by air-chemical reactions and $^1/_3$ in the soil.

Altitude profiles of the ozone concentration show that concentrations in near-ground layers can be significantly higher as well as lower as compared to the middle region of the troposphere.

Large amounts of substances leading both to ozone formation and ozone decomposition are emitted by sources on the ground or near the ground (stacks), thus leading to these considerable differences of the ozone concentrations in near-ground air layers.

However, the near-ground air layer is frequently separated from the remaining troposphere by low-level inversions or stable boundary layers. In such situations ozone newly formed below the inversion layer cannot escape into the troposphere and ozone degraded in the soil cannot be replaced from higher layers above the inversion layer.

Fig. 3.13 shows a typical diurnal course in a near-urban industrial area. In cities with high NO and HC emissions the ozone is almost completely decomposed during the night.

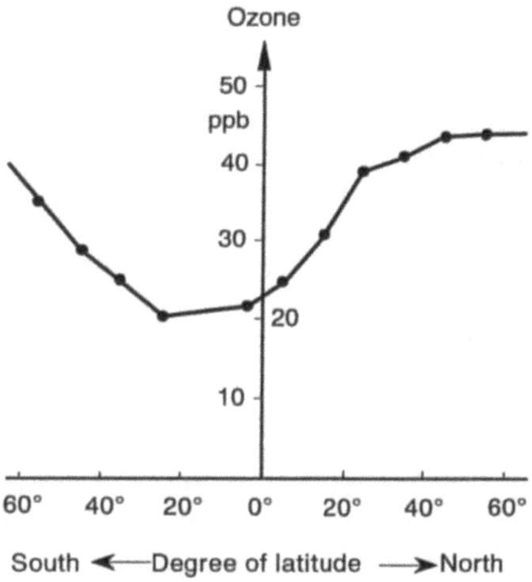

Fig. 3.12: Latitude dependency of the mean tropospheric ozone concentration

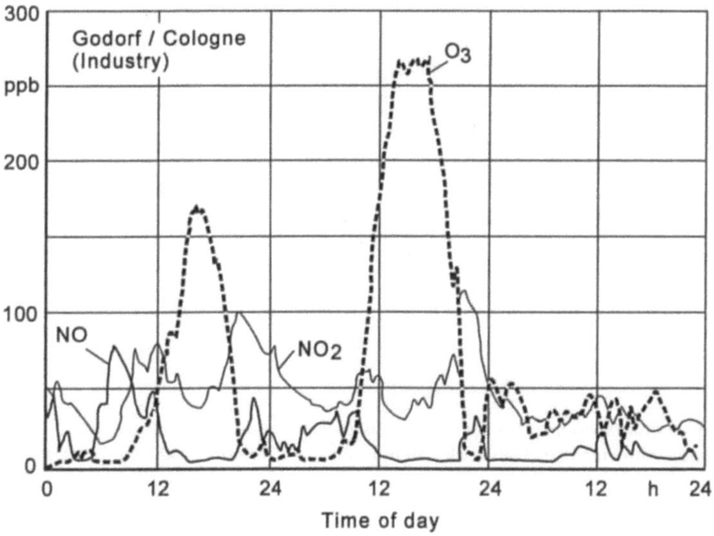

Fig. 3.13: O_3 diurnal courses over a period of 3 days with high O_3 concentrations in a near-urban industrial area. "Old" ozone in the ambient air probably contributed to the higher concentrations on the 2nd day. On the third day there are no high ozone concentrations due to a weather change

According to (3.1) and (3.2), photo-chemical ozone formation begins with the onset of solar radiation. The ozone immediately reacts with the NO and NO_2 present, though. It is only if ozone formation exceeds decomposition in the reaction acc. to (3.7) that the ozone concentration begins to rise with increasing solar radiation, reaching its peak in the afternoons. After sunset the O_3

concentration drops rapidly as the available ozone is decomposed during the oxidation of NO and other trace gases.

In areas with clean air in temperate latitudes ozone is also formed during suitable weather conditions. The emission of reactive natural hydrocarbons from forests is an important factor here. However, the maximum concentrations of approx. 70 ppb remain below those values observed in areas with high air pollution. Again, ozone is decomposed at night, with ozone concentrations of 10 to 30 ppb however remaining, depending on the season and weather conditions (s. Fig. 3.14).

It is assumed that conditions particularly favorable for ozone formation frequently exist if inversion layers are present. In the inversion layer itself, particularly high ozone concentrations exist temporarily, as the ozone can neither escape into the free troposphere nor does it reach the near-ground air layer where it could react with gases emitted by the ground or react with the soil. Fig. 3.15 shows altitude O_3 profiles during an inversion weather situation, observed in the US state of Ohio.

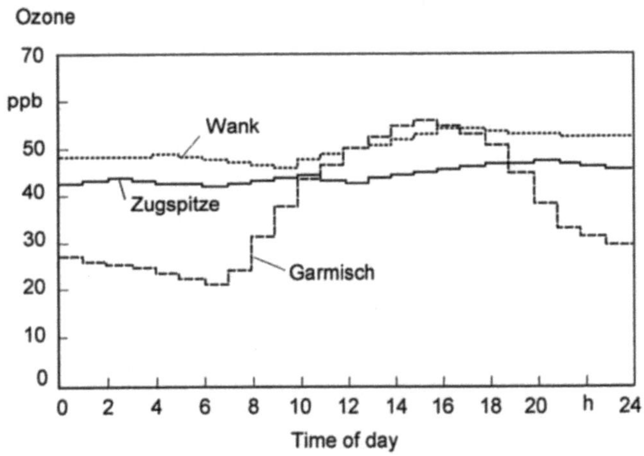

Fig. 3.14: Averaged O_3 diurnal courses for several days with sun and no precipitation from three stations in an area with clean air: Garmisch 740 m (valley), Wank top 1780 m, Zugspitze 2964 m

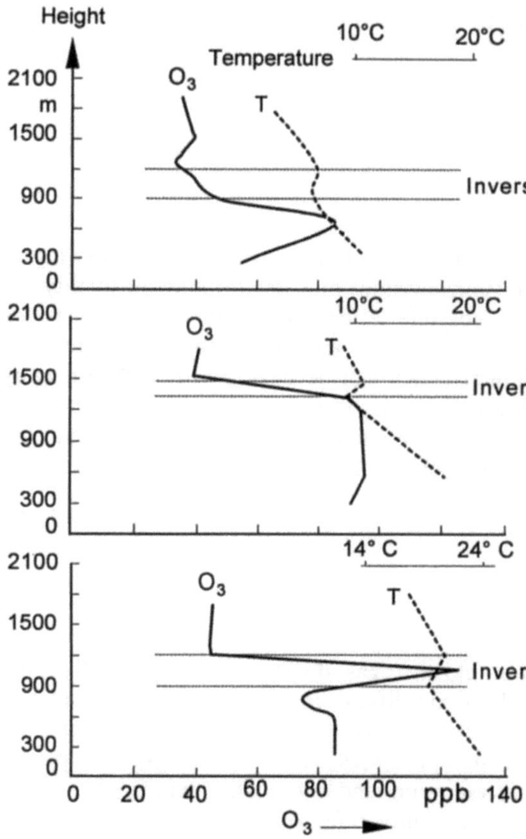

Fig. 3.15: Development of high ozone concentrations in the presence of an inversion layer, i.e., a layer where air temperature T increases with increasing altitude. The dotted lines indicate the inversion layer

4 Air Quality Control

Not the emissions themselves but rather those pollutant concentrations settling in the ambient air after their dispersal (transmission) thus determining air quality are significant for the assessment of possible effects of vehicular exhaust gas emissions on the environment, refer to Chap.1 and 5. If these pollutant concentrations cannot be measured directly, they can be determined mathematically with the help of dispersal models on the basis of emission values available. To assess their effects on human health extreme situations such as in street canyons with dense traffic are frequently drawn upon.

4.1 Dispersion Models

For the mathematical treatment of exhaust gas dispersion the generally very complex real situation is described by simplifying models where the parameters which must be taken into account are reduced to a manageable number. By mathematically simulating the dispersion processes, such dispersion models permit one to establish the link between emission and air quality within the causal relationship emisson - transmission - air quality, refer to Fig. 1.1.

Fig. 4.1 exemplifies the basic functioning of dispersion models, with the "traffic" box containing the emission behavior of vehicles and "meteorology"

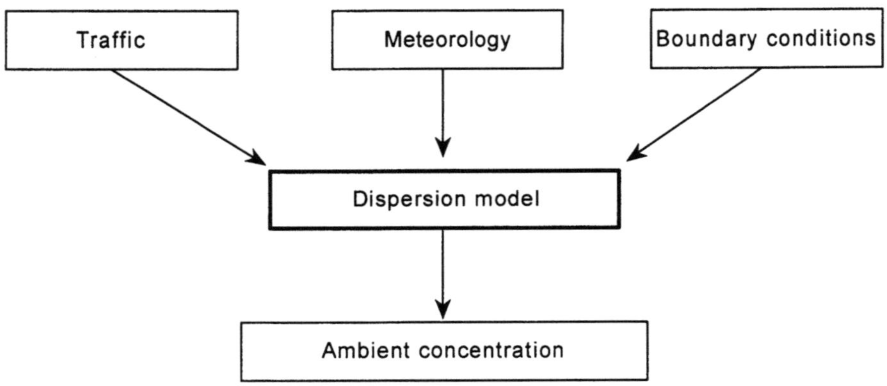

Fig. 4.1: Basic principle of dispersion models

implying parameters such as wind velocity and wind direction. "Boundary conditions" concern, e.g., type of road and the geometry of the dispersion situation at hand.

Naturally, results of model calculations can basically be only as good as the parameters available. The availability of emission data which are as representative as possible is of particular significance here.

The emission data represented in Chap. 2 mostly fulfill this decisive precondition, as emission values for common driving behavior are available for different engine concepts in vehicles. Thus a generalization for all passenger vehicles seems to be possible.

Even when all parameters and boundary conditions are known, a dispersion model can always only treat a certain dispersion situation. This is why different specific models are necessary. For the dispersion of vehicle exhaust gases two cases are mainly significant: street canyons and highways in open country.

4.1.1 Street Canyon Model

A dispersion model for street canyons developed by the Stanford Research Institute (SRI) is also used for assessing pollution situations in the United States by the EPA. In street canyons, i.e., streets with buildings on both sides of the streets and little space in-between the individual buildings, the highest pollution concentrations are to be expected. This is why this case (Fig. 4.2) is particularly important for the determination of maximum exhaust gas pollution.

In a street canyon with width *W*, lined on both sides by buildings of the same height *H*, there are one or more lanes of traffic considered as line sources with the source strength *S*. Furthermore, *x* is the distance of the reference point from the

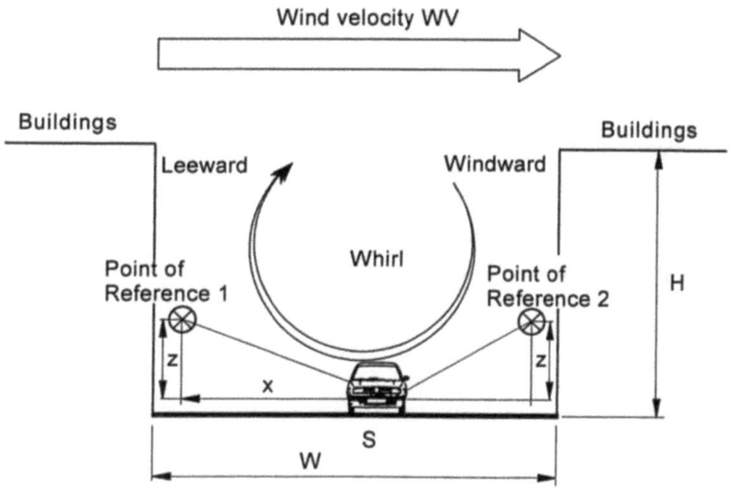

Fig. 4.2: Model of a street canyon

source and z its height above the source. When the wind direction is at right angles to the street the indicated whirl occurs inside the street canyon and causes raised pollution concentrations at reference point 1 (sides of buildings away from the wind = lee-side) as compared to point 2 (sides of buildings towards the wind = luff). When the wind direction is parallel to the street no whirls have been observed, concentrations are then the same at both points 1 and 2.

This treatment of the street canyon is an approximation, which is really only valid for the the space directly above the sidewalks. In addition, buildings lining the streets usually are not of the same height in real street canyons and adjoining buildings frequently have differently shaped roofs. This irregular profile creates more air circulation in the street space.

The model only takes into account the dilution of the exhaust gases; chemical reactions are not taken into consideration. As the atmospheric life-span of the substances examined ranges from hours to days, omission of the chemical reactions could well be considered. Despite its simplifications, the model provides usable information for generalizing observations of pollution situations in street canyons.

For the pollutant concentration C in ambient air at the reference point 1 the following applies:

$$C_1 = \frac{K \cdot S}{(WV + 0.5) \cdot \left(\sqrt{x^2 + z^2} + l_0\right)} \tag{4.1}$$

with

S emission source strength of one single lane,

K scaling factor interpreted as ratio of the above-roof wind velocity to a mean wind velocity determining the air through-put in the street,

WV above-roof wind velocity,

$\sqrt{x^2 + z^2}$ distance of reference point from source,

l_0 correction factor, taking into consideration the air mixing caused by the vehicles.

For the reference point 2 located where the wind points downward on the building fronts, the pollutant concentration is as follows:

$$C_2 = \frac{K \cdot S \cdot (H - z)}{(WV + 0.5) \cdot W \cdot H} \tag{4.2}$$

with

H height of roofs,

W width of street canyon.

This expression corresponds to a linear decrease with height z. At the roof level H $C_2 = 0$. For $z = 0$ the equation for C_1 is obtained, when empirically instead of

$\sqrt{x^2 + z^2} + l_0$ (2. Term in the denominator of C_1)

width W of the street canyon is used as the "distance". In this case $C_1 = C_2$.

These two formulas hold good for the case that the wind approaches the street at right angles. If the wind blows parallel to the street, pollutant concentrations on both sides of the street C_3 are taken to be equal to the mean value of C_1 and C_2:

$$C_3 = \frac{1}{2}(C_1 + C_2) \tag{4.3}$$

Fig. 4.3 shows lines of equal concentration calculated with the model described in a four-lane street canyon with the wind blowing at right angles to the street. As, according to definition, the model is only valid for the space directly above the sidewalks, the lines in the space above the driving lanes have been interpolated for both cases, C_1 and C_2.

Fig. 4.4 shows the street canyon during parallel wind, in this case the concentrations above the sidewalks are the same.

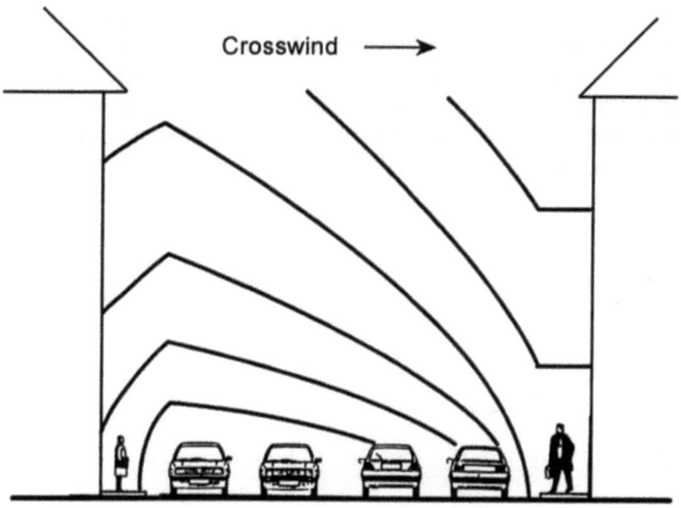

Fig. 4.3: Calculated lines of equal concentration in a street canyon with wind blowing at right angles to the street

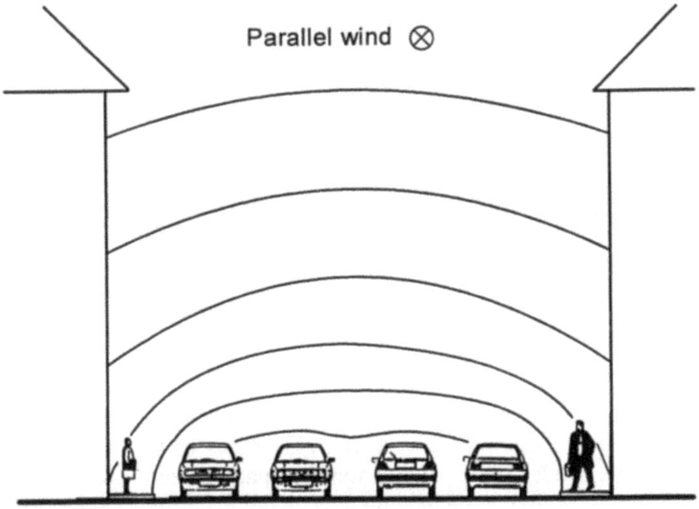

Fig. 4.4: Calculated lines of equal concentration in a street canyon with parallel wind

4.1.2 Highway Model

Apart from occurring in street canyons raised pollutant concentrations can also occur in the vicinity of highways. The highway model considered here permits calculating the course of pollutant concentrations on the leeward side from the edge of the road for distances of several hundred meters. It is assumed that the concentrations are only changed by turbulent mixing processes. Chemical reactions during transport and a possible deposition on the ground or on plants are not taken

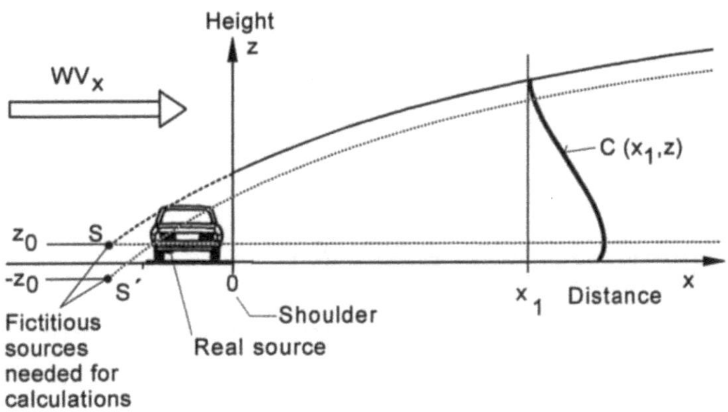

Fig. 4.5: Dispersion of exhaust gases on the leeward side of a highway (for a given distance x_1 to the highway the vertical concentration profile $C(x_1,z)$ is tilted by 90°)

into consideration. However, for the majority of the substances regarded here, except for NO_x, this is no substantial restriction. Soil and plants act as sinks for NO_x exhaust gases. This model - Fig. 4.5 shows the schematic depiction of the situation - is a one-dimensional Gaussian model. The highway is treated as an (infinitely long) line source. It is assumed that the vertical transport of the exhaust gases owing to turbulent diffusion can be neglected compared to the advective transport by wind.

In a Cartesian coordinate system, refer to Fig. 4.5, where the abscissa x runs vertically and the ordinate y runs parallel to the street, and where the applicate axis z represents the height, the concentration dispersion of the exhaust gases in the x, z plane is the solution of the continuity equation in the shape of a Gaussian distribution.

$$C(x,z)=\frac{S}{\sqrt{2\pi}(WV)_x\sigma_z}\left[exp\left\{-\frac{1}{2}\left(\frac{z-z_0}{\sigma_z}\right)^2\right\}+exp\left\{-\frac{1}{2}\left(\frac{z+z_0}{\sigma_z}\right)^2\right\}\right] \qquad (4.4)$$

with

S	emission source strength,
$(WV)_x$	component of wind vector in x direction,
z_0	level of line source above ground,
σ_z	dispersion parameter, depends on distance to source and the turbulence of the air.

Reflection by the ground is taken into consideration by introducing a virtual source S' at the point $z=-z_0$. The concentration at a certain point (x, z) is then

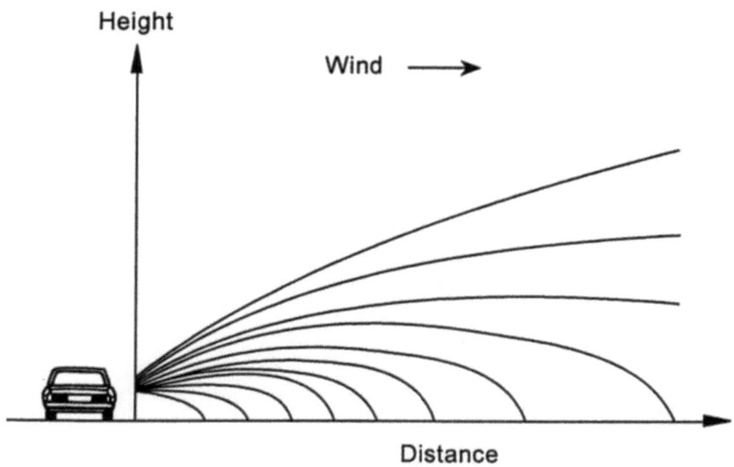

Fig. 4.6: Lines of equal concentration next to a four-lane highway, calculated with a one-dimensional Gaussian model

the result of the sum of the contributions of the real and the virtual source, for which undisturbed dispersion is assumed.

Turbulence in the driving lane space is taken into consideration by shifting the line sources used for calculating to the windward side. Just like the dependency of the dispersion parameter σ_z on the distance x, this shift must be determined empirically, e.g., with tracer experiments. The model can only be applied when wind velocity is at least 1 m/s. Fig. 4.6 shows lines of equal concentration next to a four-lane highway calculated with this method.

4.2 Scenarios for the Assessment of Air Quality Situations

Pollutant concentrations in ambient air which ultimately have an impact on man and the environment depend on a number of factors which can be assigned to the two areas of either emission or dispersion. The first group includes for instance emissions of vehicles, traffic volume, driving behavior and the type of vehicles in use, refer to Chap. 2.5 (Emission Prognosis). The area of dispersion includes for instance the source's geometry, terrain structure, meteorological parameters, possible chemical transformations, washing out and deposition. In view of the multitude of possible combinations of all these parameters it is necessary to confine oneself to the minimum necessary for obtaining distinct scenarios.

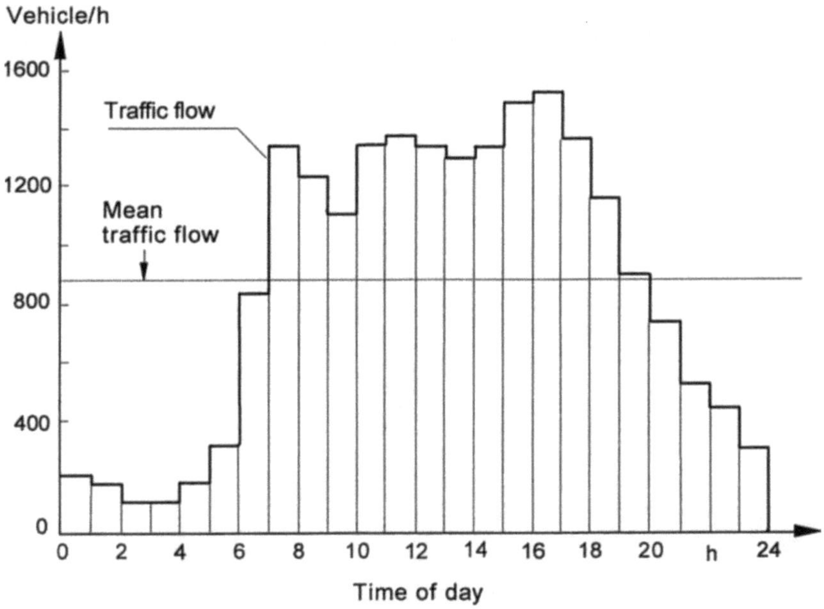

Fig. 4.7: Mean diurnal course of traffic volume on a four-lane intra-urban road. The traffic volume is restricted by local determinants such as parked vehicles blocking a lane

In practice two types of scenarios are especially significant. The "most unfavorable" case and the "average" or "normal" case. An unfavorable case would, e.g., be the coincidence of dense traffic with low wind velocity and stable air layers in the lower atmosphere (low-exchange weather situation). Each of the factors mentioned favors the formation of high pollutant concentrations.

Figs. 4.7 and 4.8 show as examples the typical diurnal courses of traffic volume averaged over longer periods of time for an intra-urban road and a four-lane highway. In both cases traffic volume recedes drastically at night. The traffic intensity averaged over 24 hours amounts to only 60 % of the maximum. The 24-hour mean value of the pollutant concentration in ambient air is then obtained by reducing the peak concentration in the maximum-to-mean traffic volume ratio.

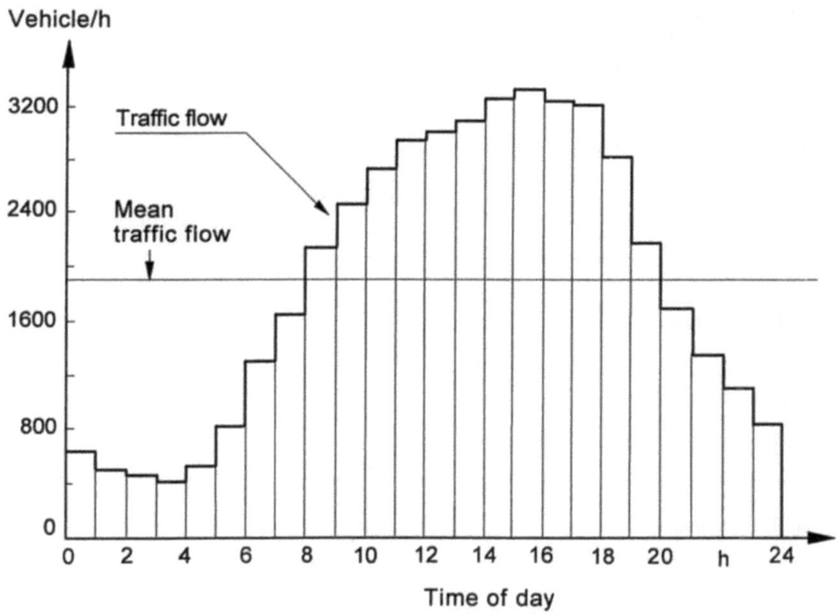

Fig. 4.8: Mean diurnal course of the traffic volume of a four-lane highway

4.3 Scenario Definitions

Measurements, refer to Chap. 2.4, have provided emission values for three groups of vehicles:

- vehicles with gasoline engines without catalytic converter,
- vehicles with gasoline engines and λ-sensor-controlled three-way catalytic converters acc. to US specifications and
- diesel vehicles acc. to US specifications.

It is possible, for instance, to calculate the air quality situations in the USA and Europe, in particular in the old states of Germany for the year 1990. However, the available data requires a number of compromises. The data for ECE vehicles without catalytic converter can be tranferred to the US scenario and, correspondingly, the values of the US catalytic converter gasoline and diesel engine vehicles to the European scenario. The vehicle composition listed in Table 4.1 is assumed as basis for passenger vehicles in the USA and in the FRG. It is based on estimates. Due to the inaccuracy of such estimates 5 % steps are used. Trucks are not taken into account in either case.

For city traffic the emission values determined acc. to the FTP-Test are used, for highways or freeways those values measured acc. to the Highway Driving Cycle (HDC) Test, refer to Chap. 8, which are included as part of the mean values in Chap. 2.4. Corresponding to the assumed total vehicle composition emission factors are calculated for each scenario from the emission values of the three vehicle groups, refer to Chap. 2.5.

Table 4.1: Composition of passenger vehicles assumed for the year 1990

Vehicle group	Proportion relative to total stock USA[a] in %	 Federal Republic of Germany in %
Vehicles with gasoline engines without catalytic converters	20	60
Vehicles with gasoline engines with three-way catalytic converters	75	25
Vehicles with diesel engines	5	15

[a] acc. to an EPA estimate

For the US and the European scenarios the street canyon is assumed to be identical: a total of four lanes with a width of 3.50 m each, sidewalks 2 m wide, i.e., a total street width of 18 m. The height of the buildings lining the streets is 15 m. Wind velocity is set at 1 m/s, calculating the two cases with wind blowing at right angles to the street and parallel to the street. Traffic density is assumed to be 3000 vehicles per hour, which represents heavy traffic in an intra-urban area. Reference points for the calculation are 1 m from the kerb and 1.5 m above the sidewalks.

As traffic and street conditions as well as meteorological conditions are assumed to be the same for the US and the European scenario, differences in the pollution concentrations calculated can only be attributed to the different total emission from the vehicles themselves.

4.4 Pollutant Concentrations in Street Canyons

As an example, the pollutant concentrations calculated for cross-wind and parallel wind are examined for the European scenario and for CO, NO_X and particles.

As a result of the linear mathematical relationships the CO, NO_X and particle concentrations calculated show the same ratio as the corresponding emission factors, apart from a few differences in rounding. The highest concentrations occur in crosswind on the leeward side of the buildings, as is elucidated by Figs. 4.9 to 4.14. In these diagrams the lines of constant concentration have been included both in mg/m^3 or $\mu g/m^3$ and in ppm (in parenthesis). It can be clearly seen how the exhaust gases are blown to one side of the street by the swirl when there are crosswinds. On the other hand, when the wind blows parallel to the street, concentrations are the same on both sides.

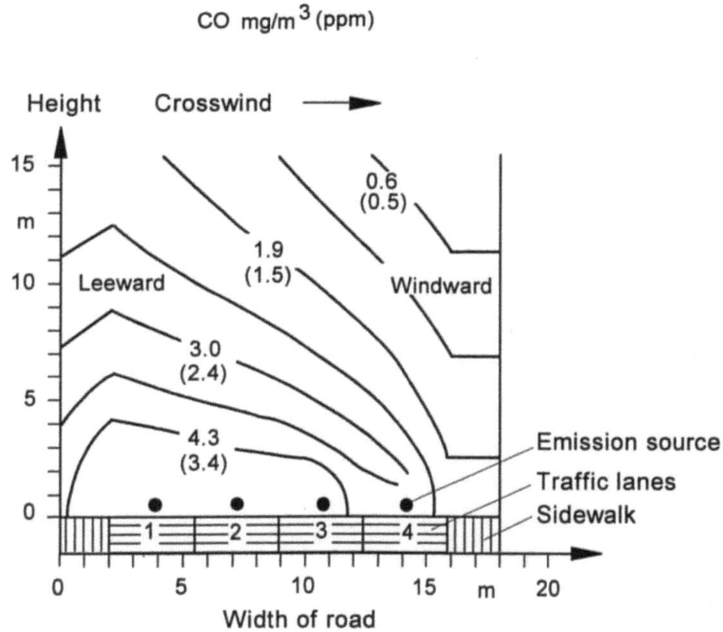

Fig. 4.9: Lines of equal CO concentration in mg/m^3 and in (ppm) in a street canyon with crosswinds (Europe scenario) for 1990

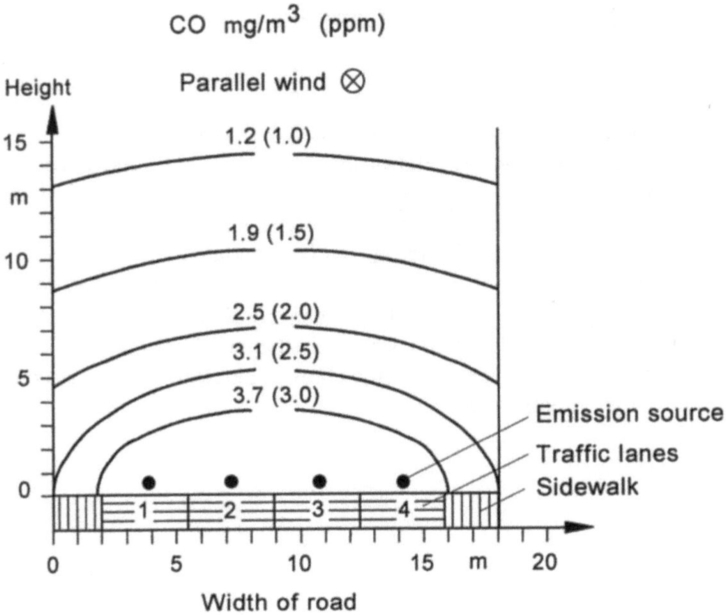

Fig. 4.10: Lines of equal CO concentration in mg/m^3 and in (ppm) in a street canyon with parallel wind (Europe scenario)

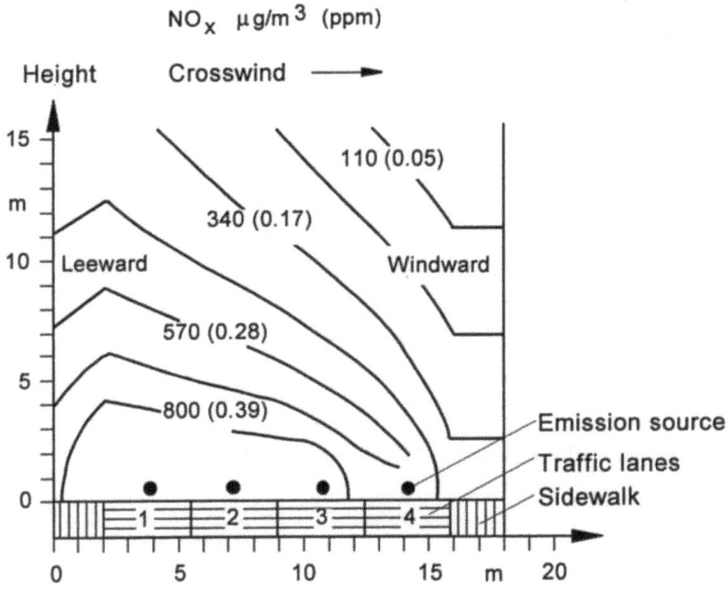

Fig. 4.11: Lines of equal NO_x concentration in $\mu g/m^3$ and in (ppm) in a street canyon with crosswind (Europe scenario)

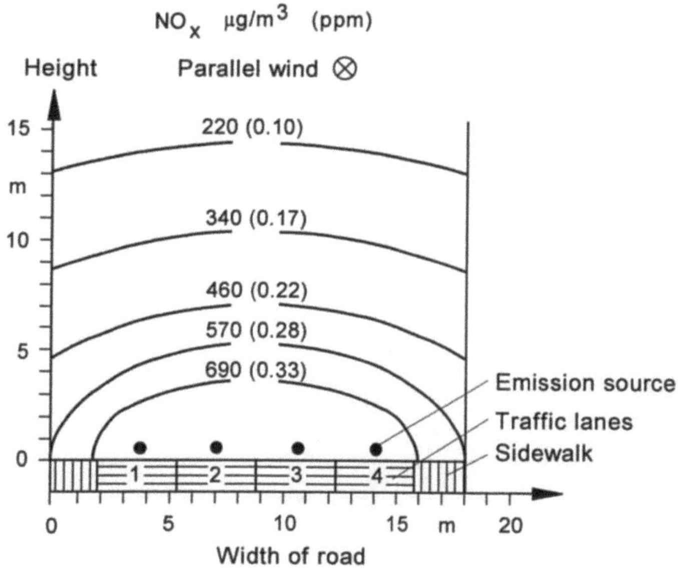

Fig. 4.12: Lines of equal NO_x concentration in µg/m^3 and in (ppm) in a street canyon with parallel wind (Europe scenario)

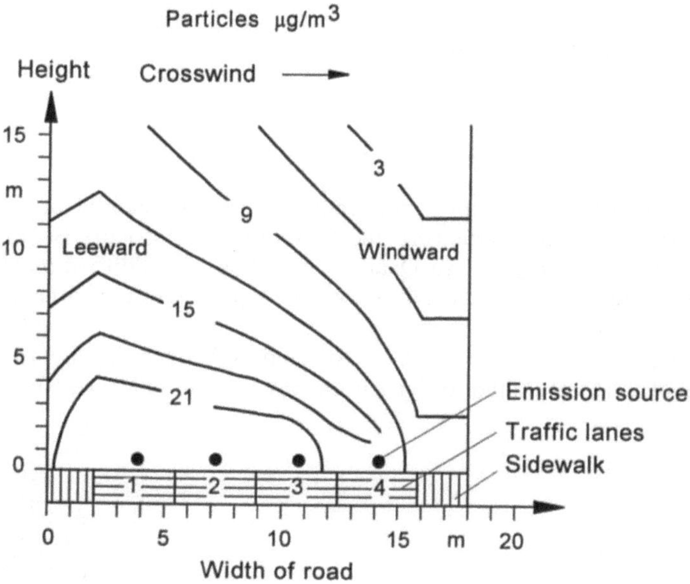

Fig. 4.13: Lines of equal particle concentration in µg/m^3 in a street canyon with crosswind (Europe scenario)

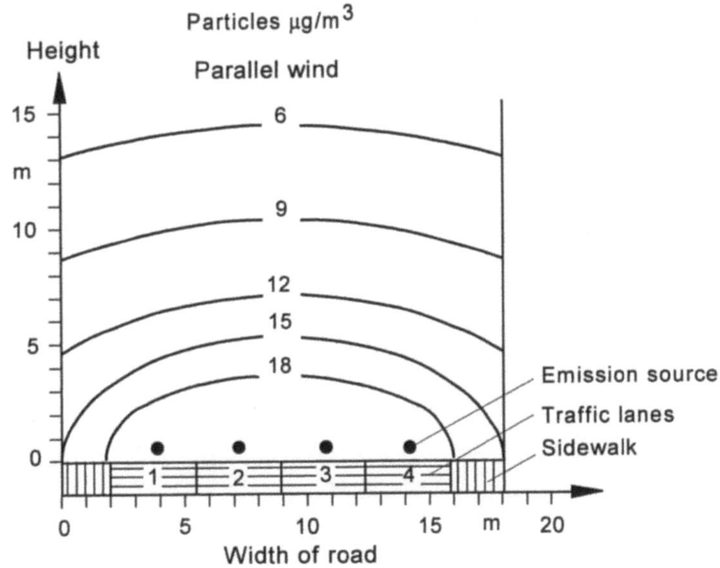

Fig. 4.14: Lines of equal particle concentration in µg/m³ in a street canyon with parallel wind (Europe scenario)

4.5 Measured Air Quality Values

4.5.1 Street Canyon

In Germany measurements of pollutant concentrations have been carried out in cities and in selected streets there. Particularly the TÜV (Technical Control Board) Rhineland and the State Authority for Air Quality Control have established many such measuring data. Fig. 4.15 schematically shows the typical set-up of a larger measuring station erected in a street canyon.

Measuring results for the components CO, NO and NO_2 recorded by two measuring stations of the State Authority for Air Quality Control NRW (LIS) are listed in Table 4.2.

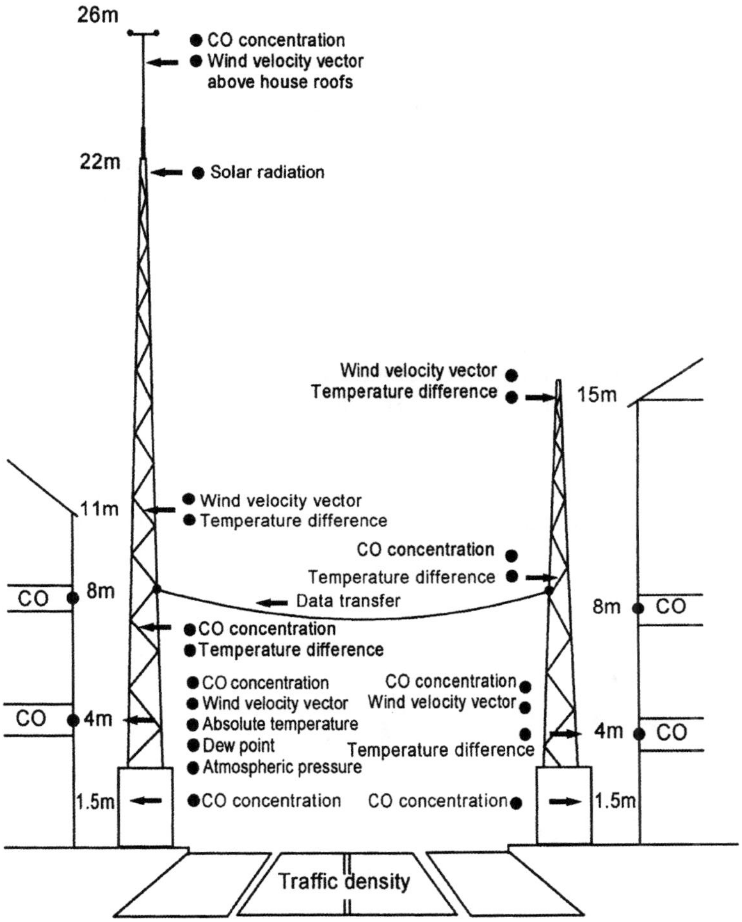

Fig. 4.15: CO pollutant measuring station (Cologne, Venlo Street, Source: TÜV Rhineland)

Table 4.2: Measured mean values of two measuring stations in street canyons (Source: State Authority for Air Quality Control, Essen)

Location of Measurement	Component	Sampling height 1.5m		Sampling height 3.5m	
		μg/m³	ppm	μg/m³	ppm
Düsseldorf	CO	3700	3.2	2900	2.5
	NO	176	0.1	153	0.08
	NO_2	70	0.04	67	0.03
	$NO + NO_2 \approx NO_x$	246	0.14	220	0.11
Essen	CO	3400	3.0	2800	2.4
	NO	194	0.1	128	0.07
	NO_2	73	0.04	62	0.03
	$NO + NO_2 \approx NO_x$	267	0.14	190	0.1

4.6 Air Quality Maps

The pollutant concentrations found over large areas, as also the ones in cities, can be inferred from the values measured by the 268 German pollution measuring stations operated by the state authorities in charge of the each area (Fig. 4.16). These values provide information on those pollutant concentrations in ambient air caused by exhaust gas components from all sources (industry, transportation, house fires etc.).

These weekly mean values, among others for CO and NO_2, are published on a weekly basis by the VDI news in the form of an environmental index map (refer to Fig. 4.17) and and as a table (refer to Table 4.3). The weekly mean value is determined as arithmetical mean value of all single values measured and, as long-term mean value, is compared with the standard for long-term exposure (IW1 of the Technical Directive on Air Pollution Control 1986). For CO this standard is 10 mg/m^3 or 0.86 ppm. To assess short-term exposures on the basis of the maximum values measured MIK (MIK - maximal pollutant concentration in ambient air) standards are used acc. to VDI guideline no. 2310, refer to Chap. 5.

Fig. 4.16: Air quality measuring stations for the recording of pollutant concentrations over large areas in Germany (Date: 1991)

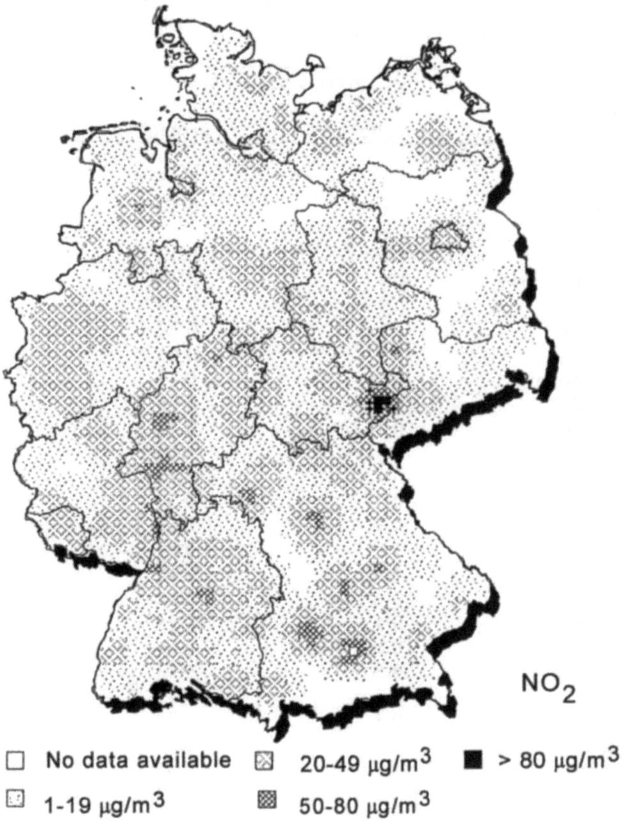

Fig. 4.17: Nitrogen dioxide pollution in Germany, mean values for the week from October 17 to October 23, 1994 (Source: Georisk)

4.7 Conclusion

If, as an example, the mean results of the CO air quality values calculated for the street canyon at a height of 1.5 m, refer to Chap. 4.4, are compared to the values measured, Chap. 4.5, there is quite a conformity as can be seen in the compilation listed in Table 4.4, with the exception of the particle concentration which can be explained since the number of diesel vehicles are lower than given in Table 4.1 (refer to Fig. 2.28).

If the decrease expected within the next years is also taken into consideration (refer to Fig. 2.36, Chap. 2.5) the values are also expected to drop by another $^2/_3$ to $^3/_4$ (Table 4.4, third column).

Table 4.3: Examples of CO measurement in the week from Sept. 12 to Sept. 18, 1994 (Weekly mean value of all 268 German stations recorded: 0.5 mg/m³ or 0.43ppm)

State	Location of Measurement	Weekly Mean Value in mg/m³	Weekly Mean Value in ppm
Schleswig-Holstein	Lübeck	1.1	0.95
Mecklenburg-Vorpommern	Rostock	0.6	0.52
Lower Saxony	Hannover	0.6	0.52
Bremen	Bremen-downtown	0.4	0.34
Sachsen-Anhalt	Halle	0.5	0.43
	Magdeburg	0.3	0.26
Brandenburg	Potsdam	0.5	0.43
Berlin	Berlin-downtown	0.5	0.43
North Rhine Westphalia	Dortmund	1.1	0.95
	Düsseldorf	0.5	0.43
	Mülheim	0.8	0.69
Hesse	Frankfurt	0.9	0.77
	Hanau	0.4	0.34
Thuringia	Gera	0.3	0.26
	Weimar	1.1	0.95
Saxony	Chemnitz	1.3	1.12
	Plauen	0.4	0.34
Rhineland Palatinate	Kaiserslautern	0.4	0.34
	Mainz	0.8	0.69
Saarland	Saarbrücken	0.5	0.43
Baden-Württemberg	Freiburg	0.2	0.17
	Mannheim	0.3	0.26
	Stuttgart	0.5	0.43
Bavaria	Augsburg	1.7	1.46
	Hof	0.2	0.17
	Munich-Stachus	2.2	1.89

Table 4.4: Compilation of calculated and recorded mean pollution values for the street canyon at a height of 1.5 m, including emission prognosis, refer to Chap. 2.5

Pollution Values / Component	Calculated	Recorded	Prognosis
	ppm	ppm	ppm
CO	≈ 3.2	≈ 3	≈ 1
NO_x	≈ 0.39	≈ 0.14	≈ 0.04
	μg/m³	μg/m³	μg/m³
Particles	≈ 20	≈ 10	≈ 0.03

The pollution standard for particle concentration in ambient air is to be initially fixed at 14 μg/m³, and later at 8 μg/m³.

5 Effects

5.1 Introduction

Since the sixties discussions on cars and traffic have increasingly dealt with the question of what kind of effect automobile exhaust gases have on humans. Later, effects on the environment were also included (forest, soil, water and on a global scale: the earth's atmosphere).

As scientific findings on the effects on man and environment are as yet insufficient, legislators and governments advocate the idea of prevention. All these discussions led to global exhaust gas legislation as has been described in Chap. 2.3.

As early as 1974, the National Academy of Science, USA, while acknowledging the prevention idea, demanded more research into effects, particularly on humans.

As a contribution to throwing more light on the facts the German and European automobile industry, represented by the Verband der Automobilindustrie (VDA) (Federation of Automobile Industry), that is by the Forschungsvereinigung Automobiltechnik e.V. (FAT) (Research Association Automobile Technology, Inc.) and by the Committee of Common Market Automobile Constructors (CCMC) started in the late seventies to support a larger number of effect research projects which were carried out by independent scientific institutes.

Moreover, in 1980, the American government, represented by the Environmental Protection Agency (EPA), and the automobile industry founded the Health Effects Institute (HEI), an institution independent of its financial backers, with the aim of starting joint effect research projects (effects on humans). Funding of these projects is shared equally by the EPA and the automobile industry.

After several years of discussions in the FRG a joint research program, with the effect on human health in the foreground, has been successfully established between the federal government, represented by the Federal Ministry for Research and Technology, and the FAT. After several years this project was discontinued on the part of the Federal Ministry for Research and Technology. The animal experiments with diesel exhaust gas and artificial aerosols could, however, be brought to a conclusion.

5.2 Legislation

Apart from having set standards for limited exhaust gas components (refer to Chap. 2) the USA also has regulations concerning non-limited exhaust gas components, refer to Chap. 7. The following excerpt, derived from the environmental law valid in the USA, the Clean Air Act, shows that the EPA passed directives in 1979 which hold automobile manufacturers responsible for the effects of all detectable exhaust gas components.

EPA directives on limiting non-limited exhaust gas components:

Installing exhaust gas purification systems or construction elements in vehicles or engines with the aim of lowering pollutant emissions may not involve unjustifiable risks to health, welfare and safety of the public.

This direction is phrased in such a general and vague manner that it can be interpreted arbitrarily. In practice, this requirement cannot be met as, according to estimates of chemical engineers, automobile exhaust gases contain a large number of different chemical compounds. However, the majority of them occur in negligibly minute concentrations and nothing is known about their effects.

The EPA's awareness of these practical difficulties contributed to the foundation of the HEI. The responsibility for researching the effects was consequently conferred to it.

For the assessment of pollutant concentrations in ambient air of automobile exhaust gas components as to their effects on human health and the environment effect-related threshold values as comparative criteria are required. Standards directed at the protection of human health and the prevention of environmental damage have been set for a series of exhaust gas components, e.g., in the USA as "National Ambient Air Quality Standards" (NAAQS), in Europe as EU directives or as recommendations with an official character, for instance the so-called MIK values (MIK - Maximale Immissionskonzentration = maximum pollutant concentrations in ambient air) in the Federal Republic of Germany. In Germany the MAK Values (Maximum Concentration Permitted in Workplaces) are special values for health protection in workplaces; they are comparable to the TLV Values (Threshold Limit Values) in the USA.

If no binding or generally acknowledged standards exist for a certain exhaust gas component, threshold values must be determined below which no negative physiological effects are to be expected. If research pertaining to the component in question has been done, results can be looked up in relevant literature. Otherwise selective effect examinations must be carried out.

MIK Values and NAAQS are directing at preventing damage to human health, in particular to high-risk groups, even when there is a long-term exposure, and at ensuring that animals, plants and objects are protected from harm.

Workplace threshold values (MAK Values, TLV) represent the highest permissible concentration of a substance in the workplace ambient air, which, according to the present state of knowledge, does not impair employees' health in general and has no disagreeable, irritating effects even after repeated, long-term and as a rule daily eight-hour exposure (5 days/week = 40 hrs).

MIK Values and NAAQS have only been fixed for a few substances which would apply when assessing possible health hazards to the population, including high-risk groups. This deficiency is to be attributed to as yet insufficient knowledge on chronic health effects of low pollutant concentrations in ambient air. For most of the non-limited components only workplace threshold values exist.

When defining a pollution-related risk range where first signs of health impairment would have to be expected ("Range of Concern") it is now possible, if no lowest pollution concentration has been published where damage symptoms can be proved, to deduce, instead, the lower effect limit from the TLV or MAK Value by including a safety factor of, as a rule, 100. The TLV and MAK Values themselves can be used as upper limit for the "Range of Concern". Maximum exposure time is assumed to be 40 hrs./week. This approach, which has been developed by the EPA, cannot, however, be applied to estimate the risks caused by genotoxic, i.e., mutagenic, teratogenic and/or carcinogenic substances, for which in most cases no standards exist either.

All in all, using the "Range of Concern" is a temporary expedient when no other suitable threshold values are available. When pollutant concentrations in ambient air are within the "Range of Concern" one should check with the help of further data whether there is an actual health hazard.

5.3 Effect on Humans, Significance of Dose

To be able to understand the effect of substances (including gases) on humans, one must start with the dose, which is defined as the product of concentration and the impact period, i.e. the exposure time:

$$\text{Dose} = \text{concentration} \cdot \text{exposure time}. \tag{5.1}$$

As can be seen in Fig. 5.1, this knowledge has long been known to physicians. This is demonstrated here by the example of a quote from Paracelsus (1493 - 1541).

In experiments with toxic substances known to be effective such as hydrogen cyanide and carbonyl dichloride quantative relationships have been set up for the concentration c as a function of exposure time, which are shown diagrammatically in Fig. 5.2. An adaptation of humans to the substance has been disregarded here.

What thing existeth that is not poison? All things are poison / and no thing is without poison / The dose alone maketh a thing wholesome or poisenous. As an instance / I give ye every food and every drink / which unto whosoever taketh more than the just quantity / it is poison / proof thereof is the outcome. But I likewise allow / that there be posion / which is poison.

Fig. 5.1: Paracelsus: dose and effect

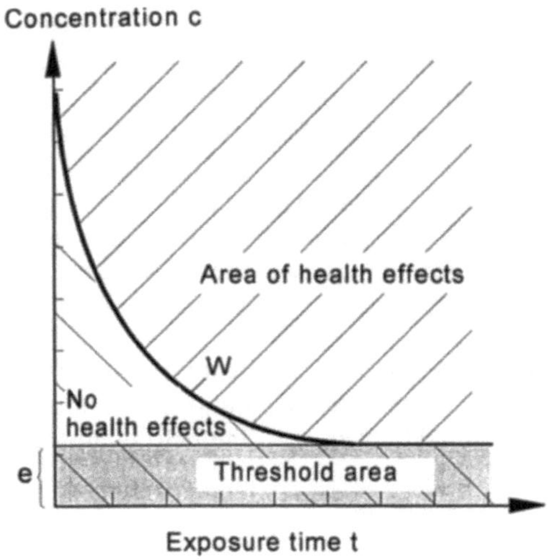

Fig. 5.2: Effect curve of toxic substances

The quantitative correlation of dose-effect relationships can be described as follows:

$$\text{Dose} = (c - e) \cdot t = W = \text{const.} \tag{5.2}$$

where W is the effect curve above which an effect occurs, e the threshold concentration, i.e. the threshold below which a substance has no effect, even over a very long exposure time up to a whole human lifetime.

One can only speak of effect then, if the concentration during life-long exposure is above such a threshold e or during a short exposure above the effect curve W. Differentiation is thus made according to acute (after short exposure) and chronic (after long exposure) effects. Carcinogenic effects are exceptions.

5.4 Using the Effect of Carbon Monoxide (CO) on Humans as an Example

The relationship shown in Fig. 5.2 is also valid for CO, one of the limited components of automobile exhaust gas.

The effect of CO is caused by the formation of carboxyhemoglobin (COHb) where the inhaled CO is attached to the red blood cells, thus preventing the formation of oxyhemoglobin (O_2Hb), the vitally important combination with oxygen (O_2). Hb has 200 to 300 times the affinity for CO that it does for O_2. The combining of CO with hemoglobin occurs as a function of time. For a certain CO concentration a corresponding COHb saturation in the blood is reached after 3 to 4 hrs. With an increasing CO concentration in the air breathed in the human organism's oxygen supply becomes more and more depleted. As a consequence physiological disorders ensue, with high CO concentrations ultimately leading to death.

Human blood has a natural COHb content, even without the influence of CO from ambient air. Table 5.1 shows some values.:

Table 5.1: Natural COHb content in human blood without CO in the ambient air

Human	COHb content %
Nonsmoker	0.8 ± 0.2
Moderate smoker	2.1 - 3.4
Heavy smoker	10

Smoking leads to a high COHb content in the blood.

Fig. 5.3 shows the effect curves of CO concentrations in the air breathed in as a function of the corresponding exposure time. The corresponding COHb amounts in the blood are included as parameter. The effects on humans are indicated above each effect curve. Below the dark line W there are no effects. The threshold concentration *e* for this is 60 ppm, refer to (5.2), and has also been included. The corresponding saturation values for COHb are reached after 3 to 4 hrs, independent of the CO concentrations. The first effect with symptoms such as slight headaches, impaired vision and slight disorientation begins at a concentration in respiratory air from 60 ppm and 3 hrs. of exposure time. At concentrations in excess of approx. 450 ppm death can occur after approx. 1.5 to 4 hrs., depending on the CO concentration.

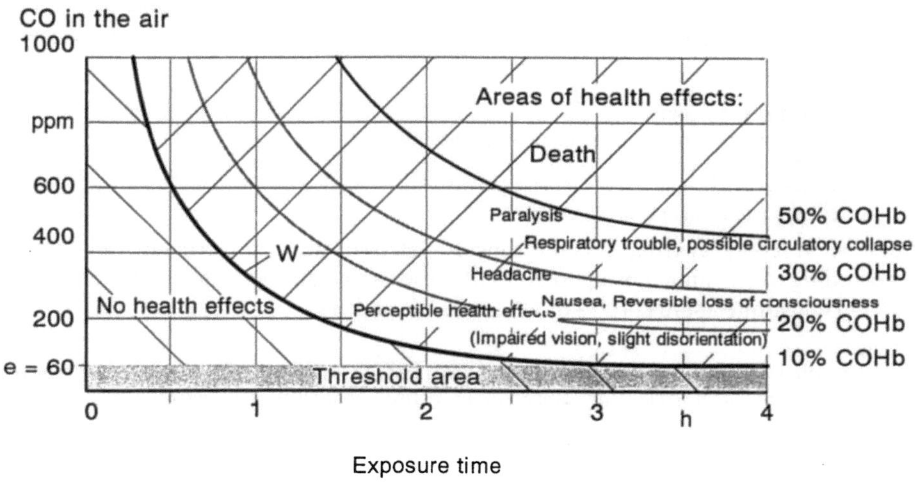

Fig. 5.3: Effect of CO in the air on humans

The traffic-caused CO mean values or COHb values determined in European cities in humans are in a range where no subjective complaints occur in humans. In street canyons with heavy traffic and in flowing traffic only CO values with a mean of 3 to 4 ppm occur, refer to Chap. 4. These CO values are equivalent to a COHb content of less than 1 %. Thus, they are no higher than the natural values for nonsmokers, refer to Table 5.1. In smokers who are exposed to this low CO concentration in a street canyon for several hours without smoking the COHb content is reduced.

Therefore, the CO concentrations in the air caused by traffic are completely harmless for healthy humans, i.e., the critical threshold "*e*" is a long way from being reached. Even for so-called risk groups (asthmatics, children etc.) 3-4 ppm are not critical, as the 24-hour MIK Value is 8.7 ppm, the half-hour MIK Value 43.5 ppm. The MAK Value used to be 50 ppm and is now 30 ppm.

With reference to the prognosticated CO concentration values from Table 4.4 the effect would be negligible.

5.5 Carcinogenic Substances

Unlike the toxic substances discussed so far, according to many scientists there are no threshold values *e* for carcinogenic substances, i.e., chemical substances which can cause cancer.

Cancer theory:

In the case of cancer it is not possible to establish threshold values for cancer-causing (inducing) or promoting chemical substances.
In the opinion of many scientists even 1 molecule can be responsible for causing or promoting cancer.

The hypothesis of effect without threshold values can hardly be proven by experiments. If one accepts this assumption nevertheless as a working hypothesis, one must estimate the possible risks to human health at low concentrations. In humans the lungs are the main target organ as far as automobile exhaust gases are concerned. Findings on the causes of lung cancer or the frequency of lung cancer are listed below in order of frequency:

Causes of lung cancer:

1. Tobacco smoke (approx. 90% of all lung cancer cases)
2. polluted workplace
3. atmospheric pollution
4. radon radiation, mainly in closed rooms.

Statistics report approx. 28,000 cancer-related deaths in Germany per year. According to examinations up to 90 % of these can be attributed to smoking which would constitute approx. 10 % of all deaths of smokers.

Note:
As these are statements based on statistics, discussions of indiviual case results or of results from a few samples are inadmissible. Individual cases do not disprove statistical statements.

The result of this evaluation is that atmospheric pollution which is only partly caused by automobiles is a small contributor to lung cancer. Accordingly, the hypothetical proportion of lung-related cancer deaths caused by automobile exhaust gases can only be minor, compared to the provenly high proportion caused by tobacco smoke and to workplace pollution.

Of automobile exhaust gas components it is mainly diesel particulate material that must be examined, as it sometimes remains in the lungs longer than gases. According to publications, the risk caused by particle concentrations occurring in street canyons, refer to Chap. 4.5, is assessed as being slight to negligible.

5.6 Methods of Effect Research

5.6.1 In-Vitro Tests

So-called in-vitro tests, i.e., examinations in Petri culture dishes with bacteria or cell cultures have been much discussed by the public in the past $2^1/_2$ decades as an alternative to animal experiments. One example is the so-called Ames Test, a mutagenicity test which was developed by B. Ames in 1972.

Ames Test

Bacteria strain: Salmonella Typhimurium (due to a mutation this special strain cannot produce the amino acid histidine required for growth).

The basis of this test is the introduction of mutagenic substances into a nutrient fluid which increase the frequency of back mutation. The bacteria which again produce histidine grow to colonies, all other bacteria die. The number of colonies is a measure of mutagenicity.

In the case of tests with automobile exhaust gas, bacteria cultures (as also cell cultures) are exposed to exhaust gas condensates or extracts which must be collected over a longer period of time, e.g., in many driving cycles, refer to Chap. 9, so that their concentrations are very high. Chemical modification of the condensate may occur during collection. For this test it must be noted that a substance which is mutagenic, meaning it can change the genotype of a cell, does not necessarily have to be carcinogenic, i.e., it does not necessarily cause cancer. Fig. 5.4 shows a photograph of Ames Test bacteria cultures. Colony growth can be seen.

The development of such in-vitro tests has not passed byond the state of fundamental research. Nevertheless, the EPA used such test and their results exclusively for many years, in particular to assess the effects of diesel exhaust gases.

As in the Ames Test, exhaust gas condensates or extracts which can change cell structures are used in the so-called cell transformation test.

Cell transformation test - Carcinogenicity studies

The basis of the test is that carcinogenic substances morphologically transform certain cells in culture.

The final state is the formation of transformed cells which can be recognized by modifications in the growth pattern when compared to normal cells.

In general it must be said that in-vitro tests permit no final conclusions as to the effect. They are useful as so-called short-term screening test, i.e., as tests with which those substances which may possibly be mutagenic or carcinogenic can be

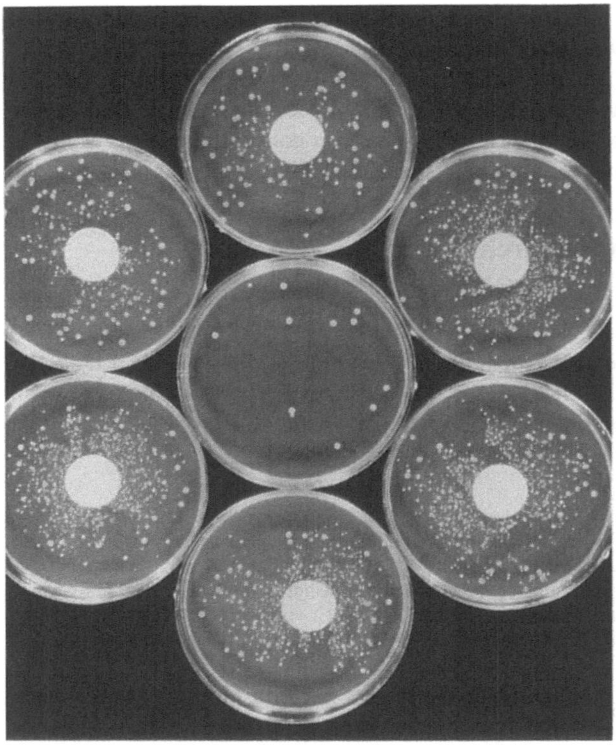

Fig. 5.4: Photograph of bacteria cultures (Ames Test)

eliminated from a large number of substances in preliminary tests. These selected substances must then be checked in animal experiments and only the results of animal experiments may be used to make decisions. This is important in the case of chemical substances as, e.g., in the USA alone approx. 50,000 new chemical substances appear on the market every year.

5.6.2 In-Vivo Tests

"In vivo" means, that studies are carried out on living beings, whereby studies using humans are restricted for ethical reasons to short-term exposures at low concentrations with acute, reversible effects or to surveys about specific groups of people (epidemiological studies). Thus, for reasons of cost, only small rodents (mice, hamsters, rats) are suitable and specially bred strains are used, e.g., "Fisher" or "Wistar" rats.

Several methods are listed below.

Toxicity and carcinogenicity tests on animals

Most frequently used methods:

a) Epicutaneous, subcutaneous and intratracheal applications with extracts, condensates, particulate material. "Epicutaneous" means application to the skin, "subcutaneous" injection below the skin and "intratracheal" injection into the lung tract. For the case in question extracts, condensate or particulate matter from automobile exhaust gases have been used.

b) Inhalation tests with highly concentrated exhaust gases for acute and chronic effects, i.e. the laboratory animals inhale certain exhaust gas concentrations for a certain period of time almost up to the end of their life-span. A control group is exposed to normal air.

Fig. 5.5 shows a photograph of inhalation chambers. Diluted exhaust gas and normal air for the control group flows horizontally through the chambers, evenly distributed by punched plates. The animals are accommodated in cages.

The animals are well cared for. In all experiments carried out so far the animals' normal life-span was not affected. In these types of experiments the animals must be kept in separate cages to prevent mutual injury and cannibalism.

One feature of in-vivo experiments are the already mentioned epidemiological examinations of specific exposed groups of people, e.g., gas station attendants or automobile mechanics, in comparison to unexposed control groups. Normally, this would be the most suitable method if it were possible to prove effects. The difficulties of epidemiological examinations are outlined below:

Difficulties regarding epidemiological examinations of specific population groups:

There is no single group with a defined exposure situation and hardly any control groups free of exposure. Interfering factors such as smoking predominate.

All epidemiological examinations carried out the world over so far have failed. According to many toxicologists, effects could not be proven. Animal experiments are therefore indispensable.

Fig. 5.5: Inhalation chambers

5.6.3 Problems of In-Vitro and In-Vivo Examinations

Fig. 5.6 illustrates the problems of the tests. In this diagram the effect is depicted in relation to the dose.

The condensate used in in-vitro tests represents the maximum dose possible (uppermost part of curve in Fig. 5.6). Inhalation tests are carried out with low doses (lower part of curve in Fig. 5.6), with the concentrations being even higher than the concentrations in a street canyon by a factor of up to 10,000.

These relatively high concentrations in inhalation tests are chosen to find out whether there are any effects at all. The number of animals is also restricted for reasons of cost. Accordingly, the dose must be increased to attain statistically usable results.

Time *t* is given due to the normal life-span of the animals. As it is much shorter than human life-spans results are obtained within a few years time.

In the middle of the curve of Fig. 5.6 "pre-treatment" has been included. A special test model is marked here in which animal groups are pre-treated with highly carcinogenic substances (e.g., nitrosamines), so that a certain number of tumors develop. The inhalation of automobile exhaust gases is to determine whether they promote or inhibit cancer. This model had to be chosen originally as it was assumed that inspite of the high doses applied no tumors would be detectable.

The S-shaped curve in Fig. 5.6 illustrates the problem of these test methods in so far as it is practically impossible to extrapolate from the effects of a high dose to the effects of a low dose, with the concentration in a street canyon with an average human exposure time of 2 hrs./day being considerably lower than in the inhalation experiments.

Moreover, transferring the results of animal experiments to humans is fundamentally problematic, although it is general practice among toxicologists.

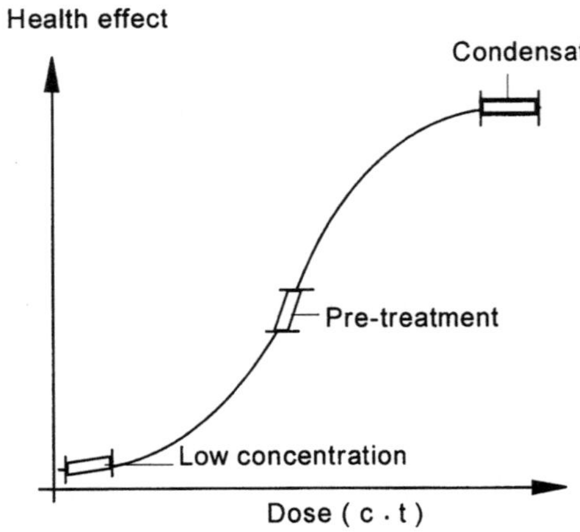

Fig. 5.6: Dose - effect relationships (c - conzentration, t - exposure time)

5.7 Research Projects and Results in Chronological Order

Research projects which in the past have mainly been subsidized by the automobile industry and their results are described below. Only inhalation experiments are introduced as only they permit direct inferences.

5.7.1 Experiments with Gasoline Engine Exhaust Gas (a UBA/FAT Project)

The most important data of a gasoline engine project jointly sponsored by the Forschungvereinigung Automobiltechnik (FAT) (Research Association Automobile Technology, Inc.) and the Umweltbundesamt (UBA) (Federal Environmental Agency of Germany) are as follows.

FAT/UBA Project:

Institution:	FhG/ITA Münster-Roxel
Program:	gasoline engine
Animals:	rats, hamsters and mice
Period:	end of 1979 to the end of 1985, i.e. 6 years
Cost:	1.8 mio. German marks

(FhG: Fraunhofer - Gesellschaft e.V., ITA: Institute of Toxicology and Aerosol Research)

Results of the FAT/UBA Project:

a) Rats and hamsters without tumor-inducing pre-treatment:

After exposure to relatively high concentrations of gasoline engine exhaust gas no increase of the spontaneous tumor rate was found. After two years of exposure to the highest exhaust gas concentration, physiological changes were found to be only in rats, which were partly reversible after concluding the exposure.

b) Laboratory animals with tumor-inducing pre-treatment:

Rats:	decrease in malignant tumors in the lung increase in benign tumors in the upper respiratory tract
Adult mice:	decrease in benign lung tumors increase in malignant lung tumors
Hamsters / newborn mice:	no significant changes of the induced tumor rate

The animals' life-span was not shortened. Only animals breathing the highest selected exhaust gas concentration with a CO content of approx. 300 ppm show physiological changes. If the concentration is only slightly lower, practically no effect is detectable. Automobile exhaust gases from gasoline engines caused no tumors in these experiments.

In the animals pre-treated with carcinogenic substances contradictory evidence appeared, part of which has not been statistically verified.

5.7.2 Test with Diesel Engine Exhaust Gas (a Volkswagen/Audi Project)

The most important data of a project sponsored by VW/Audi is listed in the following overview. It is important, that exhaust gases both with and without particulate matter were used here. The respirable particulate material was filtered in the second case.

VW/Audi Project:

Institution:	FhG/ITA Hannover
Program:	Diesel engine (with and without soot particles in the exhaust gas)
Animals:	Rats, hamsters and mice
Period:	Beginning of 1980 to the end of 1985, i.e. 6 years
Cost:	3.5 mio. German marks

Results of the VW/Audi Project:

- After exposure to relatively high concentrations of filtered diesel engine exhaust gas (without particulate material) carcinogenic or chronic-toxic effects were not to be found in any of the animal species.
- After exposure to unfiltered exhaust gas with soot particles cytological, functional and morphological changes in the respiratory tract were to be found in all three animal species.
- After exposure to unfiltered exhaust gas (with soot particles) rats showed a lung tumor rate of approx. 16%.
- After exposure to both filtered and unfiltered exhaust gas mice showed an increased lung tumor rate of approx. 13% (spontaneous tumors) to approx. 30%. It seemed necessary to repeat this test to verify the results.
- Hamsters showed no increase in the spontaneous tumor rate.
- Generally, no cocarcinogenic effect of diesel engine exhaust gas was to be found in the pre-treated animals.

Table 5.2 shows the tumor rates of rats after exposure to exhaust gas with and without particulate matter and those of the control group (exposure to normal air). After exposure to unfiltered exhaust gas with particulate material 17 tumors were diagnosed in the 95 animals, with one of the tumors being malignant.

Table 5.2: Number of histological results in rat lungs after an exposure of 140 weeks in the VW/Audi project

Histological evidence	Control	Exhaust gas without soot particles	Unfiltered exhaust gas (with soot particles)
Number of rats examined	96	92	95
Broncho-alveolar adenoma	0	0	8 (8 benign)
Tumor in the squamous epithelium	0	0	9 (8 benign, 1 malignant)

After exposure to unfiltered exhaust gas, i.e., with soot particles, an increased lung tumor rate was found in the rats. However, the dose was very high. No tumors were to be found in hamsters. The evidence for mice was contradictory.

Fig. 5.7 shows as an example a section of a rat lung with a pronounced tumor.

Surprisingly for the high dose applied during the inhalation experiments with unfiltered diesel exhaust gas (with soot particles) statistically verified tumors developed only in one single animal species, the rat. The rats used in this experiment were so-called Wistar rats.

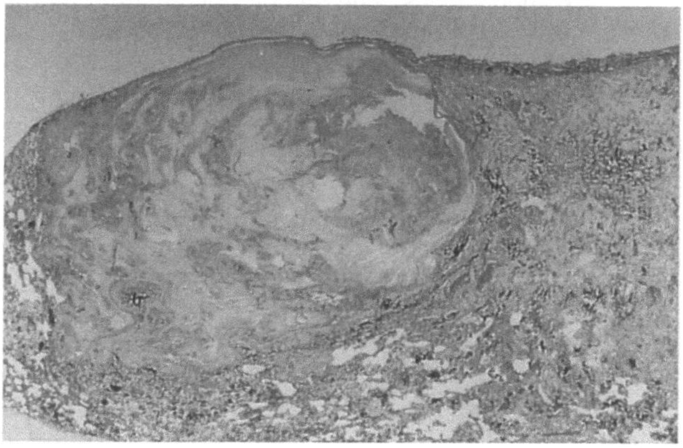

Fig. 5.7: Tumor in the squamous epithelium, rat lung

5.7.3 Experiments with Gasoline and Diesel Engine Exhaust Gas, Gasoline Engine with and without Three-Way Catalytic Converter, Diesel Engine Exhaust with and without Soot Particles (a CCMC Project)

The most important data of the CCMC project (Committee of Common Market Automobile Constructors) are shown in the overview below.
As laboratory animals hamsters as well as male and female "344-Fisher" rats (a different breed than the Wistar rats) were used. This project, the largest worldwide, supplemented the two projects described above.

CCMC project:

Institution:	Battelle, Geneva
Program:	Gasoline engine with and without three-way catalytic converter, diesel engine with and without soot particles in the exhaust gas
Animals:	Rats, hamsters
Period:	Beginning of 1980 until the end of 1985, i.e., 6 years
Cost:	15 mio. German marks

Results of the CCMC project:

1. No increased lung tumor rate after exposure to exhaust gas from:
   - Gasoline engines with and without catalytic converter,
   - Diesel engines (exhaust gas filtered),
   - Diesel engines (exhaust gas unfiltered) with soot particle concentrations 350 times higher than those particle concentrations to be found in large cities.
2. From soot particle concentrations in diesel engine exhaust gas of at least 1000 times higher than those found in large cities lung tumors occurred only in rats.

In this project three different concentrations of unfiltered diesel exhaust gas (with soot particles) were used. In hamsters, generally no tumors occurred. In rats, lung tumors were found after exposure to the medium and the highest dose.

Table 5.3 shows the number of lung tumors classified acc. to male and female rats and the total number of rats with tumors. The data suggests a dose-effect relationship. It is only from medium doses and higher concentrations that the lung tumor rate is higher than the low spontaneous lung tumor rate in the control animals.

Table 5.3: CCMC project: lung tumor rate of rats after exposure to unfiltered diesel exhaust gas, i.e., with soot particles, with different doses

Dose	Number of dissected rats	Male with tumor	Number of dissected rats	Female with tumor	Number of dissected rats	Total with tumor
low	72	1*	71	0	143	1*
medium	72	3	72	11	144	14
high	71	16	72	39	143	55
Control	134	2*	126	1*	260	3*

*spontaneous rate

5.7.4 Worldwide Projects

In 1986 a congress took place in Tsukuba near Tokyo where, among others, reports were given on all important globally carried out inhalation experiments with diesel exhaust gases.

Table 5.4 shows an overview of all projects including the diesel engines used, highest doses and animal species as well as the tumor rates found.

The corresponding proportional tumor incidence in relation to the total dose (standardized cumulatively over exposure time times particle concentration in $mg \cdot h/m^3$) are shown in Fig. 5.8.

As can be seen from Fig. 5.8 in the case of inhalation there is a strong dependency of the tumor rate upon the total dose. The suggestion of a threshold value is of particular importance. Fig. 5.9 shows this result in a different form, where the tumor rate taken from the experiments of Battelle and Lovelace is shown against the calculated amount of particles deposited in the lungs. At the point of intersection of the regression line with the x-axis a threshold value can be seen which is, however, not statistically verified due to the great dispersion.

If, in contrast to the above mentioned cancer hypothesis, such a threshold value exists below which no tumor occurs, one speaks of a so-called *epigenetic* effect. Its mechanism has not been explained yet.

Test results show that lung function is impaired when there is exposure to high concentrations of particulate material. If the particles are not expelled from the lungs by natural mechanisms as is normally the case they accumulate in the lungs. If the hypothesis of a threshold value were verified by further tests, then this would indicate that cancer formation is not to be expected at a low particle dose. The scientific directors of the Tokyo conference thus recommended that further tests be carried out. If, however, the hypothesis were verified that a biological-chemical carcinogenesis were involved, then the risk for humans would have to be assessed.

Table 5.4: Long-term inhalation studies with diesel engine exhaust gas (data and results)

Laboratory/ country	Vehicle/ operation	Exposure h/d	Exposure d/w	Highest dose mg h/m³	Animal species	Tumor rate, with malignant tumors in parentheses %
Niosh Japan	7 l CV (mining) stationary	7	5	6,720	Rats Monkeys	0 0
SWRI USA	5.7 l PV stationary	20	7	12,600	Rats Mice	not conclusive 0
GMR USA	5.7 l PV stationary	20	55	15,840	Rats Pigs	0 0
Lovelace USA	5.7 l PV FTP 72	7	5	29,400	Rats Mice	13.1 (7.7) not analyzed
FhG/ITA Germany	1.5 l PV FTP 72	19	5	53,200	Rats Mice Hamsters	17 (1.0) not conclusive 0
Battelle Switzerland	1.5 l PV FTP 72	16	5	50,688	Rats Hamsters	38.5 (32.2) 0
Saitama Japan	0.3 l small engine idling	4	4	6,656	Rats Mice	0 12.4 (4.1)
Kyushu Uni Japan	11 l CV 1.8 l PV stationary	16 16	6 6	49,920 24,960	Rats Rats	6.5 not conclusive
Matsuyama Japan	2.4 l CV stationary	8	7	28,537	Rats	42.1 (26.3)

FTP 72 ≙ Part of driving cycle acc. to US-75 Test, refer to Chap. 8, CV = commercial vehicle, PV = passenger vehicle

Quotation from "Conclusions and Recommendations" by Dr.R.McClellan:

"There is also a need for studies to clarify the relative roles of epigenetic versus genetic mechanisms of carcinogenesis. For example, it would be useful to conduct studies in which rats and other laboratory animals were exposed to carbon black and other particulate material devoid of organic compounds to see if such exposures induce lung tumors.
A logical follow-on would be studies with specific chemical compounds adsorbed onto fine particles such as carbon black."

At the Tokyo conference, E.L.Wynder, a well-known U.S. epidemiologist, drew the following conclusion. According to him, cancer risk as a result of diesel exhaust gas would be extremely low even in the hypothetical case that no threshold value existed.

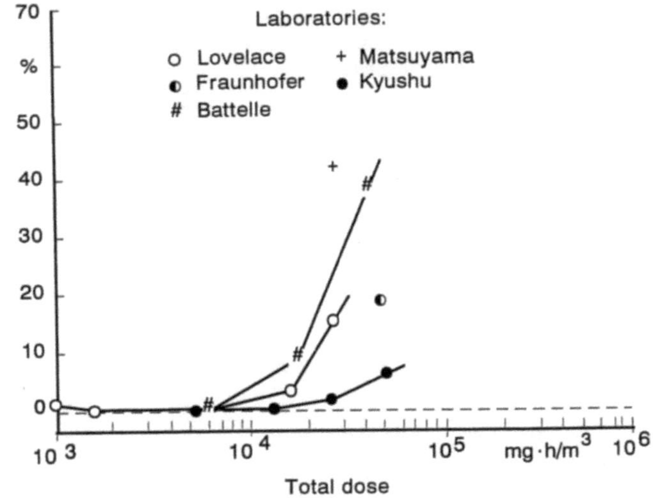

Fig. 5.8: Proportional tumor rate as a function of the total dose (without Niosh, SWRI and GMR, where tumor rates ≈ 0%, refer to Table 5.4)

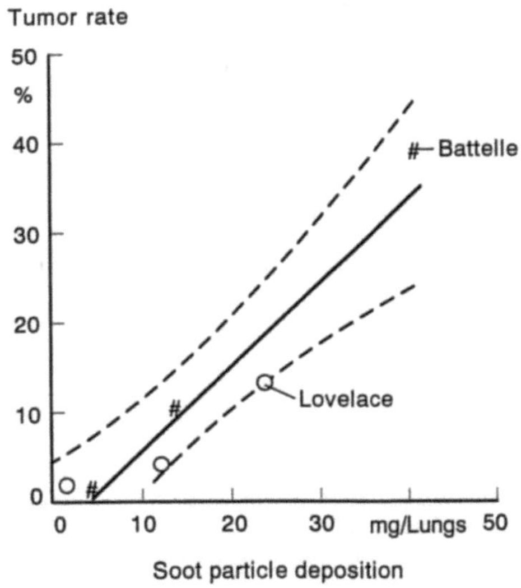

Fig. 5.9: Lung tumor incidence as a function of the soot particle amounts deposited in the lungs (only Battelle and Lovelace)

Risk assessment of Dr.med.E.L.Wynder:

In a concluding American plenary lecture the epidemiologist Dr.med.E.L.Wynder expounded, that the risk of "passive smoking", i.e., involuntarily inhaling cigarette smoke in poorly aired rooms must be small compared to the risk of smoking, but that the risk of inhaling diesel exhaust gas from the roadside must remain behind, by orders of magnitude, the even small risk of passive smoking.

5.7.5 Joint Research Program (JRP) in the Federal Republic of Germany

In 1986 the Federal Minister for Research and Technology (BMFT) and the German automobile industry (FAT) agreed upon initiating an extensive joint research program. An excerpt from the announcement in the Federal Gazette of Feb.1, 1986, reads:

FAT/BMFT - JRP:

Announcement for the promotion of research projects within the framework of a joint research program (Jan. 22, 1986):

"Effects of automobile exhaust gases on health and environment"

This joint research program "Effects of automobile exhaust gases on health and environment" has been divided into the following four focal points:

1. Studies on the dispersion and transformation of automobile exhaust gas components
2. Clinical and epidemiological studies
3. Animal experiments and in-vitro tests
4. Studies on the effects of automobile exhaust gases on ecosystems

Originally, a total of 30 million German marks was to have been spent in five years on effect research, with 50 % contributed by the FAT, i.e., the automobile industry, and 50 % by the BMFT, i.e., the German government. Later on, the program was restricted to 10 million German marks on part of the BMFT and then discontinued.

These 10 million marks were spent mainly on the inhalation-effect research program described below.

Animal experiment-based inhalation studies on the tumor-inducing effect of diesel engine exhaust gases and of two particulate test materials (a FAT/BMFT - JRP Project)

FAT/BMFT - JRP:

Institution:	FhG, ITA Hannover
Program:	Diesel engine exhaust gas with soot particles, aerosol with artificially generated soot particles (Printex), aerosol with titanium dioxide particles
Animals	Rats, mice (NMRI and C57BL/6N species), for the mice: only diesel engine exhaust gas
Period:	July 1987 to August 1992, i.e. 5 years
Cost:	6.1 mio. German marks

The special feature of these experiments was the detection of the mass of soot particles deposited in the lungs.

Rats

In the studies carried out here rats were exposed to three different concentrations of unfiltered diesel engine exhaust gas (exhaust gas dilution 1:9, 1:27 and 1:80) approx. 18 hrs. per day and five days per week over a maximum of two years in special chambers with a volume of 12 m^3. These exhaust gas dilutions contained 7.5; 2.5 and 0.8 milligrams of diesel particles per cubic meter of air.

As a control two other test groups were exposed to titanium dioxide particles (P 25, Degussa Company, Frankfurt) or technical soot (Printex 90, Degussa Company, Frankfurt) with concentrations of 7.5 milligrams particles per cubic meter of air each. The technical soot consisted of pure carbon, PAH were not adsorbed to it.

After concluding the particle inhalation the laboratory animals were kept in clean air for a maximum of another 6 months, so that the total test time for rats amounted to 2.5 years.

Mice

The results described in Chap. 5.7.2 showed a possible carcinogenic effect in the lungs of mice of a NMRI breed when using filtered and unfiltered diesel engine exhaust gas. Due to these results mice of this breed were chosen as the second species of laboratory animal for reasons of comparability. Mice of the NMRI breed have a relatively high spontaneous lung tumor rate.

In addition to the NMRI mice a second breed of mice (C57BL/6N) was used. This C57BL/6N breed of mice has a very low spontaneous lung tumor rate. It can be assumed that these animals react less sensitively to carcinogenic substances in the lungs. Mice of the C57BL/6N and NMRI breed were, in the test group, exposed to diesel engine exhaust gas with a soot particle concentration of 4.5 milligrams per cubic meter air and, in a control group, to the filtered particle-free exhaust gas.

In mice no tumors were found.

Results for rats:

Tables 5.5 and 5.6 show test results for rats.

The tests were concluded in 1992. The results suggest that the deposition of the inert dusts titanium dioxide and pure soot (Printex) indeed leads to comparable tumor incidence in the lungs of rats. PAH apparently play only a small role, if at all, quite in contrast to the decades of discussions, particularly by the EPA, about the decisive role of particle-attached PAH in the formation of tumors. Moreover, lung damage occurs only from concentrations of 0.8 mg/m^3 and more.

Table 5.7 shows the average mass deposited in the rat lungs for the different particle concentrations in the inhalation chambers

After just 3 months of inhaling, damage to the natural clearing mechanisms of the lungs occurs, even at the lowest diesel particle concentration.

At 7.5 mg/m^3 lungs of rats are already heavily damaged. It is probable that the tumors form due to mechanical damage to the cells - one indication that it is an epigenetic effect.

Hamsters and mice thus show no carcinogenic effects.

Table 5.5: Tumors in the lungs of rats after 24 months of exposure time and subsequent exposure to clean air of up to 6 months

		Exposure atmosphere		
			Diesel soot particles	
Tumors	Clean air/ control	Dose 61.7 $g/m^3 \cdot h^{+)}$	Dose 21.8 $g/m^3 \cdot h^{+)}$	Dose 7.4 $g/m^3 \cdot h^{+)}$
Type	Number of tumors x relative to the number of test animals y, i.e., x/y or $\frac{x}{y} \cdot 100\,\%$			
Squamous epithelium tumor (B)	0/217	13/100*** 13.0%	7/200** 3.5%	0/198
Squamous cell carcinoma (M)	0/217	3/100* 3.0%	0/200	0/198
Adenoma (B)	0/217	4/100** 4.0%	2/200 1.0%	0/198
Adeno-carcinoma (M)	1/217 0.5%	5/100* 5.0%	1/200 0.5%	0/198
Bronchiolar papilloma (B)	0/217	0/100	1/200 0.5%	0/198
Hemangioma (B)	0/217	1/100 1.0%	0/200	0/198
Animals with tumors, total Percent, total	1/217 0.5%	22/100*** 22.0%	11/200**a 5.5%	0/198

	(B) = benign tumor (M) = malignant tumor
*	P: ≤ 0.05 (with 95% certainty);
**	P: ≤ 0.01 (with 99% certainty)
***	P: ≤ 0.001 (with 99.9% certainty) [compare clean air group]
a	P :≤ 0.001 (compare with high diesel concentration)
+)	7.5; 2.5 or 0.8 mg/m^3 diesel particles over 24 months
g/m^3 h	To be able to compare different exposure atmospheres with each other, even though particle concentrations were changed in part during the test, the dose as particle concentration in grams per cubic meter multiplied with the total exposure time in hours(h) was used.

Note: In the animals different tumors can occur simultaneously. This explains the total percentages of the tumors.

Table 5.6: Tumors in the lungs of rats after 24 months of exposure time and subsequent exposure to clean-air of up to 6 months

Tumors	Clean air/ control	Exposure atmosphere (Refer to Table 5.5) Diesel soot particles Dose 61.7 g/m³·h a)	Printex 90 Dose 102.2 g/m³·h b)	Titanium dioxide Dose 88.1 g/m³·h c)
Type	Number of tumors x relative to the number of test animals y			
Squamous epithelium tumor (B)	0/217	13/100 13.0%	20/100 20.0%	20/100 20.0%
Squamous cell carcinoma (M)	0/217	3/100 3.0%	4/100 4.0%	2/100 2.0%
Adenoma (B)	0/217	4/100 4.0%	13/100* 13.0%	4/100 4.0%
Adeno-carzinoma (M)	1/217 0.5%	5/100 5.0%	13/100* 13.0%	13/100* 13.0%
Bronchiolar Papilloma (B)	0/217	0/100	0/100	0/100
Hemangioma (B)	0/217	1/100 1.0%	0/100	0/100
Animals with tumors, total	1/217	22/100	39/100**	32/100
Percent, total	0.5%	22.0%	39.0%	32.0%

(B) = benign tumor (M) = malignant tumor

* P: ≤ 0.05 (with 95% certainty);

** P: ≤ 0.01 (with 99% certainty) [compare diesel exhaust gas group]

a) 7.5 mg/m³ diesel particles over 24 months

b) 7.5 mg/m³ over 4 months, subsequently 12 mg/m³ over 20 months Printex 90 - soot

c) 7.5 mg/m³ over 4 months, subsequently 15 mg/m³ over 4 months, afterwards 10 mg/m³ over 16 months titanium dioxide

g/m³·h: standardized particle concentration (s. Table 5.6)

Note: In animals different tumors can occur simultaneously. This explains the total percentages of the tumors.

Based on Fig. 5.8, Fig. 5.10 lists the tumor rate for Printex, titanium dioxide and diesel exhaust gas at high doses. The values fit in with those of the previous experiments.

This supports the conjecture that the formation of lung tumors is to be attributed solely to the overdose of particles (epigenetic effect) which is necessary in experiments for the generation of effects which can be evaluated statistically, and that a threshold value could exist. This would mean that given the low diesel particle dose in the environment carcinogenesis in humans is not to be expected or is at least highly unlikely.

Table 5.7: Deposited particle mass in mg in rat lungs

Type of particles	Concentration mg/m^3	Deposited mass mg
Diesel soot particles	7.5	62
	2.5	24
	0.8	6
Pure soot particles (Printex)	at first 7.5 then 15 then 12	44
Titanium dioxide particles	at first 15 later 10	39

Nevertheless, the senatorial commission of the Deutsche Forschungsgemeinschaft (German Research Association) for the examination of harmful occupational materials has classified diesel exhaust gas into category III-A2, i.e., as a carcinogenic substance in animal experiments. The technical standard in workplaces (TRK Value) has been fixed at 200 $\mu g/m^3$.

Inhalation experiments have thus shown, that diesel exhaust gas with higher soot concentrations is carcinogenic in rats. The precise mechanism of effect leading to

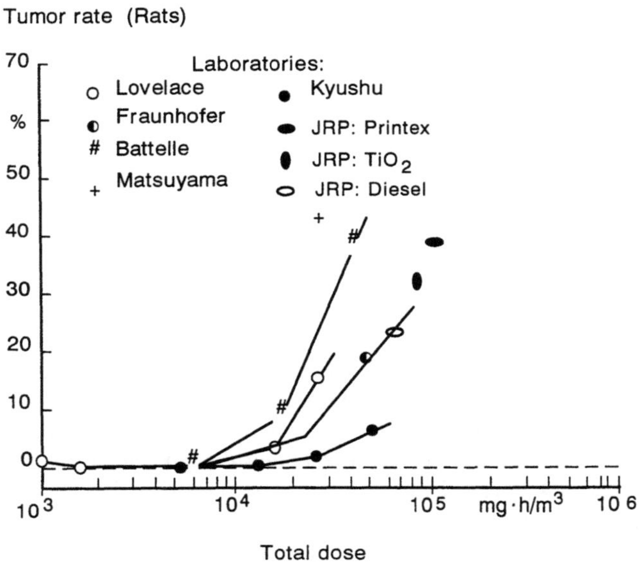

Fig. 5.10: Tumor rates for rats in inhalation experiments of the joint research program (JRP) as additional data to Fig. 5.8

the formation of tumors has not been totally clarified. Furthermore, other poorly soluble fine particulate material such as technical soot and titanium dioxide, administered in comparably high concentrations, caused tumors in rat lungs after having been inhaled with the air. It is therefore *not* a problem specific to diesel soot.

According to the results available so far, the carcinogenic effect is not to be attributed to the polycyclic aromatic hydrocarbons attached to the diesel soot, as was believed up to this point, but to the inner part of the diesel soot particle, the so-called soot core.

The tumor-inducing effect of the soot core has only been statistically verified at higher concentrations. At a particle content of 0.8 milligram per cubic meter of air no tumors were found even in rats. Whether these experiments indicate that diesel soot causes tumors only from a certain concentration onwards, meaning that a so-called effect threshold does exist, could not be determined with absolute certainty.

Inhalation experiments with mice and hamsters show that after these animals were exposed to the same diesel exhaust gas soot particle concentrations as rats there was no tumor formation. It therefore remains unexplained whether this particle effect is specific to rats.

Even if these studies with animal experiments permit no consistent inferences on the question of the carcinogenic effect of diesel soot, it should be said with regard to preventive health protection that a very small risk of cancer caused by fine particulate material in higher concentrations can at present not be excluded for humans, but it cannot be proven either. It can at the very most be a hypothesis.

To sum up it may be said that so far there are no reliable, sensible data indicating a cause-effect relationship between automobile exhaust gases and the diseases in question. It can only be proven that certain disorders increase only when concentrations are high. These limits, however, vary and are considerably higher than the present air pollution values.

5.8 Forest Damage

5.8.1 General Remarks

Forest damage and the term "acid rain" or expressed more accurately "acid precipitation" were discussed vehemently in the public years ago. In the early eighties the public was alarmed by reports that forests were supposed to be dying. These reports continue.

Unprecedented damage, such as high needle losses and other growth defects, mainly to spruce and firs, but also to pine trees and deciduous trees, were reported. Some researchers made the prognosis that gradually all tree species would be affected. They predicted a large-scale dying of forests and disappearance of tree species, e.g., spruce in the Black Forest, in a few years. The scenarios sketched by the media projected that forest owners would suffer tremendous economic losses and that mountain areas would become inhospitable.

The cause given was the complex interaction between rising concentrations of air pollutants from industry and traffic. Photographs of the fir afforestations on the ridges of the Ore Mountains dying on a large scale since the seventies were shown as gloomy scenarios.

The doubts voiced by experienced forest scientists about the occurrence of a new type of epidemic, fell mostly on deaf ears.

Ministries and their subordinate forest research institutes began to trace the development with annual "forest damage inventories", but their results are controversial. A great number of research projects were started to explain the causes.

5.8.2 The Terms "Acid" and "Alkaline"

The terms "acid" and "alkaline" for an aqueous solution are defined by the pH value (negative Brigg's logarithm of the hydrogen ion concentration). The acid scale is from pH 0 to pH 7, with a pH value of 7 indicating a neutral solution. The smaller the pH value, the more acidic the solution.

With regard to the above hypothesis of “acid rain“ it must now be said that meteorological studies have shown that the mean pH value of precipitation in Germany has not changed for 50 years.

Statement:

The mean value of the acid content of depositions (rain, snow, particulate material), expressed with the pH value, has not changed in the past 50 years.

This means: in connection with its effect on vegetation the term **"acid rain"** is an incorrect term.

Seen from a chemical standpoint the following is meant by the term "acid rain" or acid precipitation:

Acid transformation products are acid anhydrides, acids or their salts. These dissolve in fog and raindrops or form aerosols, with considerably higher acid content occurring in fog than in rain. The acid content is mainly caused by sulfur dioxide emissions, as approx. 80 % of them can be retraced in dry and wet depositions.

Of the nitrogen oxides emitted, however, just under 20 % reappear in depositions. The whereabouts of the majority of nitrogen oxides remains unexplained. Due to the type of chemical transformation, the proportion of nitrogen oxide compounds in the total deposition of all acidifiers only amounts to approx. 12 %. The rise of nitrogen oxide emissions in the last decade has been mainly attributed to road traffic. However, at the clean air stations of the German Federal Environmental Agency nitrogen oxide emissions have not increased to the same extent as the increased traffic would lead one to expect.

These data suggest that only a small proportion of the nitrogen oxides emitted by automobiles at ground level get into the atmospheric long-range transport. A possible cause of this is the absorption of nitrogen oxides possibly by plants and soil (fertilizer effect) near the emission source automobile.

Fig. 5.11 shows the course of deposition of exhaust gas components into the environment.

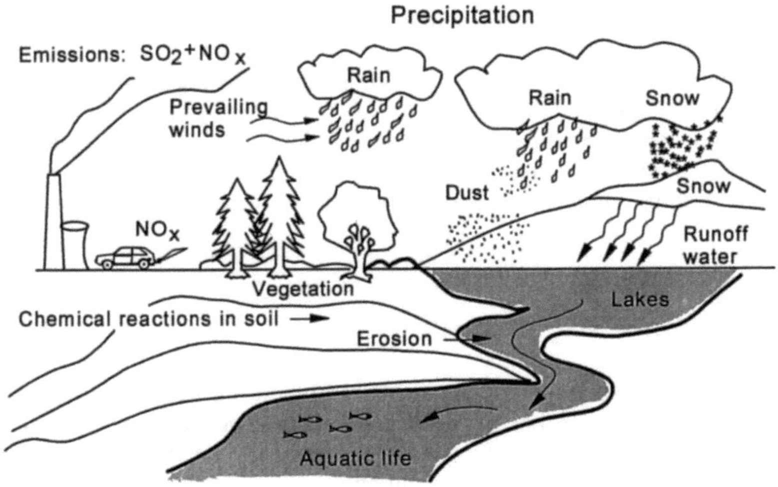

Fig. 5.11: Deposition of exhaust gas components into the environment

Accumulation of substances in soil and lakes formed by further chemical reactions is feared in the long run.

That the problem of "acid rain" is not a new one is shown by Fig. 5.12 with the cover page of an extensive and extraordinary work from the year 1872. In this work the expression "acid rain" was already used.

5.8.3 Hypothesis for the Explanation of Tree Diseases

There are large gaps in our knowledge of the causal correlations of forest damage. These correlations, however, seem to be much more complicated than was generally assumed. All serious attempts at explanation are based on the assumption that there is not only one polluter, but that a series of factors interact with each other.

The following is an overview of some of the most important hypotheses on forest damage (in the meantime there are up to 250 hypotheses)

1st hypothesis: air pollution

- Ambient air concentrations caused by various harmful substances such as SO_2, NO_x (→ acid precipitation);
- Strong reduction of predominantly alkaline dust emissions in the exhaust of stacks in the past years (elimination of acid-neutralising air components and soil depositions).

2nd hypothesis: climatic stress

- Dry spells;
- Strong frost, sudden freezing temperatures in active growth phases of trees (late and early frost)

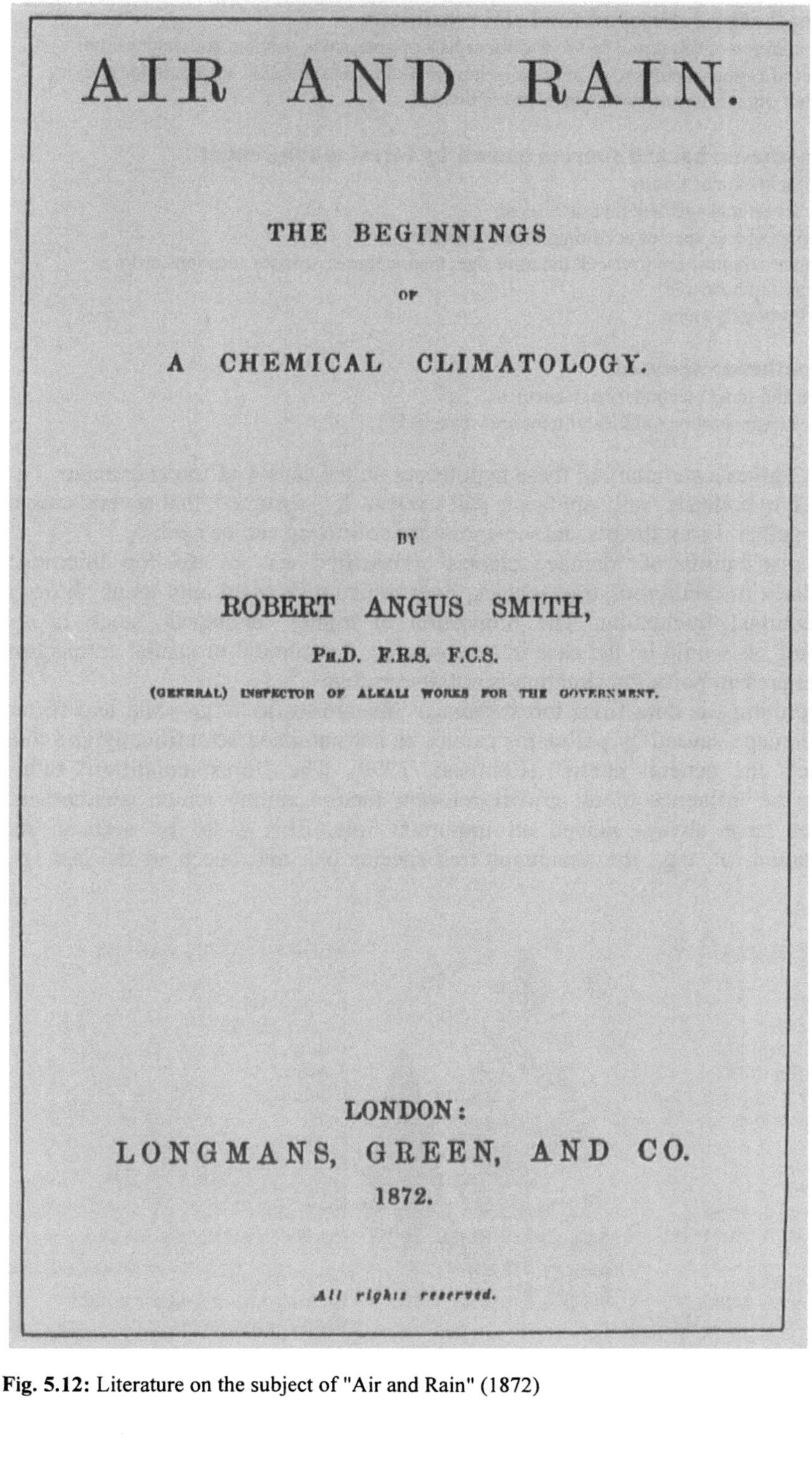

AIR AND RAIN.

THE BEGINNINGS

OF

A CHEMICAL CLIMATOLOGY.

BY

ROBERT ANGUS SMITH,

PH.D. F.R.S. F.C.S.

(GENERAL) INSPECTOR OF ALKALI WORKS FOR THE GOVERNMENT.

LONDON:

LONGMANS, GREEN, AND CO.

1872.

All rights reserved.

Fig. 5.12: Literature on the subject of "Air and Rain" (1872)

3rd hypothesis: deterioration of soil conditions

- Poor condition of forest soil (no fertilization: lack of potassium, calcium and magnesium)
- Soil acidification: mobilisation of toxic metals (in particular aluminum accumulation), dying of higher organisms and microorganisms in the soil

4th hypothesis: hazard sources caused by forest management

- Insufficient care of forests
- Faulty forest management measures as to
  * choice of tree species according to soil condition
  * timber composition (trees all the same age, monocultures, inferior locations as to nutrient household)
  * Browsing by game

5th hypothesis: diseases

- Insects and fungi (secondary parasites)
- Viruses (regeneration abilities of trees are weakened)

Fig. 5.13 gives a summary of these hypotheses on the causes of forest damage.

Which hypothesis really applies is still unclear. It is assumed, that several causes work together. From the present viewpoint the following can be said:

The distribution of "damage classes" (classified acc. to tree-top thinning), particularly in coniferous tree species, over registration years and areas shows a nondirectional fluctuation. The proportion of highly "damaged" stock is not increasing, as would be the case in an epidemic. A temporal or spatial connection with the present pollution situation is not discernible.

Interpreting the data from forest damage inventories as large-scale and recent forest damage caused by pollutants cannot be substantiated scientifically and thus "misleads the general public" (Rehfuess, 1990). The "forest condition" rather reflects the influence of all growth-relevant factors among which weather and parasites have always played an important role. This is to be seen in the development of, e.g., the deciduous tree species oak and beech in the last few years.

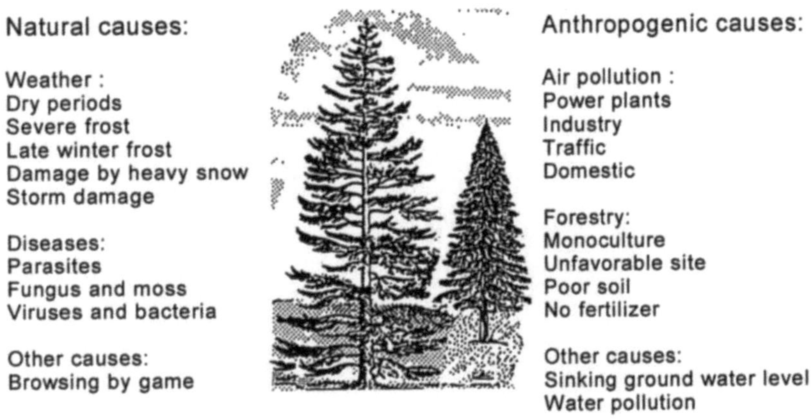

Fig. 5.13: Summary of hypotheses on forest damage

The increased damage to these tree species, above all to oaks in northern Germany, is represented as a new pollution-related catastrophe in some reports. Nevertheless, in none of the examinations of the oak stock in Western, Central and Eastern Europe "air pollutants were identified as relevant damage factors, neither as regards to their effect on leaves nor as regards to mediate effects above ground". The cause complex here is weather stress, insect and fungus damage. Damage to a similar extent are known from earlier decades (mainly from the twenties) and always alternated with periods of recovery.

5.9 Global Environmental Pollution

5.9.1 Natural Greenhouse Effect

Only the natural "greenhouse effect" makes human life possible on earth. Many of the gases present in the air, such as CO_2 etc. are trace gases, refer to Chaps. 2 and 3. They have the property of absorbing the sun's light in certain spectral ranges. The spectrum of incident sunlight is shown in Fig. 5.14. The sunlight is partly absorbed by the ground, warming it up. As a consequence the ground radiates light in the infrared wavelength range. This infrared radiation is absorbed by the gases in the air, refer to Fig. 6.9 in Chap. 6.1.2.6, so that part of the energy remains in the atmosphere and the air is warmed up.

In a greenhouse (or also in an automobile with its windows closed) materials are warmed by absorbing light and they radiate infrared light. This infrared radiation is

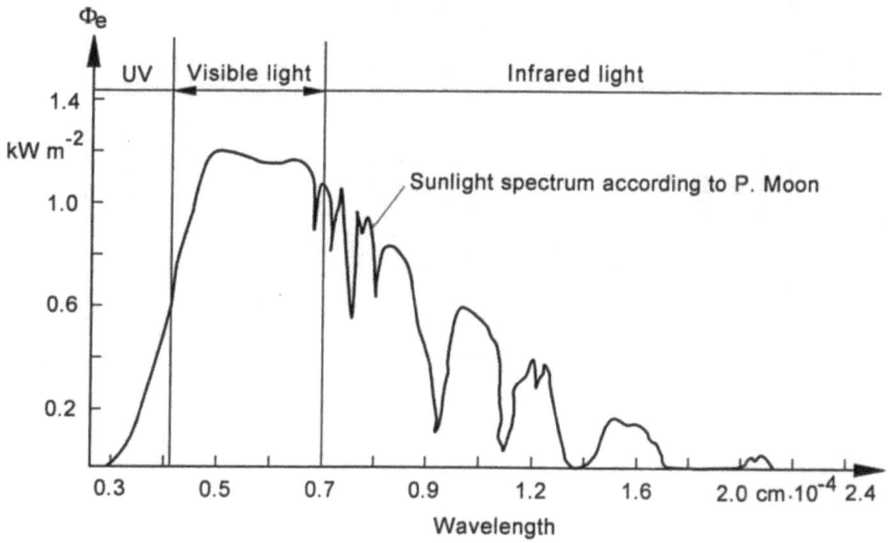

Fig. 5.14: Sun spectrum acc. to P. Moon

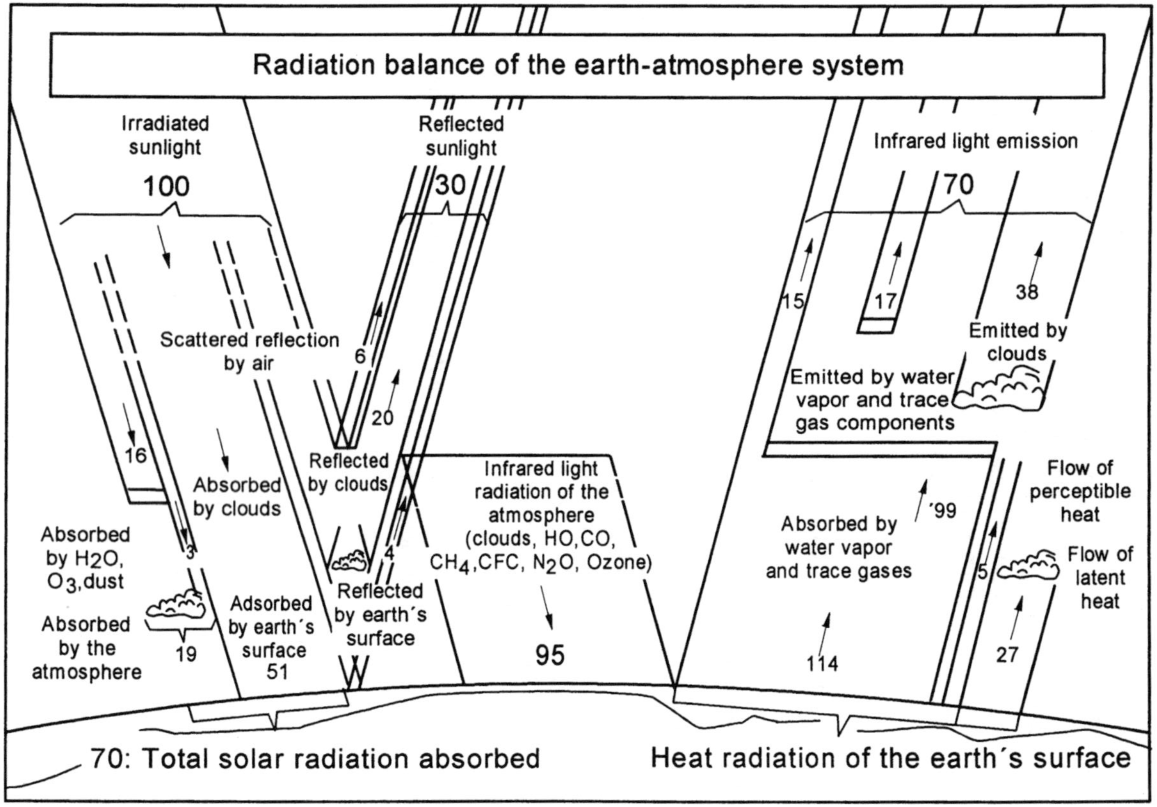

Fig. 5.15: Radiation in the earth - atmosphere system. Reference parameter is the incoming solar radiation. (Source: Interim report of the Enquete Commission of the 11th German Bundestag, 1988)

reflected by glass, heating up the interior. This is a similar effect, thus the name "greenhouse effect" for this process. (However, heat convection and heat conduction should also be taken into account.)

Fig. 5.15 shows the distribution of incoming solar radiation on ground and atmosphere. The balance sums are formed at the outer edge of the atmosphere and on the ground respectively. The left side describes the short-wave radiation flows, the right side the long-wave radiation flows of earth and atmosphere. Reference figure, i.e., the value 100, is the solar radiation coming in at the outer edge of the atmosphere. Changes in the natural concentrations of greenhouse gases lead to changes of temperatures in the earth's atmosphere.

Table 5.8 gives an overview of the most important natural gases and trace gases and their contribution to the crucial warming effect of the earth's atmosphere.

Table 5.8: Contribution of the most important trace gases to the natural greenhouse effect

Trace gas	Atmospheric concentration	Warming effect K
Water vapor (H_2O)	2 ppm bis 3%	20.6
Carbon dioxide (CO_2)	350 ppm	7.2
Ozone, ground level (O_3)	0.03 ppm	2.4
Dinitrogen monoxide (N_2O)	0.3 ppm	1.4
Methane (CH_4)	1.7 ppm	0.8
other	-	approx. 0.6
Total		approx. 33

If the most important trace gases contributing to this greenhouse effect were removed from the earth's atmosphere, then the world mean temperature of the ground-level air layers would drop by approx. 33 K and would then be at -17 °C instead of at the present +16 °C. Life on earth would not be possible.

5.9.2 Anthropogenic Contribution

Man's manifold activities such as energy consumption, agricultural and industrial production, traffic and private household consumption lead to the increasing concentrations of certain trace gases in the earth's atmosphere. However, one must differentiate between trace gases with toxic effects (environmental pollution in a narrower sense due to environmentally relevant trace gases) and those with climatic effects. A trace gas is then termed climatically effective if it has a relatively long "residence time" in the atmosphere (i.e., at least several years) so that it can disperse globally in the atmosphere independent of its emission location,

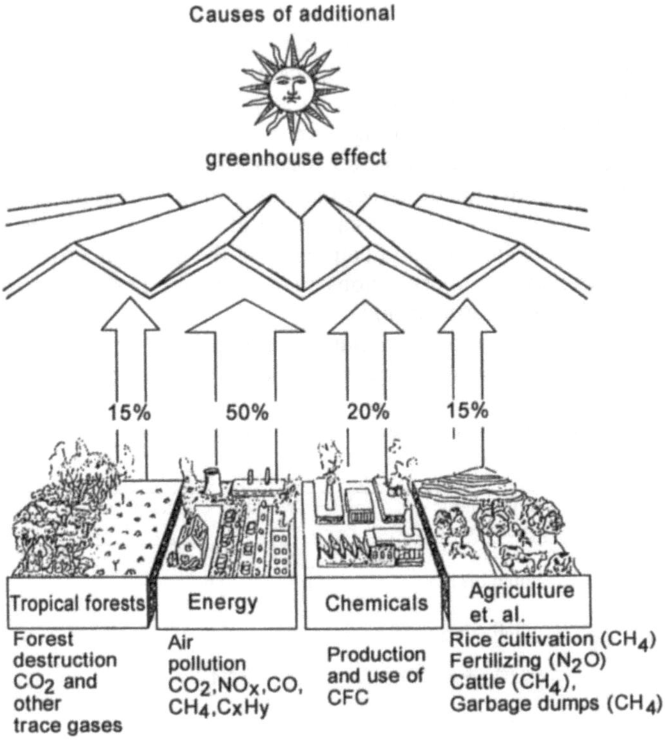

Fig. 5.16: Causes of the additional anthropogenic greenhouse effect. Rounded off percentages for different polluters. (Source: Interim report of the Enquete Commission of the 11th German Bundestag, 1988)

and if it has certain properties of light absorption (solar radiation and radiation by the earth) which can lead to a further warming of the lower atmosphere.

Fig. 5.16 schematically shows the main polluters and their percentages.

As some of the life-spans of greenhouse gases (refer to Table 3.1 in Chap. 3.1) are very long, long-term effects can be noticed.

Table 5.9 lists the properties of "greenhouse" gases.

It can be inferred from this that CFCs (fluorochlorinated hydrocarbons) play a decisive role. Reducing these greenhouse gases thus has first priority.

Fig. 5.17 shows the energy balance between the years 1900 and 1990. Here a mean temperature increase of the earth's surface by approx. 1 K has been recorded.

Table 5.9: Properties of greenhouse gases (Source: Interim report of the Enquete Commission of the 11th German Bundestag, 1988)

Greenhouse gas	CO_2	CH_4	N_2O	Ozone	CFC 11	CFC 12
c (in ppm)	354	1.72	0.31	0.03	0.00028	0.00048
t (in years)	120	10	150	0.1	60	130
$\Delta c/\Delta t$ (in % /year)	0.5	1.0	0.25	0.5	5	3
rel. GP (mol)	1	21	206	2,000	12,400	15,800
rel. GP (kg)	1	58	206	1,800	3,970	5,750
Proportion in %	50	13	5	7	5	12

c Concentration in ppm
t Residence time in the atmosphere and biosphere in years
$\Delta c/\Delta t$ Increase in percent per year
GP (Mol.) Relative greenhouse potential, relative to the same volume of CO_2, in mol
GP (kg) Relative greenhouse potential, relative to the same mass of CO_2, in kg
Proportion Proportion of individual greenhouse gases to the additional anthropogenic greenhouse effect in the eighties of this century

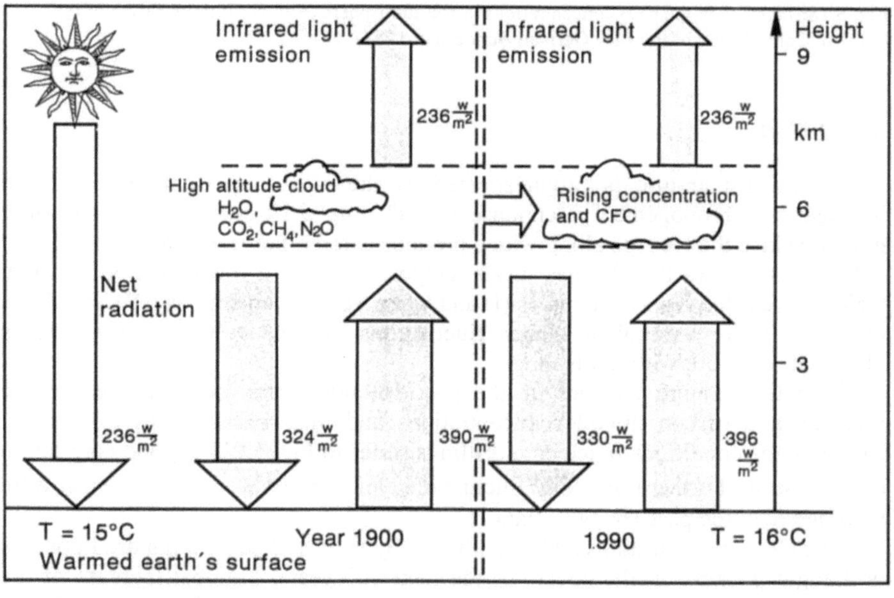

Fig. 5.17: Energy balance of the years 1900 and 1990 (Source: Interim report of the Enquete Commission of the 11th German Bundestag, 1988)

5.9.3 Increase in the CO_2 and Methane Concentrations in the Atmosphere

The practically classic CO_2 pollution measuring results have been explained in Chap. 3.2.3, refer to Fig. 3.7. The rising values of CO_2 concentrations in the atmosphere from the 18th century until today are shown in Fig. 5.18 without the seasonal variations.

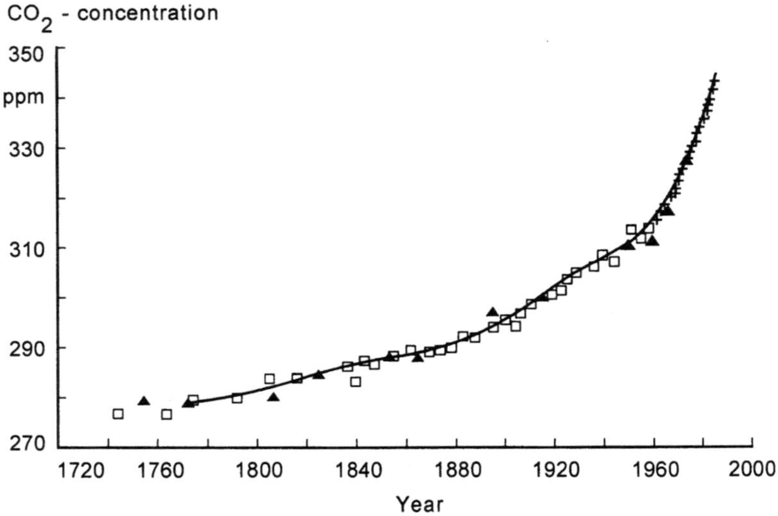

Fig. 5.18: Increase in the CO_2 concentrations in the atmosphere (Source: Interim report of the Enquete Commission of the 11th German Bundestag, 1988)

5.9.4 Effects

A rise in temperature is prognosticated as the most important effect of the increasing anthropogenic "greenhouse effect". Fig. 5.19 shows the temperature increase since the Ice Age along with a prognosis, refer, however, to Fig. 5.20.

Fig. 5.20 shows the changes in air temperature over a longer period of time. Cold periods (Ice Ages) entered as (I) and warm periods, entered as (W) have been included. There have always been fluctuations of several K. The fluctuations observed so far are within this range.

If the temperature changes of the past 160,000 years are compared to the methane and carbon dioxide concentrations and their uncertainties determined from isotope conditions in ice core drillings (refer to Fig. 5.21), a good correlation can be seen. Furthermore, big fluctuations, of e.g. CO_2 concentrations, were recorded over the past 160,000 years.

If the CO_2 concentrations additionally rise by values of several hundred ppm, as is thought possible in the course of the next 100 years, a global increase of the mean annual temperatures in the troposphere by approx. 1-2 °C is prognosticated (Fig. 5.19). The prognosticated increase will be especially pronounced in the higher latitudes, temperatures in the tropics will change minimally. Apart from strong regional climatic changes, such as a shifting of dry zones, such a temperature increase is also expected to lead to a rise of the world oceans by up to approx. 5 m as a result of melting glaciers and particularly of a part of the antarctic ice masses. However, a certain prognosis of the course of CO_2 concentrations expected in the long run und the climatic changes connected with it is not possible at present.

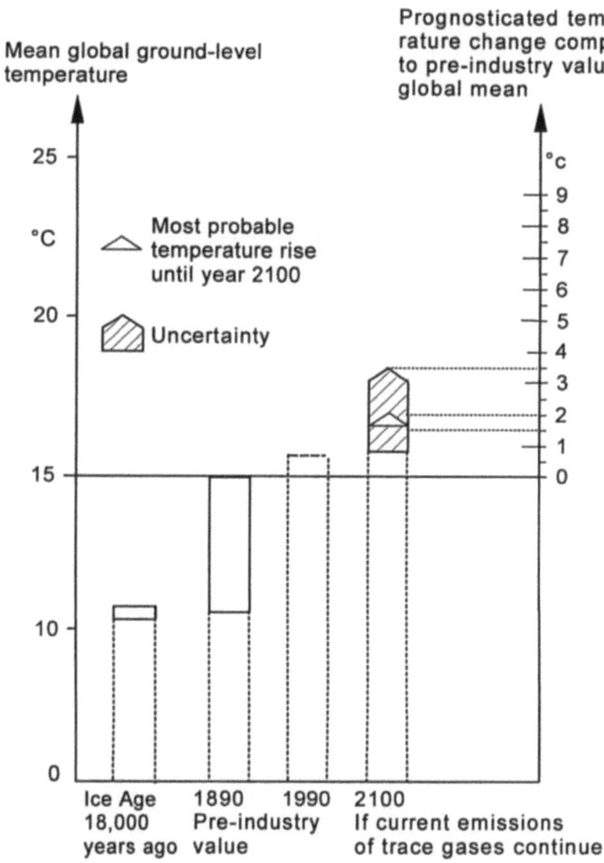

Fig. 5.19: Temperature increase since the Ice Age (since 18,000 years ago) and the prognosticated temperature increases compared to the preindustrial value with continuing trend of the present rates of trace gas emissions until the year 2100. (Acc. to more recent model calculations.)

But the actual problem is the dramatic population development, refer to Fig. 5.22, combined with the increased industrial development of the Third World countries.

Feeding a larger number of people alone causes the methane concentration to rise correspondingly, refer to Chap. 3.2.1. Fig. 5.23 shows the temporal trend of the methane content in the atmosphere together with a curve for world population growth. A good correlation can be seen.

This practically renders the situation hopeless. Although the attempts of the Federal Republic of Germany to lower CO_2 *emissions by 25 to 30 % until the year 2005 are commendable, their effects are insignificant. In terms of road traffic such a decrease has an effect of merely parts per thousand on a global scale and is more than completely neutralized by the consequences of the earth's population growth.*

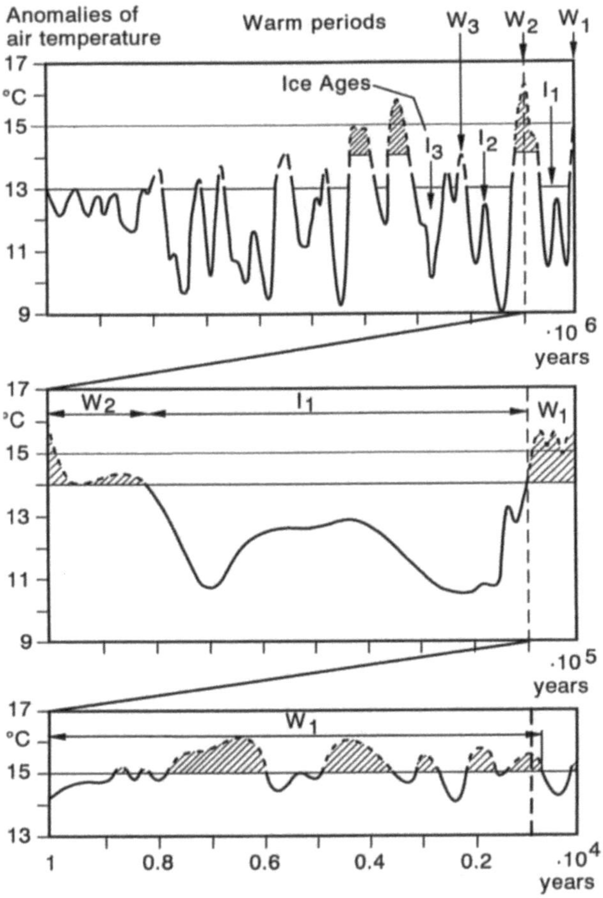

Fig. 5.20: Change of air temperature over a longer period of time with an extended time axis (Source: Interim report of the Enquete Commission of the 11th German Bundestag, 1988)

5.9.5 Critical Remarks

The explanations given in Chap. 5 so far represent current opinion of the scientific community.

Recently, some scientists have published critical papers on the greenhouse effect. E.g., Gerlich (Technical University Braunschweig) points out that there are no computations founded on physics permitting the determination of the mean surface temperature of a heavenly body. This puts the frequently mentioned 33 K temperature difference of the greenhouse effect into the realm of pure fiction. The increase of the mean global temperature used as a basis for the common climatic model computations is made plausible in literature as a consequence of the greenhouse effect. This explanation, however, is equivalent to the description of a

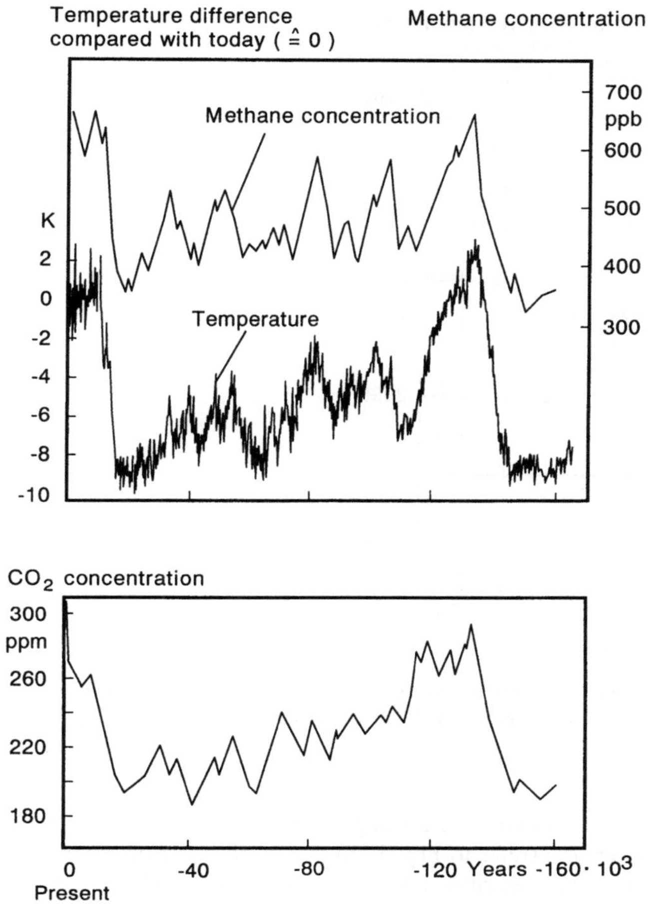

Fig. 5.21: Temporal variation of CH_4 and CO_2 concentrations and temperature during the past 160,000 years (Source: Interim report of the Enquete Commission of the 11th German Bundestag, 1988)

perpetuum mobile of the 2nd kind, i.e., to a violation of the 2nd law of thermodynamics (the earth's surface heats up due to the transport of thermal energy from a colder gas layer at an altitude of 6 km down to the warmer earth, thus defying principle). Gerlich draws the conclusion that such a statement is only possible because in the atmospheric models thermal conductivity, which ought to include all types of thermal transfer (conduction, convection, condensing vapors and radiation), is equated to zero and radiation included in the models is not considered a part of it. As a result of this, convection in particular is left unconsidered when looking at the earth's energy balance.

W. Thüne (of the Ministry for Environmental Protection of the German state Rhineland-Palatinate) also criticizes these points and adds that it is scientifically

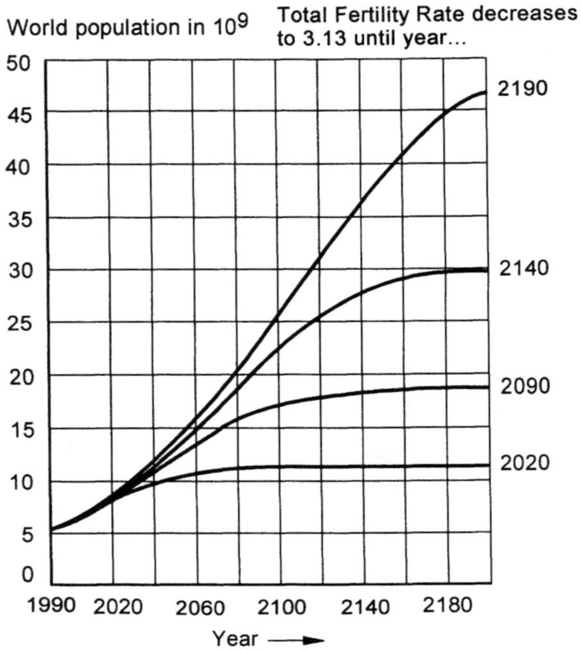

Fig. 5.22: World population development in relation to the speed of an assumed decline in the birth rate (called fertility rate) (in the reference year 1990: total fertility rate = 3.4)

questionable to deduce a causal relationship between an increased CO_2 concentration in the atmosphere (refer to Fig. 5.18) and an increase of the earth's mean temperature by 0.6 K observed since 1860 (refer to Fig. 5.19 and 5.21) and to establish a climatic catastrophe on the basis of this.

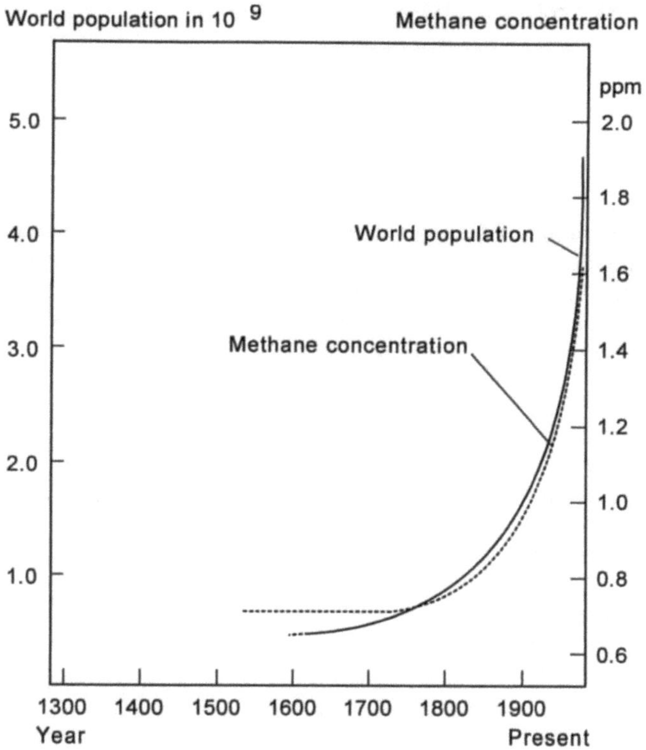

Fig. 5.23: Temporal trend of the methane content in the atmosphere as compared to world population (Source: Interim report of the Enquete Commission of the 11th German Bundestag, 1988)

6 Measuring Methods and Measuring Instruments

For a quantitative determination of the individual limited and nonlimited gaseous components of automobile exhaust gas measuring methods and measuring instruments based on the different physical or physico-chemical properties of the molecules of the individual gas types are used. The most important methods are listed below.

Measuring methods:

Absorption spectroscopy
Chemiluminescence methods
Ionisation methods
Oxygen measurement methods

Separation and measuring methods:

Chromatography
Mass spectrometry

6.1 Absorption spectroscopy

6.1.1 Introduction

Absorption spectroscopy is the most comprehensive method for analysing gas mixtures. It is used in a wavelength range from ultraviolet to microwaves, but mainly in the infared range. The measuring effect is brought about by an interaction of the electromagnetic radiation with interior, discrete energy states of the molecule or its atoms.

If one generally disregards the non-discrete translation energy of the atoms or molecules, the energy of an atom consists of, apart from its nuclear energy, the energy of its valence electrons. The energy of a molecule, however, is composed of three types (if one discounts the interaction of the individual energy types):

- E_e, the discrete energy of the valence electrons,
- E_V, the discrete energy of the vibrations of the atoms bound in the molecule around a position of equilibrium and
- E_r, the discrete energy of the rotations of the whole molecule,

i.e., a first approximation of the total energy E is

$$E \approx E_e + E_v + E_r. \tag{6.1}$$

A more accurate approximation shows that the energy levels are influenced by the interaction of the three energy types. In the case of strong rotation lines, e.g., next to the main lines which are related to the vibrational ground state, one can observe another one or more weaker rotation lines related to the excited vibrational states.

An overview of the excitation types in molecules by electromagnetic radiation in each of the spectral ranges from microwave to ultraviolet range is shown in Fig. 6.1. The three types of energy are specified at the bottom.

In spectroscopy the wave number $\tilde{\nu} = 1/\lambda$ is used instead of wavelength λ, which is also included. With the known correlation

$$\lambda \nu = c_0 , \tag{6.2}$$

where λ is wavelength, ν frequency and c_0 speed of light, and where

$$c_0 \approx 3 \cdot 10^{10} \text{ cm/s} \tag{6.3}$$

is to be substituted, the following wave number is obtained

$$\tilde{\nu} = \frac{1}{\lambda} = \frac{\nu}{c_0} \ . \tag{6.4}$$

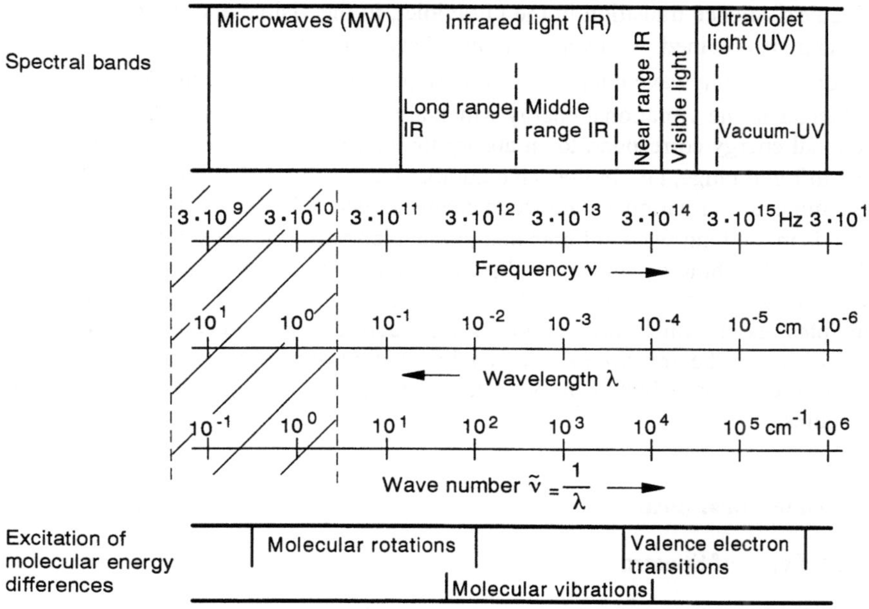

Fig. 6.1: Energy types in molecules for each of the spectral ranges (wavelength ranges)

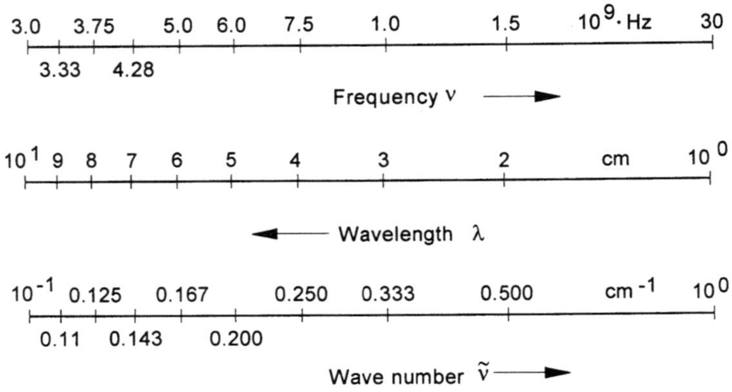

Fig. 6.2: Correlation between λ and $\tilde{\nu}$ for one decade each, refer to Fig. 6.1 (shaded area)

The relationship between $\tilde{\nu}$ and λ is thus not linear. This non-linear correlation is shown in Fig. 6.2 for one decade of the wavelength λ or the wave number $\tilde{\nu}$ of Fig. 6.1 (shaded area) whereby the logarithmic measure in Fig. 6.1. must be taken into consideration.

If the energy state of the valence electrons changes, with the vibrational and rotational states of the molecule also changing as a rule, the spectrum of the molecule's valence transition in the ultraviolet, in the visible and the near infrared (IR) range is obtained, with wavelengths λ between approx. $3 \cdot 10^{-4}$ and $2.3 \cdot 10^{-6}$ cm and corresponding wave numbers $\tilde{\nu}$ of between approx. $3.3 \cdot 10^{3}$ and $4.3 \cdot 10^{5}$ cm^{-1}.

Changes in the vibrational energy which are usually connected with a change in rotational energy correspond to an absorption or emission in the middle and long-wave infrared range, i.e., in the wave number range of $\tilde{\nu} \approx 7.1$ cm^{-1} to $1 \cdot 10^{4}$ cm^{-1}. If the molecule's rotational state alone changes, then only the spectrum is obtained which is in the long-wave infrared and in the microwave range ($\tilde{\nu} < 10^{2}$ cm^{-1}).

Just by absorbing or emitting radiation a molecule cannot change from a given energy state into any other energy state. For these transitions certain selection principles exist which are based in and derived from quantum mechanics. Moreover, for the transitions allowed by quantum-mechanics certain transition probabilities exist which can be calculated in the simplest case of diatomic molecules.

6.1.2 Theoretical Basis

6.1.2.1 Types of Spectra

Depending on the type of generation and observation of the spectrum one differentiates between emission and absorption spectra.

An emission spectrum is created when atoms or molecules excited by radiation absorption or high temperatures or an electron impact release their excessive energy in the form of radiation.

Fluorescence
If atoms or molecules reach excited energy states by absorbing radiation and return to lower energy states by emitting radiation, one speaks of fluorescence. The normal life of excited electron states is around 10^{-8} seconds. This means that the moment the excitation radiation stops, the fluorescence practically ceases

Phosphorescence
There are cases, however, where emission outlasts excitation by up to several seconds. This is called phosphorescence. This long decay period is only possible if long-lived (so-called metastable) excitation states exist.

Absorption
The second type of spectra are absorption spectra. As, acc. to Kirchhoff's law, a body absorbs the type of radiation it emits when excited, spectra can also be observed via absorption. A dissolved or gaseous substance submitted to a continuous spectrum quasi absorbs the wavelengths characteric to it or weakens their remaining intensity. The remaining residual spectrum is measured.

In exhaust gas measuring technology the absorption spectra are of interest. Instead of measuring absorption, transmission *D* is frequently measured, or, to use another designation, transmittance *T*. When reflection can be neglected, absorption is defined as

$$A = \frac{I_0(\tilde{\nu}) - I(\tilde{\nu})}{I_0(\tilde{\nu})} = 1 - \frac{I(\tilde{\nu})}{I_0(\tilde{\nu})} = 1 - T(\tilde{\nu}), \tag{6.5}$$

where $I_0(\tilde{\nu})$ is the incident initial light intensity and $I(\tilde{\nu})$ the light intensity transmitted. Both are dependent on the wave number.

Thus, transmission or the equivalent transmittance is

$$T(\tilde{\nu}) = \frac{I(\tilde{\nu})}{I_0(\tilde{\nu})} . \tag{6.6}$$

The following applies

$$0 \leq A \leq 1 , 0 \leq T \leq 1 , \tag{6.7}$$

where $T(\tilde{\nu})$ is complementary to $A(\tilde{\nu})$ and:

$$A(\tilde{\nu}) + T(\tilde{\nu}) = 1 . \tag{6.8}$$

Multiplied by 100, *A* or *T* are indicated in %.

6.1.2.2 Energy - Frequency Correlation

The basis of optical spectroscopy is the Bohr-Einstein frequency relation:

$$\Delta E = E_2 - E_1 = h\nu . \tag{6.9}$$

It links discrete atomic or molecular energy states E_i with the frequency ν of electromagnetic radiation. The proportionality constant h is Planck's elementary quantum of action ($6.626 \cdot 10^{-34}$ Js or $6.626 \cdot 10^{-27}$ erg s). As it is common practice in spectroscopy to use the wave number $\tilde{\nu} = \frac{1}{\lambda}$ instead of the frequency ν, the result of (6.9) with (6.4) is

$$\Delta E = E_2 - E_1 = h\tilde{\nu}\, c_0 \; . \tag{6.10}$$

Absorbed or emitted radiation of the frequency ν or of the wave number $\tilde{\nu}$ can thus be assigned to certain energy differences or certain term differences:

$$\tilde{\nu} = \frac{\Delta E}{hc_0} = \frac{E_2}{hc_0} - \frac{E_1}{hc_0} = T_2 - T_1 \tag{6.11}$$

with

$$T_i = \frac{E_i}{hc_0} \qquad (i = 1,2) \tag{6.12}$$

as definition of a term.

It follows from the definition of the term that acc. to the SI System[1] it has the dimension m^{-1}. General practice is still the usage of a dimension in cm^{-1}. As the wave number has been indicated in cm^{-1} exclusively in literature published so far, it will be used here as well. For the conversion $1\ cm^{-1} = 100\ m^{-1}$ applies.

The magnitude of the excitation energy acc. to (6.9) or the term differences acc. to (6.11) determine the position of the spectrum in question within the range of the electromagnetic radiation.

6.1.2.3 Electron Band Spectra

From the ultraviolet to the visible part of the electromagnetic radiation spectrum primarily electronic transitions are excited in the molecules which are overlapped by vibrational and rotational transitions. The spectra thus formed are, with the exception of a few cases, not sufficiently structured to permit their application in a selective gas analysis. Absorption-spectroscopic methods working with radiation

---

[1] Système International d'Unités (International System of Units)

in the ultraviolet and visible range of the spectrum therefore cannot be used for gas analysis, apart from a few exceptions.

The formation of the individual spectral lines can be best visualized by means of a level or term diagram (Fig. 6.3).

When radiation is absorbed the electron makes a transition from an orbit closer to the nucleus to one further out in Bohr's atomic model. During emission it drops back from the outer to the inner orbit. Thus, the atom has different energy states in the excited and non-excited state. In the term diagram the energy stages the excited electron of the molecule can enter are represented one above the other by horizontal lines. The transitions occurring during emission or absorption are marked by arrows pointing in the direction of the transition.

The example in Fig. 6.3 is the simplest case of a term diagram for the energy states of the orbital electrons of an atom, a so-called singlet system. Vibrational and rotational conditions have not been included.

Of the transitions included in the term diagram only the transition represented schematically in Fig. 6.3, Transition I, is recorded in normal absorption spectroscopy. Transition II is two-step transition where primarily T_1 must be

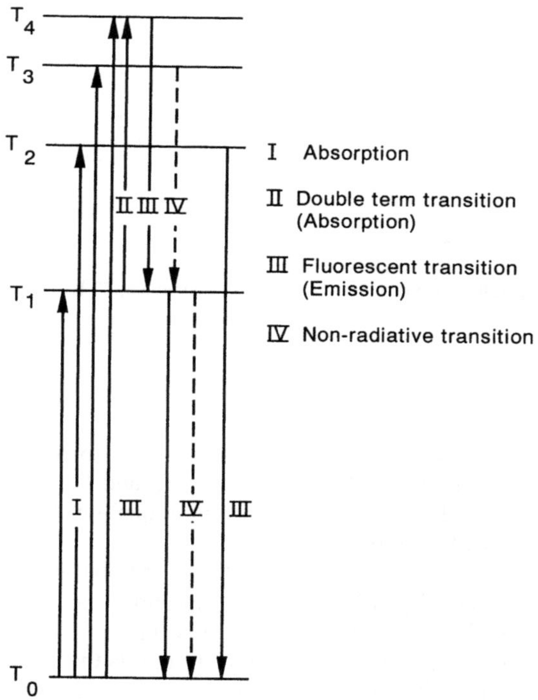

Fig. 6.3: Term diagram of a singlet system, I absorption (transition from ground state T_0 to T_1, T_2, T_3 or T_4), II two-step transition (absorption), III fluorescence transitions (emission), IV two-step radiationless transition to ground state T_0

excited. Transitions III are fluorescence transitions and IV is a radiationless transition, which is called internal rearrangement.

As a result of the influence of the electron spin, the influences of electric and magnetic fields and of the nuclear spin a splitting of the individual energy levels occurs, i.e., the number of discrete spectral lines increases. One then speaks of doublet, triplet, quadruplet and generally of multiplet systems.

The example of an electronic spectrum for benzene is shown in Fig. 6.4.

The right side of Fig. 6.4 shows the laterally reversed course of fluorescence versus absorption with the dashed line forming the line of reference

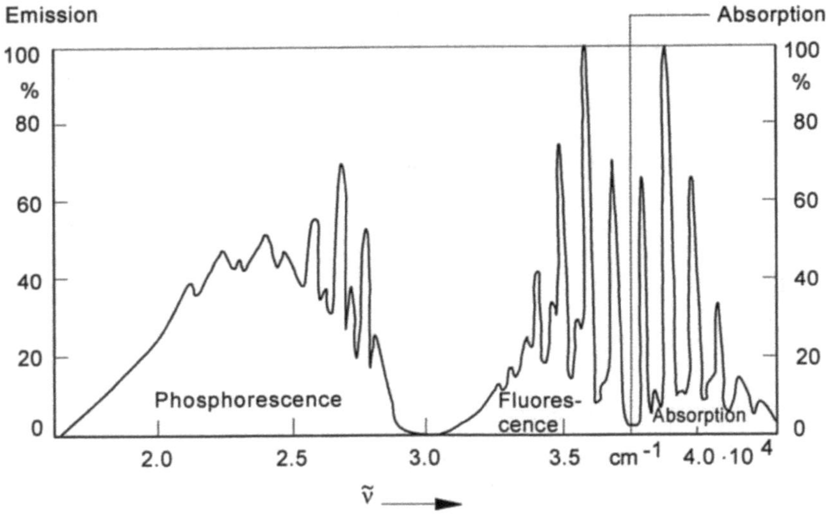

Fig. 6.4: Absorption, fluorescence and phosphorescence electron spectra of benzene at 77 K

6.1.2.4 Vibrational Spectra

The term diagram shown in Fig. 6.5 illustrates the influence of the molecular vibrations on emission and absorption. It shows the ground state T_0 and the first excited state T_1 of a valence electron singlet system with their vibrational levels S_0 to S_5 respectively and the corresponding transitions.

Hence, the fluorescence spectrum provides information on the vibrational states as related to the ground electron state T_0, whereas the absorption spectrum provides the vibrational states as related to the excited electron state T_1.

The more extended transition of Abb. 6.5 is called a 0.0 transition, it is the same both in absorption and fluorescence. The absorption bands stretch towards higher, the fluorescence bands towards lower wave numbers $\tilde{\nu}$. Virtually mirror-symmetrical absorption and fluorescence bands are created, as is implied in the lower part of the diagram.

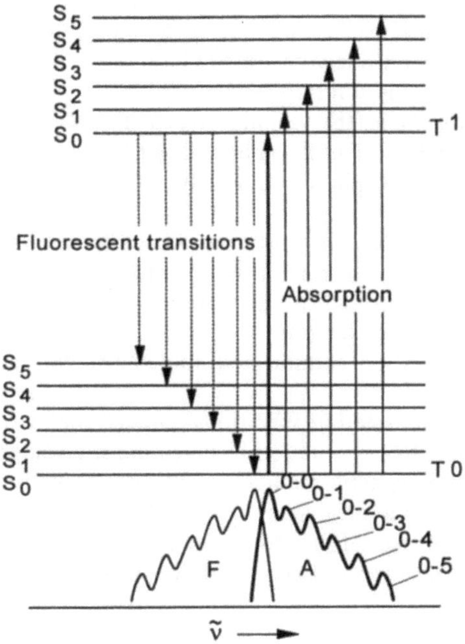

Fig. 6.5: Term diagram with vibrational transitions S_1 to S_5 in the absorption (A) and fluorescence spectrum (F) of a singlet system

Pure molecular vibrations

From a simple viewpoint molecules could be considered an accumulation of mass points, which are kept in their equilibrium positions by elastic but massless springs. Theory shows that a molecule of n atoms can execute barely (3n-6) different vibrations, a linear molecule merely (3n-5).

Example:

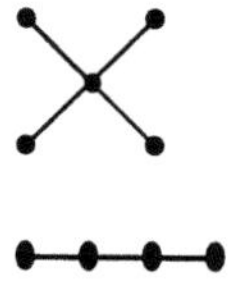

three-dimensional molecule

linear molecule

Each atom can move independently in the direction of three vertical axes of a coordinate system. It thus has three so-called *degrees of freedom of movement*. The atoms of a molecule with n atoms therefore have 3n dregrees of freedom of movement. In a non-linear molecule three of those degrees of freedom are ascribed to the rotations around the three main inertial axes. In linear molecules, however, there are only two degrees of rotational freedom vertical to the molecule's axis. Three more degrees of freedom must be figured as the three translation degrees of freedom for the mass center of the whole molecule. In the remaining (3n-6) or (3n-5) degrees of freedom of movement the distances between the atoms in the molecule change, hence there are (3n-6) in non-linear or (3n-5) vibrations in linear molecules. During this, the mass center remains at rest.

These vibrations are called *normal vibrations*, the vibration types are described by *normal coordinates*. They are independent of each other and are thus able to superimpose with arbitrary amplitudes and phases (linearity of differential equations).

The sum of all normal vibrations, the *vibration spectrum*, characterises the whole molecule. Frequency and intensity of the normal vibrations in the infrared spectra observed are determined by the mass and type of the atoms, the elasticity of the linkages and by the linkage angle and linkage lengths.

Nowadays, even vibrational frequencies of complicated molecules can be calculated.

In the infrared spectrum, absorption or emission of vibrational energy quanta of a normal vibration can only be observed if this vibration changes the molecule's dipole moment. Only then can an electromagnetic radiation field interact with a molecule, i.e., transfer energy to the molecule or absorb energy from it. Infrared intensity is proportional to the square of change of the molecular dipole moment μ with the normal coordinate q, with which the deflection of the atoms during vibration is described:

$$I_{IR} \approx \left(\frac{\partial \mu}{\partial q}\right)_0^2 \tag{6.13}$$

Based on the condition that the vibrations must coincide with dipole moment changes to become IR-spectroscopically effective, it follows that no IR spectra are to be observed in symmetrical molecules such as H_2, N_2, and O_2. Such compounds can, however, be examined with the help of Raman spectroscopy. One therefore speaks of infrared-active or -inactive and raman-active or -inactive compounds. Both methods are complementary. Raman spectroscopy will not be treated in detail here.

Vibration frequencies of molecules

The vibration frequency of a diatomic molecule can be determined approximately with the help of a model of a double-mass vibrator with a massless spring (Fig. 6.6).

To begin with let us consider a mass point m, which is attached to a helical spring. Deflection is described by the x coordinate. The vibration of this mass point can be described, if one uses Newton's law, as
force = mass times acceleration:

$$F = m\frac{d^2x}{dt^2} \tag{6.14}$$

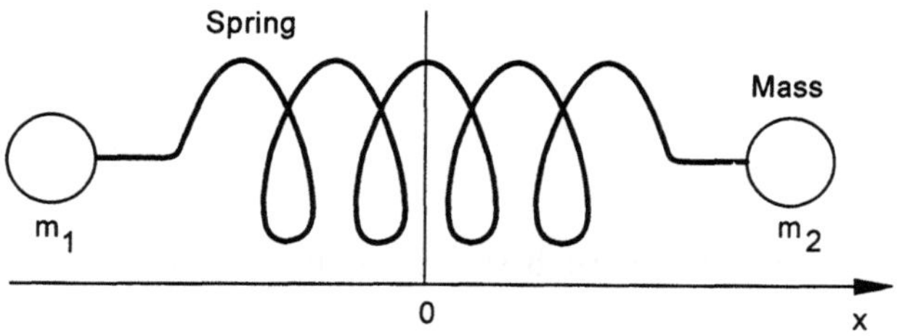

Fig. 6.6: Model of a diatomic molecule with massless spring

as also Hooke's law as:

$$F = -f \cdot x \tag{6.15}$$

According to Hooke's law, the rebounding force is proportional to the deflection x and the spring constant f of the spring. The negative sign is explained by the fact that the rebounding force acts in the opposite direction of the deflection.

The two equations can be combined:

$$m\frac{d^2x}{dt^2} = -f \cdot x\,. \tag{6.16}$$

If the movement is harmonious, such a differential equation has the following equation as solution

$$x = x_0 \cos 2\pi \nu t \qquad (2\pi\nu t = 2\omega t). \tag{6.17}$$

If the second derivative with respect to time is formed from this equation, one gets:

$$\frac{d^2x}{dt^2} = -4\pi^2 \nu^2 x_0 \cos 2\pi \nu t = -4\pi^2 \nu^2 x\,. \tag{6.18}$$

This expression is substituted into (6.16) and one gets:

$$4\pi^2 \nu^2 m = f \tag{6.19}$$

and from this:

$$\nu = \frac{1}{2\pi}\sqrt{\frac{f}{m}}\,. \tag{6.20}$$

This leads to ν as basic vibration frequency of a mass attached to a massless elastic spring. m can also be understood as *reduced mass* of a diatomic molecule with the atomic masses m_1 and m_2, expressed by:

$$\frac{1}{m}=\frac{1}{m_1}+\frac{1}{m_2}. \tag{6.21}$$

With this, the following is obtained for the basic vibrational frequency of a diatomic molecule

$$\nu=\frac{1}{2\pi}\sqrt{f\left(\frac{1}{m_1}+\frac{1}{m_2}\right)} \quad . \tag{6.22}$$

Acc. to (6.4) what follows for the wave number is

$$\tilde{\nu}=\frac{1}{\lambda}=\frac{\nu}{c_0}=\frac{1}{2\pi c_0}\sqrt{f\left(\frac{1}{m_1}+\frac{1}{m_2}\right)} \quad . \tag{6.23}$$

As during the vibration of a molecule its center of mass in principal does not move, the vibration amplitudes of a diatomic molecule are inversely proportional to the masses acc. to the center-of-mass theorem:

$$\frac{x_1}{x_2}=\frac{m_2}{m_1}. \tag{6.24}$$

Light atoms show the largest amplitudes. The vibration amplitudes of hydrogen atoms are in the magnitude of 10^{-12} m.

Example:

trans - dichloroethylene

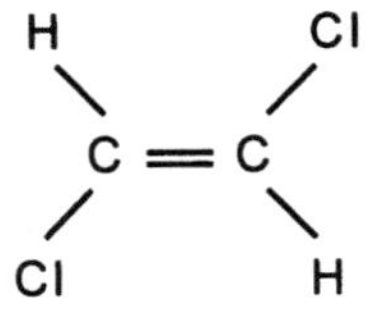

Frequency ν or the wave number $\tilde{\nu}$ of the vibrations of the C=C bond is to be calculated:
Spring constant: $f = 9.4$ N/cm $= 9.4\cdot 100$ N/m $\quad m_1 = m_2 = 12\cdot 1.66\cdot 10^{-27}$ kg .
With (6.23) we get

$$\tilde{\nu}=\frac{1}{2\pi\cdot 3\cdot 10^{10}}\sqrt{9.4\cdot 100\cdot\frac{2}{12}\cdot\frac{1}{1.66\cdot 10^{-27}}}=1630\ cm^{-1}$$

The measured value is $\tilde{\nu} = 1588$ cm^{-1} and corresponds well to the above calculations, if the rough model is considered.

In high-molecular hydrocarbons with many C-atoms, the masses become so great that the vibrational spectra become indiscernible.

Spring constants are almost proportional to the bond order. For carbon-carbon single, double or triple bonds $f = 4.5$; 9.4 or 15.7 N/cm are to be found. A formula with which the spring constant of arbitrary single bonds between atoms x and y can be estimated was given by Siebert:

$$f_{xy} = 7.2 \frac{\xi_x \cdot \xi_y}{n_x^3 \cdot n_y^3} . \tag{6.25}$$

In this case ξ_i is the nuclear charge and n_i the principal quantum number of the valence electron shell of the atom i.

For more complicated molecules the vibration frequency cannot be expressed with closed formulas. Newton's Law is then replaced by the Lagrangian equation of the 2nd order:

$$\frac{d}{dt}\left(\frac{\partial L}{\partial \dot{q}_i}\right) - \left(\frac{\partial L}{\partial q_i}\right) = 0 , \tag{6.26a}$$

with

$$L = T - U = E_{kin} - E_{pot} . \tag{6.26b}$$

In this case the Langrangian function L is given by the difference between kinetic energy T and potential energy U of the molecule in dependency of the i atoms' deflection in generalised coordinates q_i and their temporal change to

$$\dot{q}_i = \frac{d q_i}{d t} . \tag{6.27}$$

The solution of the Langrangian equation of the 2nd order provides the vibration modes and vibration frequencies.

Modes of vibration, example: CO_2 as a linear molecule

The CO_2 molecule is linear and has n =3 atoms. It can therefore perform 3n - 5 = 4 normal vibrations. Fig. 6.7 shows all four vibrations and the diagrams of their dipole moments, which are changed depending on the bond angles and lengths. Two normal vibrations of the CO_2 are deformations of the O-C-O-angle. During this, the O-atoms move in a parallel direction to each other and both together in a vertical direction to the molecular axis with two possibilities respectively, Fig. 6.7 left side. They have the same frequencies, but are independent of each other. The vibrations are then said to be degenerated. This term stems from the fact that the two deformation vibrations can overlap with any phase. As a result of this

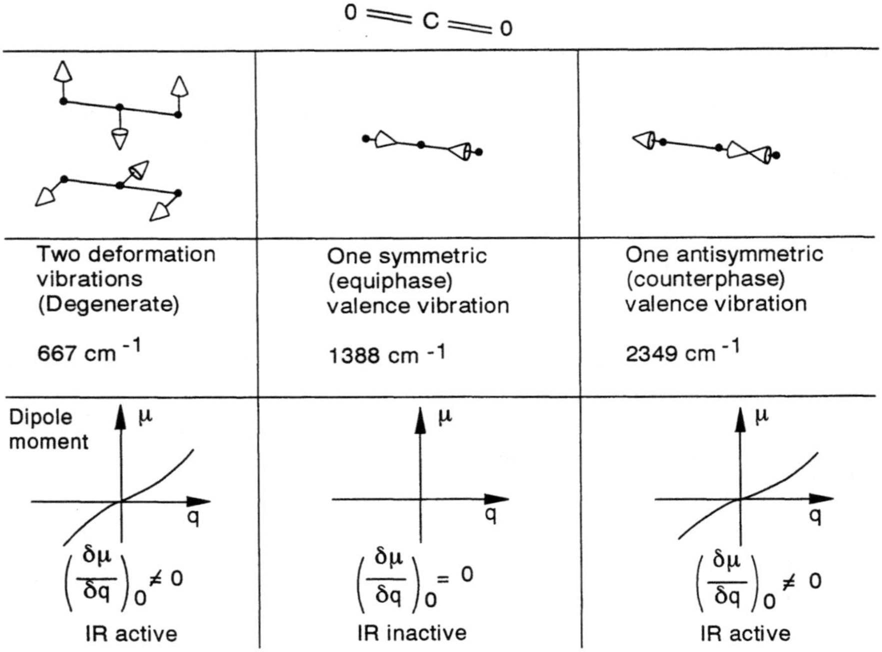

Fig. 6.7: Modes of vibration of the CO_2 molecule and modification of its dipole moment $\frac{\partial \mu}{\partial q}$

overlapping, elliptical movements have been observed in mechanical models, in contrast to the movements on linear paths in non-degenerated vibrations.

Moreover, there is a symmetrical and an antisymmetrical valence vibration. These vibrations, where bonds vibrate with different relative phases respectively are called equiphase and counterphase vibrations.

The CO_2 molecule has no dipole moment when in an equiphase condition, i.e.,. $\left(\frac{\partial \mu}{\partial q}\right)_0 = 0$ is valid. In this state the molecule is infrared-inactive.

However, the two extreme positions of the atoms both during degenerated deformation vibrations and during antisymmetrical valence vibrations have a dipole moment, $\left(\frac{\partial \mu}{\partial q}\right)_0 \neq 0$, of different signs respectively. Hence, these vibrations are infrared-active. The findings are, of course, also true of other linear symmetrical triatomic molecules.

6.1.2.5 Pure Rotational Spectra

Pure rotational spectra have wave numbers of between 0.1 and 10^2 cm^{-1} or frequencies between $3 \cdot 10^9$ Hz and $3 \cdot 10^{12}$ Hz (refer to Fig. 6.1). They can also be used for gas analysis.

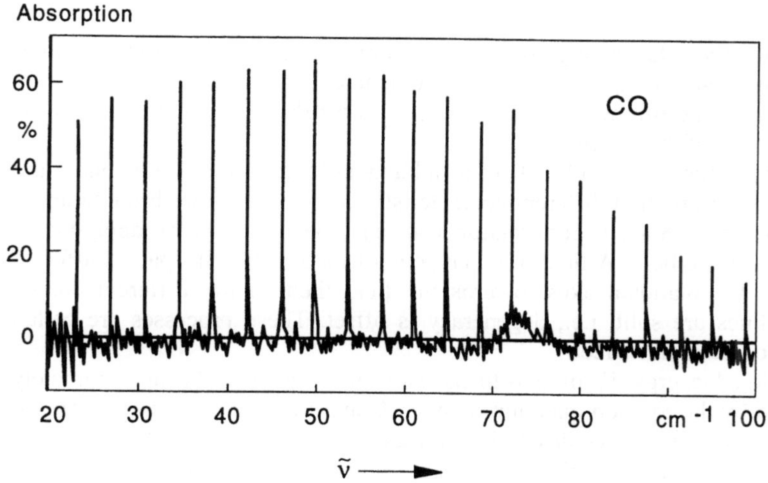

Fig. 6.8: Part of a pure absorption-rotation spectrum of CO, absorption plotted as function of the wave number

As an example Fig. 6.8 shows a part of the pure absorption-rotation spectrum of CO with equidistant lines as functions of the wave number $\tilde{\nu}$.

As a result of this interaction, the rotational states of the molecules change, electron and vibration states remain unchanged. As a rule, absorption spectra are measured.

Rotational energy of a linear molecule, e.g. of CO or HCl, is obtained as solution of the Schroedinger equation of the simple rigid rotator. The solution is given by

$$E_r = J(J+1)\left[\hbar^2 / 2I\right] \qquad J = 0,\ 1,\ 2,\ \ldots \tag{6.28}$$

where

E_r Rotational energy

J Rotation quantum number, being zero or a positive whole number,

$\hbar =$ $h/(2\pi)$,

h Planck's constant,

I Moment of inertia of the molecule with

$$I = \sum_i m_i r_i^2, \tag{6.29}$$

where

m_i Mass of the atom i and

r_i Distance from the molecule's center of mass.

Any energy state with a certain J value is degenerated $(2J+1)$ times; the individual states with the same J value can be differentiated by the directions of the rotational axes.

Degeneracy

In quantum theory states having different quantum numbers (and thus representing different states) but having the same characteristic energy value are called jointly degenerated. The occurrence of degeneracy is, as a rule, linked with some symmetrical property of the system.

Frequently, degeneracy can be offset by adding an "perturbation", i.e., under the influence of a perturbation different characteristic energy values are brought about for different states. Such a perturbation can, e.g., be produced by applying an electric or magnetic field. A field of this type marks a direction in space. Different positions of the rotational axes towards the field then signify different energy states. The lines are split, i.e., degeneracy is offset. These processes are called Stark effect or Zeemann effect.

The rotational energy E_r of non-linear molecules can also be approximately characterized by the rotation quantum number J, and it can be expressed in closed form for symmetrical molecules (with at least a triple symmetry axis such as CH_3Cl, NH_3). For asymmetrical molecules, however, E_r must be determined numerically from the spectra. It generally holds true for molecules of any symmetry, that only those transitions between rotational energy levels correspond to the lines in the spectrum where J changes by ± 1 (or not at all), +1 in absorption, -1 in emission.

Acc. to (6.28) the spectrum of linear molecules is a group of equidistant lines:

$$\Delta\, E_r = \left[\hbar^2 / 2I\right] \sim \tilde{\nu}\,. \tag{6.30}$$

In fact, the spectra of molecules of arbitrary symmetry observed consist of a group of approximately equidistant, inter-structured lines (refer to Fig. 6.8). The distance between the purely rotational lines of a gas is inversely proportional to the moment of inertia. In molecules of high mass (high moment of inertia) the fine structure may disappear and turn into a continuum.

In molecules with the shape of a symmetrical gyroscope (with triple or multiple symmetry axis) two of the principal moments of inertia are the same. Series of almost frequency-equidistant lines, which are generally multiple degenerated, are yielded as absorption spectra. Lines deviating from an equidistant series stem from centrifugal distortions, a result of the not completely rigid molecular structure.

A special case is represented in linear molecules, which are completely described by one single moment of inertia. In this case non-degenerated but equidistant lines have been observed.

In an asymmetrical gyroscope (with only a bidentate symmetrical axis or even lower symmetry) there is no rigid molecular rotational axis left, i.e., the inertia ellipsoid has no rotational symmetry. Circumstances are much more complicated here than in the symmetrical gyroscope. The number of lines is much higher. It is not possible to detect a simple pattern any longer. Lines are distributed irregularly over the entire microwave range reaching as far as the long-wave infrared.

Vibrational fine structure

Dividing energies of an isolated molecule into electron, vibrational and rotational energy cannot be done accurately. An approximation of a higher order shows that the rotational levels are influenced by the vibrations and the electron movement. Hence, in the case of strong rotation lines, one or several other weaker lines caused by the excited vibration conditions can be observed next to the main lines.

Hyperfine structure

It is induced by the rotational momentum of the individual atoms. One nuclear spin of $I_K = 0$ does not have an effect on the spectrum, but with a nuclear spin of $I_K > 1/2$ a weak magnetic interaction is possible. Due to the electric nuclear quadrupole moment a nuclear spin of $I_K > 1/2$ leads to considerably stronger interaction with the molecular rotation. The hyperfine structure observed during this is caused by the different possible settings between the nuclear spin and the intrinsic spin of the molecular rotation. From the point of view of chemical analysis the hyperfine structure complicates the microwave spectrum, which, as a rule, becomes highly complex if more than one atom has a nuclear spin of $I_K > 1/2$.

Stark effect and Zeemann effect

The Stark effect is highly characteristic of rotation transitions. Applying an external electrical field to the gas molecules leads to a line splitting. As the intensity of this splitting depends on the electric dipole moment of the molecule, the Stark effect permits a highly accurate determination of the electric dipole moment.

The Zeemann effect is mainly significant in parmagnetic molecules such as NO, NO_2 and O_2, where a relatively wide Zeemann splitting is obtainable in an external magnetic field.

6.1.2.6 Rotational-Vibrational Spectra under Normal Conditions

What is meant by rotational-vibrational spectra under normal conditions are complex molecular spectra, resulting from the overlapping of a molecule's rotational and vibrational spectra, hence being an expression of both rotational and vibrational excitation. One speaks of vibrational bands.

Rotational-vibrational spectra occur in the wavelength range of infrared spectroscopy, namely in the wave number range of approx. 10^3 to $5 \cdot 10^3$ cm^{-1}. As can be seen in Fig. 6.9., the vibrational excitation leads to absorption bands in this range, i.e., to regularly occurring line accumulations in the spectrum.

At a higher resolution these bands show a rotational fine structure in the form of a multitude of individual equally spaced absorption lines, as is shown in the lower half of Fig. 6.9 with the example of CO.

Spectral position, form and fine structure are parameters specific to each gas, i.e., each gas absorbs within sharply defined frequency intervals or wavelength intervals.

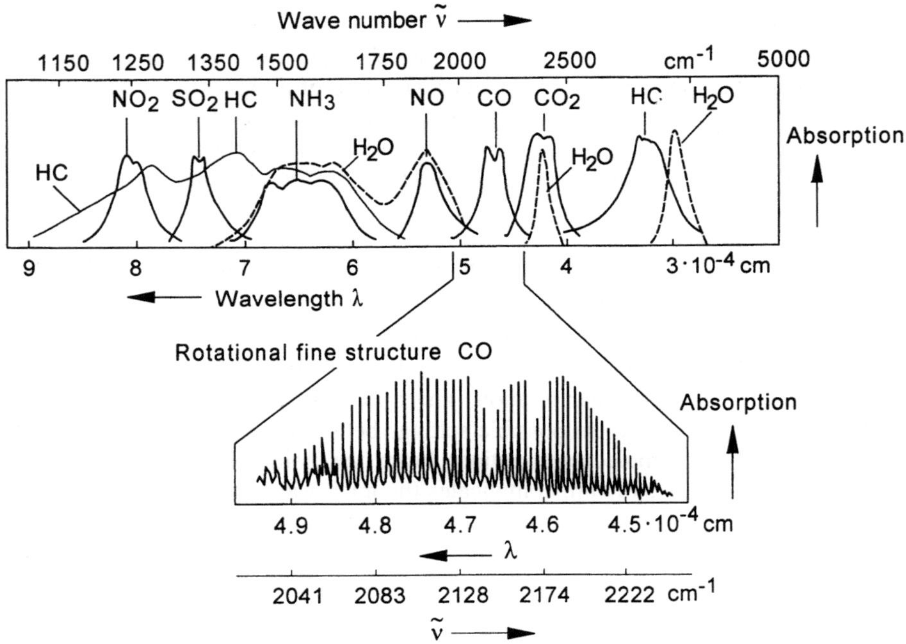

Fig. 6.9: Rotational-vibrational spectrum, vibration bands of different gases and below, as a pull-out, rotational fine structures of the CO bands

This is why radiation absorption in the intermediate infrared range is predominantly used for gas analysis.

As can be seen from the (schematically represented) gas spectra in the upper half of Fig. 6.9, although the spectra of the individual gases are separate, overlapping is also to be observed. Overlapping H_2O bands are particularly disturbing.

The lower section of Fig. 6.10 shows two rotation lines "pulled out" of the CO fine structure in their theoretical form and their theoretical distance.

Pressure broadening

The line form is determined by the collisions of the molecules and depends on gas pressure and gas temperature. Half-band width $\Delta\tilde{\nu}$ and intensity at the mean wave number of $\tilde{\nu}_o$ change in dependency of gas pressure, where the area below the line remains approx. constant. The mean half-band width $\Delta\tilde{\nu}$ of gases with simple molecules such as CO, NO, CO_2, CH_4 etc. is 0.1 cm^{-1} under atmospheric pressure conditions, the mean line distance approx. 4 cm^{-1}.

Doppler effect

The influence of the molecules' movements is small compared to pressure broadening and only responsible for the line form if gas pressure is very low (vacuum).

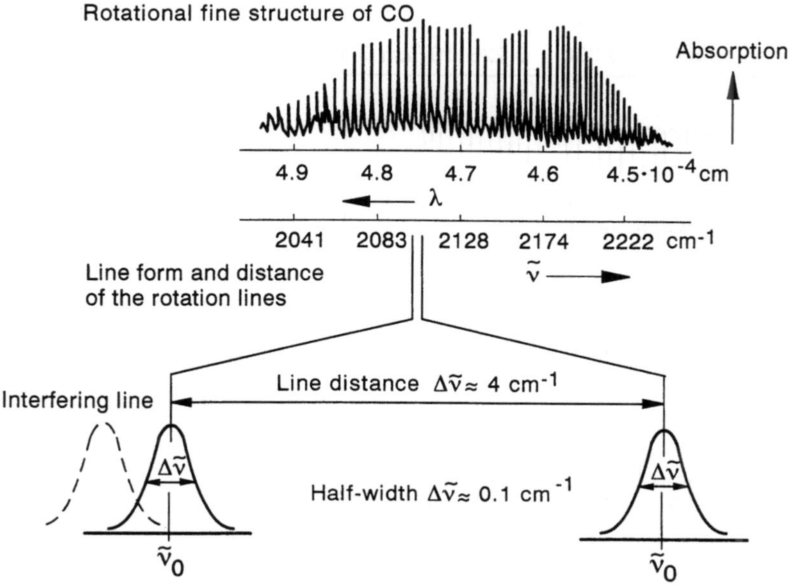

Fig. 6.10: Form of and distance between two CO band rotation fine structure lines at atmospheric pressure, lower section

Cross sensitivity

An interfering line of another gas is indicated in the left part of the lower half of Fig. 6.10 with a dashed line. This overlaps with the line to be measured and falsifies measurement. It is therefore important to find spectral ranges in which such an interference does not occur or is very small.

Special conditions are to be found when looking at hydrocarbons. For this, let us look again at Fig. 6.9, upper half.

The upper half of this illustration indicates two absorption ranges for hydrocarbons (HC) (thin line on the left and a line on the right).These wide absorption bands represent a greater number of different hydrocarbons.

A fine structure, similar to the one depicted for CO below, occurs in, e.g., methane (CH_4). In the higher hydrocarbons the complicated molecular structure and the higher mass result in high rotation moments of inertia. The rotation's moments of inertia determine, as has been explained above, the distance between the rotation lines (small moment of inertia meaning greater distance and vice versa). The rotation lines of the heavier hydrocarbon molecules move into such close proximity that a quasi-continuum is formed. Fig. 6.11 shows an example, the upper half for methane, the lower half for propane. Transmittance is shown here, i.e., the lines are pointing downward.

Methane (CH_4) shows a dissolved line structure. It has a small moment of inertia. Even in propane (C_3H_8) a fine structure can only be recognized indistinctly due to the greater moment of inertia.

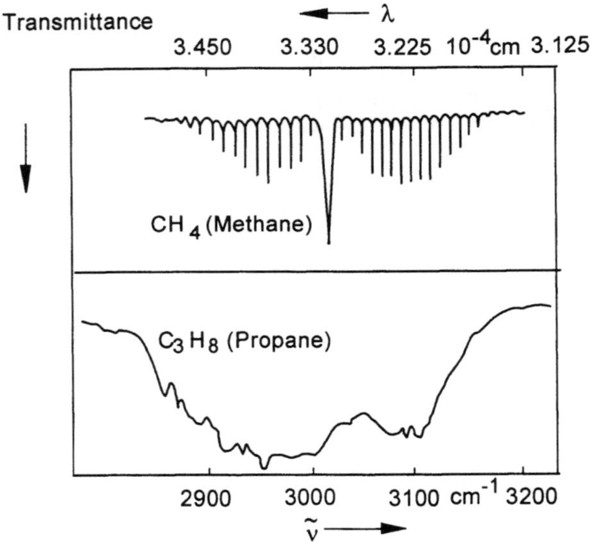

Fig. 6.11: Transmittance for bands of methane and propane

6.1.3 Lambert-Beer law

When measuring absorption in a medium, e.g., in a gas, reduction of light intensity is measured after passing through the medium. The Lambert-Beer law holds true under the important precondition that *pressure and temperature of the gas are constant in space*

$$I = I_0 \, e^{-\varepsilon(\tilde{\nu}) c_i d} \tag{6.31}$$

or for absorption $A(\tilde{\nu})$ and transmittance $T(\tilde{\nu})$, refer to (6.5) or (6.6),

$$A(\tilde{\nu}) = 1 - e^{-\varepsilon_i(\tilde{\nu}) c_i d} \quad \text{and} \quad T(\tilde{\nu}) = e^{-\varepsilon_i(\tilde{\nu}) c_i d}, \tag{6.32}$$

or for extinction $E(\tilde{\nu})$

$$E_i(\tilde{\nu}) = -\ln T(\tilde{\nu}) = \ln \frac{I_0(\tilde{\nu})}{I(\tilde{\nu})} = \varepsilon_i(\tilde{\nu}) \, c_i \, d \tag{6.33}$$

where

$I(\tilde{\nu})$ Light intensity after passage through the medium,

$I_0(\tilde{\nu})$ Initial light intensity (neglecting reflection),

$\varepsilon_i(\tilde{\nu})$ Molar decadic extinction coefficient of the gas component,

c_i Concentration of the gas component i,
d Thickness of gas layer = absorption path length.

Extinction $E(\tilde{\nu})$ is the measuring parameter proportional to the concentration c_i of the absorbing gas component *i* and its path length *d*. The proportionality constant, the molar decadic extinction coefficient $\varepsilon(\tilde{\nu})$, is a substance-specific parameter depending on the wave number $\tilde{\nu}$.

Strictly speaking, the law applies when the line width of the radiation absorbed is small compared to the line width of the absorption line (which is only possible with monomode laser light), then integration is possible using the line widths, or when $I(\tilde{\nu})$ and $\varepsilon(\tilde{\nu})$ are constant over a large range of $\tilde{\nu}$.

If these preconditions are not fulfilled, the law only applies approximately. The curves of dependency of the absorbed intensity on the gas's density and on the absorption path length are not linear due to the exponential dependency. Moreover, absorption saturation can occur when gas density is high and path length is correspondingly long, so that a further increase of gas density at a constant path length practically does not change the measuring signal.

If integration (6.33) is carried out with the wave number, the following is obtained

$$\int_{\tilde{\nu}_1}^{\tilde{\nu}_2} E_i(\tilde{\nu})\,d\tilde{\nu} = \int_{\tilde{\nu}_1}^{\tilde{\nu}_2} \ln I_0(\tilde{\nu})\,d\tilde{\nu} - \int_{\tilde{\nu}_1}^{\tilde{\nu}_2} \ln I(\tilde{\nu})\,d\tilde{\nu} = d \cdot c_i \int_{\tilde{\nu}_1}^{\tilde{\nu}_2} \varepsilon_i(\tilde{\nu})\,d\tilde{\nu}\,. \tag{6.34}$$

Using the so-called oscillator strength *S*

$$S = \int_{-\infty}^{+\infty} \varepsilon_i(\nu)\,d\nu = c_0 \int_{-\infty}^{+\infty} \varepsilon_i(\tilde{\nu})\,d\tilde{\nu} \approx \int_{\tilde{\nu}_1}^{\tilde{\nu}_2} \varepsilon_i(\tilde{\nu})\,c_0\,d\tilde{\nu}\;, \tag{6.35}$$

(c_0 = speed of light)

with *S* being dependent on line form and thus on gas temperature and gas pressure,

$$\int_{\tilde{\nu}_1}^{\tilde{\nu}_2} E_i(\tilde{\nu})\,d\tilde{\nu} = \int_{\tilde{\nu}_1}^{\tilde{\nu}_2} \ln I_0(\tilde{\nu})\,d\tilde{\nu} - \int_{\tilde{\nu}_1}^{\tilde{\nu}_2} \ln I(\tilde{\nu})\,d\tilde{\nu} \approx \frac{S}{c_0}\,d c_i \tag{6.36}$$

is obtained for the concentration c_i of the component i from (6.34) and ultimately

$$c_i \approx \frac{c_0}{S_i\,d}\left\{\int_{\tilde{\nu}_1}^{\tilde{\nu}_2} \ln I_0(\tilde{\nu})\,d\tilde{\nu} - \int_{\tilde{\nu}_1}^{\tilde{\nu}_2} \ln I(\tilde{\nu})\,d\tilde{\nu}\right\} = \frac{c_0}{S_i\,d} \int_{\tilde{\nu}_1}^{\tilde{\nu}_2} E_i(\tilde{\nu})\,d\tilde{\nu}\,. \tag{6.37}$$

In equations (6.36) or (6.37) the symbol $\approx$ is indispensable as the oscillator strength S is only approximately determined by the definite integral between the limits $\tilde{\nu}_1$ and $\tilde{\nu}_2$. The degree of approximation depends on the interval width $(\tilde{\nu}_1 - \tilde{\nu}_2)$.

If S_i is known and $I(\tilde{\nu})$ and $I_0(\tilde{\nu})$ can be measured, then the concentration c_i with given d can be determined in all routine measurements by the difference of the integrated logarithmic light intensities I and I_0 acc. to (6.37).

Deviations

One distinguishes between true and seemingly true deviations from the Lambert-Beer law. True deviations are deviations which can, in a wider sense, be attributed to chemical changes of the absorbing substance, if the partial pressure (in gases) or concentration is changed. Among these are all changes having a perceptible influence on the absorption coefficient of the gas, such as pressure broadening of the absorption lines.

In the case of *pressure broadening* a linear dependency between the absorption and the product $c_i \cdot d$ approximately exists, e.g., in the case of short path lengths d, if the exponential function is expanded in one series and approximated with the 1st term of the series, resulting in a linear term,

$$A(\tilde{\nu}) \approx c_i \cdot d \tag{6.38}$$

and in the case of long path lengths d there is an approximate root dependency

$$A(\tilde{\nu}) \approx \sqrt{c_i \cdot d}\,, \tag{6.39}$$

if the series development of the radical expression is taken into consideration.

Apparent deviations occur if, when applying the Lambert-Beer law, the radiation used is not as strictly monochromatic as is required by the problem to be examined and the accuracy of the measuring method used. Hence, such deviations have physical reasons.

Measurements with "monochromatic" radiation are, in practice, always (except in the case of monomode laser light) measurements with a spectral line isolated by a filter or monochromator. In addition to this main line ($\tilde{\nu}_1$) adjacent side lines ($\tilde{\nu}_2 \ldots \tilde{\nu}_n$) with smaller intensity always slip through, as filters almost always have a relatively large spectral width and monochromators always transmit some stray radiation of other wave numbers.

If ε_1 is the molar extinction coefficient of the main line and ε_n the one of the sidelines there are basically three possibilities:

1st $\varepsilon_n >> \varepsilon_1$.

The sidelines are weakened more by the absorbing substance than the main lines. Hence, the medium itself has an additional filtering effect for the elimination of sidelines. As these, as assumed, have a relatively small intensity as compared to the main line the error ensuing from their complete elimination is not very serious.

2nd $\varepsilon_n \approx \varepsilon_1$.

This is the case when the substance for the spectral section concerned is approximately "grey" (constant absorption). This state is largely achieved if measurements are carried out in the range of a flat absorption maximum, which is always especially advantageous. The relative error then becomes very small and can frequently be left practically unconsidered. This also applies when using a closely spaced spectral line group for measuring. If, however, measurements are carried out in the section of a steeply declining band, the extinction coefficients frequently differ by several percent despite the closely spaced wavelengths, so that the relative error can take on perceptible values in the case of higher extinctions E.

3rd $\varepsilon_n << \varepsilon_1$.

It is chiefly the main line which is absorbed. The sidelines are negligibly weakened and thus always yield an excessively high total intensity after leaving the absorbing layer. This case leads to considerable errors in measurement.

Fig. 6.12 illustrates these three cases. Fig. 6.13 shows the relative error as relative to the extinction $E(\tilde{\nu})$ with $(\varepsilon_n / \varepsilon_1)$ as parameters.

In practice, direct determination of $\varepsilon(\tilde{\nu})$ or of $E(\tilde{\nu})$, refer to (6.33), is frequently not possible. It is necessary to work dispersively, i.e., with a spectral resolution in closely spaced spectral ranges up to the spectral line, as is common practice with spectrometers.

If this is not possible, calibration must be carried out with a calibration gas of component i (refer to Chap. 8.2.8) with the length of the measuring cell known,

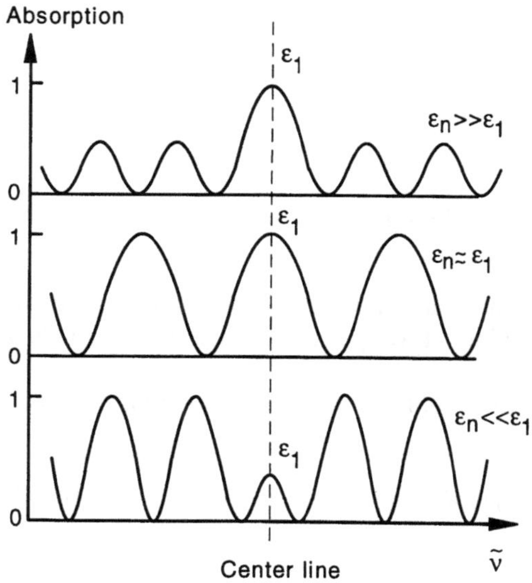

Fig. 6.12: Main and sidelines when IR radiation is not strictly monochromatic

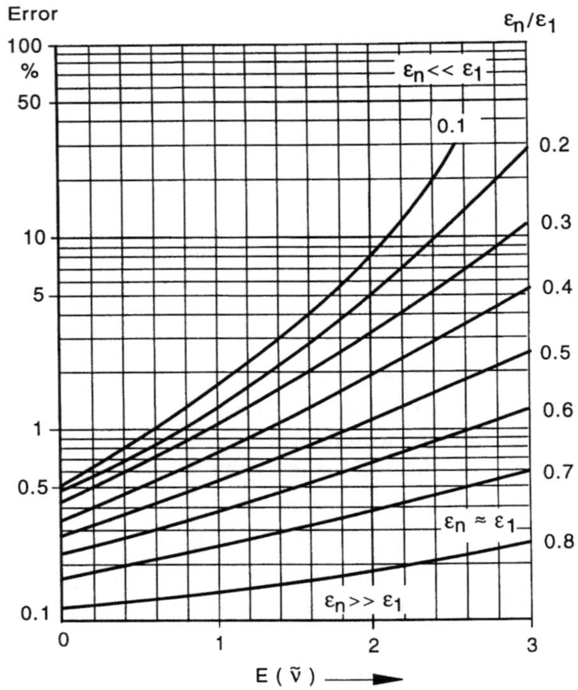

Fig. 6.13: Relationship between relative error of extinction measurement and extinction $E(\tilde{\nu})$ as a result of sidelines at a line intensity ratio of 0.6% and with $\frac{\varepsilon_n}{\varepsilon_1}$ as parameter

and the following is obtained

$$c_i = \frac{1}{f_i} \ln \left(1 - A(\tilde{\nu})\right) = \frac{1}{f_i} \ln T(\tilde{\nu}) \tag{6.40}$$

where f_i is the calibration factor including the sign.

Lambert-Beer law for gas mixtures

The law also applies to gas mixtures and can, thus formulated, also be used for multicomponent gases, if no interactions take place between the individual components.

Hence, one can generally formulate for n components of type i in j ranges of $\tilde{\nu}$:

$$E_{j\,total}(\tilde{\nu}_j) = \sum_{i=1}^{n} E_i(\tilde{\nu}_j) = \sum_{i=1}^{n} \varepsilon_i(\tilde{\nu}_j)\, c_i\, d\ . \tag{6.41}$$

$$E = \sum_{j=1}^{m} E_{j\,total}(\tilde{\nu}_j) = \sum_{j=1}^{m} \sum_{i=1}^{n} \varepsilon_i(\tilde{\nu}_j)\, c_i\, d\ . \tag{6.42}$$

These equations are the basis for analytical application of absorption spectroscopy for multicomponent gases. m linear equations can be set up for n components, if at m different wave numbers j the extinctions $E_i(\tilde{\nu}_j)$ have been measured.

Together with (6.36) the following is obtained

$$E = \frac{d}{c_0} \sum_{j=1}^{m} \sum_{i=1}^{n} S_{ij}\, c_i \tag{6.43}$$

6.1.4 Nondispersive Infrared Measuring Method (NDIR)

6.1.4.1 Operating Principle of the NDIR Measuring Instrument

The principle of the nondispersive NDIR instrument, i.e., a measuring instrument which does not work with spectral resolution, is to be seen in Fig. 6.14 showing a diagram of the set-up (left) and the spectra used (right). In the NDIR measuring instrument the infrared beam emitted by the emitter as wide bands, almost as a continuum, passes through the measuring cell and the identically set up reference cell. The latter is filled with a gas, e.g., nitrogen, which does not absorb in the spectral range concerned. The light leaving the two cells reaches a detector which consists of two chambers separated by a membrane. The symmetrical set-up of

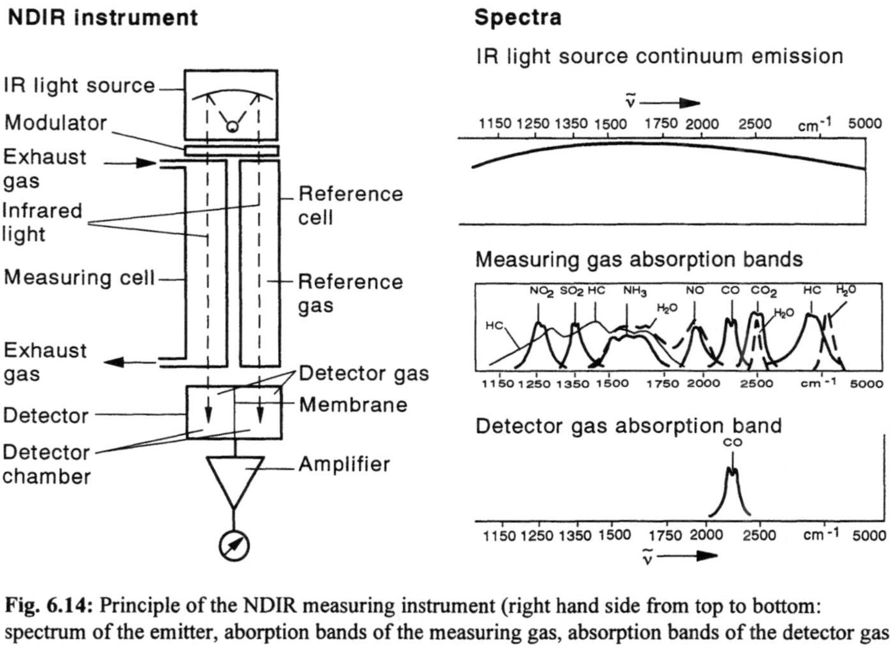

Fig. 6.14: Principle of the NDIR measuring instrument (right hand side from top to bottom: spectrum of the emitter, aborption bands of the measuring gas, absorption bands of the detector gas CO)

both optical paths with the cells' comparable optical properties simplifies evaluation.

The basic idea of this method is based on the fact that the two separate detector chambers below the measuring and reference cell are filled with the gas component to be measured, e.g., CO. CO in the detector chambers absorbs selectively within its absorption band and heats up. This heating is a measure of the light intensity in this spectral range of the CO bands. Thus, the detector is adjusted selectively to the CO bands. This method is called *gas correlation.*

Both detector chambers are heated differently. Under the precondition that further bands of other molecules do not overlap CO bands, the intensity leaving the measuring cell and entering the detector chamber (shown in Fig. 6.14, left-hand side) is, in the spectral range concerned (in this case CO bands), reduced to the extent of the radiation absorbed in the measuring cell by the CO in the exhaust gas when compared to the intensity entering the right detector chamber from the reference cell, as nitrogen does not absorb in the spectral range of interest here. As a consequence, the gas in the left detector chamber below the measuring cell heats up less than on the right below the reference cell. These different degrees of heating or the related pressure differences between the two chambers are measured by a sensor. The deformation of a membrane between the two chambers can, e.g., be used for this purpose.The measuring gas component's concentration (here CO) is determined from the measuring signal. Beforehand, the instrument is calibrated with a reference gas of known concentration (here CO).

Acc. to (6.31) the concentration is dependent on absorption acc. to an exponential function. As the exponential function of smaller exponents can be approximated by linear term, the measuring cell's length should be as short as possible.

If the membrane is, e.g, part of a condenser measuring the pressure difference via the change in capacity, the approximate values of Table 6.1 result at full deflection.

Table 6.1: Data of a membrane-condensor-receiver

Temperature change	ΔT	≈	$2 \cdot 10^{-4}$ K
Pressure change	Δp	≈	$3 \cdot 10^{-2}$ Pa
Deflection of membrane	Δs	≈	10^{-5} mm
Capacity changes	ΔC	≈	10^{-3} pF

The modulator between emitter and cells alternately releases light incidence into the two cells. This generates an alternating voltage or an alternating current at the sensor exit. This creates an independence from zero point drifts and one can work with an alternating voltage amplifier.

Ideally only one gas component is detected with this selective method. If, however, the spectra of several measuring gas components overlap, measuring results could be falsified (cross sensitivity). Frequently, rotation lines also overlap, refer to Fig. 6.10. Remaining cross sensitivity can be eliminated by narrowing the spectral range with the help of interference filters which are mounted between the two cells and the pertinent detector chamber.

The whole set-up is accommodated in a temperature-stable housing, with gas pressure in each of the cells kept constant, refer to Chap. 6.1.3.

According to regulations, CO and CO_2 (as control parameter) are measured with NDIR instruments. Measuring ranges of the different components are between ppm (for CO) and Vol.-% (for CO_2).

6.1.4.2. Principle of Multicomponent Measurement acc. to the Gas *Filter* Correlation Method

The NDIR method can also be used for multicomponent measurements. Fig. 6.15. shows an illustration of a set-up for multicomponent measurement acc. to the gas filter correlation method, which is also suitable for single component measurement, refer to Chap. 6.1.4.1.

Several measuring cells and filter cuvettes are set up in a circular pattern. The gas to be measured flows through the measuring cells set up in series. The emitter's beam is passed through all measuring cells and filter cuvettes. The advantage of this multiple arrangement is that for each measuring component a cell adjusted in length to suit its individual concentration can be installed, thus preventing an overamplification or a saturation.

Behind the corresponding measuring cell sealed filter cuvettes are installed for each component to be measured, containing the measuring component in question, e.g., CO, or a nonabsorbing gas (inert gas), e.g., nitrogen. The filter cuvettes filled with the measuring component thus provide the corresponding spectrum of reference. The measuring gas concentration in the filter cuvette has been selected to *completely* absorb the measuring component's corresponding radiation

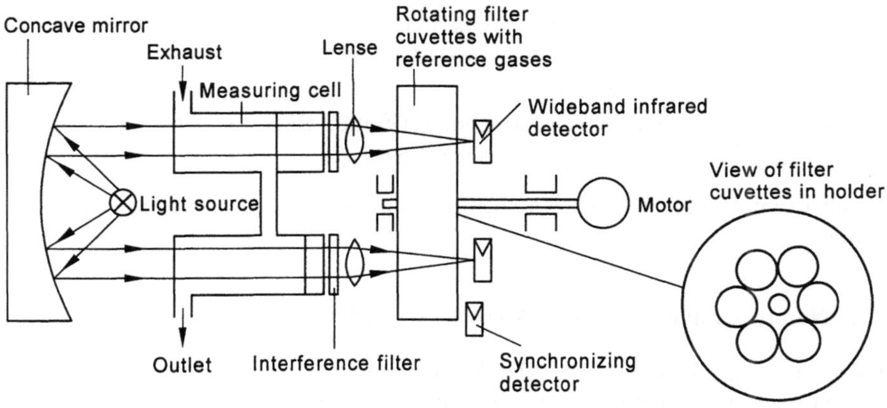

Fig. 6.15: Set-up for multicomponent measurement acc. to the gas filter correlation method

proportion, of e.g., CO. A wideband infrared detector has been included behind each filter cuvette. The measuring effect is periodically brought about by pivoting the filter cuvettes with the gas component to be measured into and out of the optical path. When the filter cuvette with the gas component is swung into the optical path no relevant radiation reaches the detector. When it is swung out of the optical path and the the filter cuvette containing the inert gas is swung in, the radiation reaching the detector has been weakened by the quantity of the measuring component absorbed in the measuring cell. The detector registers the alternating intensity. The difference or the quotient of the detector signals indicates the concentration of the measuring component. Interfering components in the measuring gas, both with and without measuring gas in the filter cuvette, result in the same intensity and become zero when the difference of the signals is formed.

In practice a rotating device is used, containing filter cuvettes for each measuring component and the corresponding cuvettes filled with inert gas. Alignment of the measuring gas filter cuvettes and inert gas cuvettes with the appropriate optical path and receiver is carried out by a synchronizing signal. Pre-filtering is carried out by preliminary interference filters for the spectral ranges of the individual gas components. This prevents interfering overlapping by neighboring spectral ranges. Cross sensitivity must always be guarded against. Calibration with calibration gases is necessary.

6.1.4.3 Electronics

Modern analyzers are microprocessor controlled and linked to an external control unit by standardized interfaces. A selection of the possibilities provided by microprocessor electronics is listed below:

- Convenient user surface with alphanumerical indication;
- Digital display of values measured, sometimes also quasi-analogous display on screen;
- Linearisation of characteristic lines, correction of interference due to cross sensitivity, drift correction;
- Mean value formation over arbitrary periods of time;
- Selection of measuring range and automatic switch of measuring range;
- Pressure and temperature correction;
- Error recognition and status information (operation and error status);
- Open-ended due to bidirection interface (RS 232).

As an example Fig. 6.16 shows such a system configuration by using a block graph.

6.1.5 Nondispersive Ultraviolet Analyzer (NDUV) for NO

The measuring principle of the NDUV analyzer is based on the specific radiation absorption of nitrogen monoxide in the spectral range of around 226 nm, i.e., in the near- ultraviolet spectral range. An absorption line of NO is in this range.

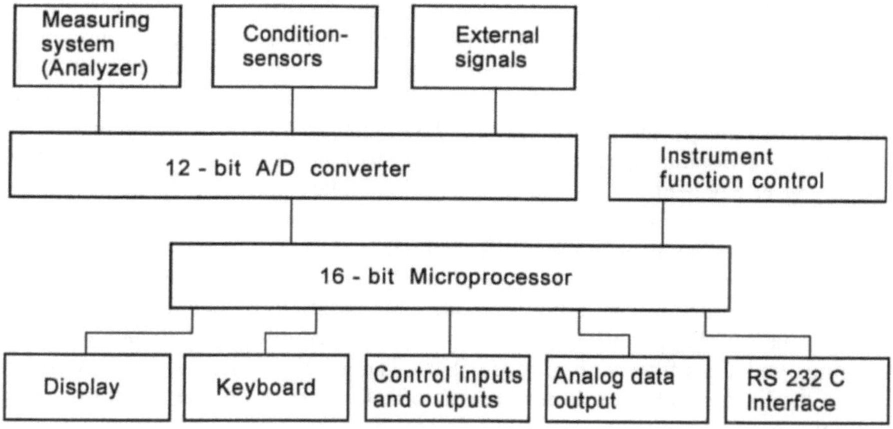

Fig. 6.16: Block graph of the configuration of a microprocessor gas analyzer (Siemens AG)

The actual control panel of an NDIR gas analyzer is shown in Fig. 6.17

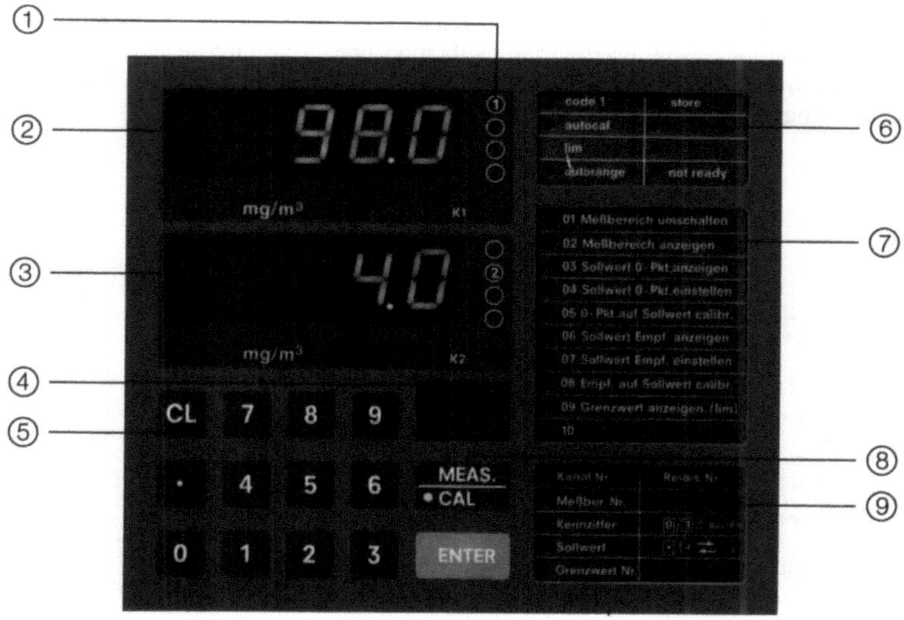

Fig. 6.17: Control panel of an NDIR gas analyzer with 2 measuring channels (Ultramat 5 E, Siemens AG)

1	Measuring range	5	Keyboard
2	Channel 1	6	Operating conditions
3	Channel 2	7	Table of codes
4	Function code	8 + 9	Control panel

Excited NO molecules are formed during a glow discharge in a hollow cathode lamp filled with nitrogen and a small proportion of oxygen at reduced pressure. During the subsequent transition to their ground state these molecules emit an NO-specific radiation in the UV range of around 226 nm. This radiation is absorbed by the NO molecules in the measuring cell. The radiation source is thus selective, it provides an emission spectrum corresponding to the absorption spectrum of the measuring component; a "resonance absorption" takes place. As far as radiation absorption is concerned (gas properties, concentration, cell length) the same interrelationships apply as in the NDIR method.

Fig. 6.18 shows the function diagram of a UV resonance absorption analyzer.

The beam emitted by the hollow cathode lamp passes through a modulation unit whose rotating chopper wheel contains a disc transmitting UV radiation as an opening, and a gas filter with nitrogen (NO). The gas filter filled with NO absorbs part of the lines which then do not have an effect on the NO in the measuring gas. After the opening is pivoted into place all lines pass through. After passing through a collimating lense and an interference filter the beam reaches a partially transmitting mirror used as a beam splitter. One part of the beam passes through the mirror to reach a reference detector, whereas the other part of the beam reaches the measuring detector, after having passed through the measuring cell and been possibly weakened by NO. The measuring signal is determined from the four detector signals by forming a dual quotient. This dual quotient formation also permits calibration by inserting the calibration unit, a vacuum-sealed, NO-filled gas cell with a known NO concentration. Gas temperature and pressure inside the measuring cell are kept constant.

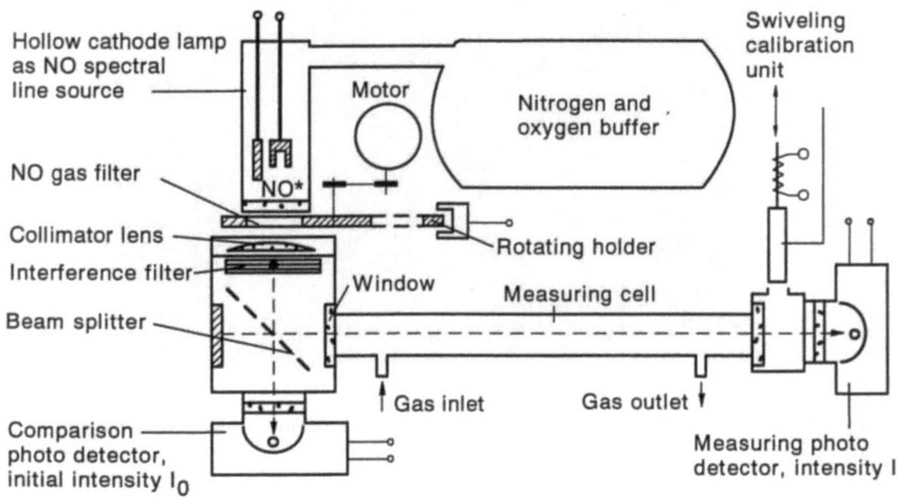

Fig. 6.18: Diagram of an NDUV measuring instrument (RADAS, Hartmann & Braun AG)

6.1.6 Microwave Gas Analysis

Microwave spectrometry used as an industrial measuring instrument for the purpose of gas analysis has only just begun. In contrast to conventional spectroscopy, frequences and wavelengths are used to describe the processes examined. Due to the relatively low energy of the molecules' rotational transitions microwave radiation can be used for excitation. The energy differences of these transitions are small compared to the energy differences in the UV, VIS and IR spectrum. Owing to the extremely discrete structure of pure rotational spectra and to the extremely narrow frequency interval in which microwave sources emit, microwave spectrometry is a highly suitable means in science for elucidating the properties and structure of molecules.

Microwaves require metallic wave guide systems with predominantly rectangular or sometimes even circular or oval cross-section, unless a "stripline" method is used.

Wave guide systems are restricted to limited frequency ranges, as otherwise efficiency losses are too high. Table 6.2 lists the frequency and wavelength ranges of some characteristic microwave frequency ranges (bands) and the corresponding internal dimensions of rectangular wave guides.

Table 6.2: Frequency ranges of microwave bands and wave guide dimensions

Band	Frequency range GHz (10^9 Hz)	Wavelength range mm	Wave guide dimensions mm x mm
X	8.6 - 10.0	35 - 30	25.40 x 12.70
Ku	12.4 - 18.0	24 - 17	15.80 x 7.90
K	18.0 - 26.5	17 - 11	10.64 x 4.32
Q	26.5 - 40.0	11 - 7.5	7.02 x 3.15
R	40.0 - 60.0	7.5 - 5.0	4.57 x 2.18

For microwave spectrometry for gases frequencies of between approx. 10 GHz (excitation of rotational transitions of larger organic molecules) and 150 GHz (excitation of rotational transitions of smaller molecules, e.g., CO excitation at approx. 115 GHz) are suitable.

The selected pressure in the measuring cell must be so low that the width of the absorption lines is as narrow as possible. In practice, pressures are set between 1 Pa and 100 Pa with line widths in the range of approx. 100 kHz to 10 MHz. The low number of absorbing molecules in the measuring cell, combined with the low absorption coefficients of the rotational transitions, leads to a very low total absorption.

Stark effect modulation is used for modulation, where an electrical field with a voltage of up to 2000 V is applied to an electrode installed inside the measuring cell. If the field is modulated with a certain frequency (10 to 100 kHz) e.g., in a rectangular shape, the absorption line in field-free modulation time cycle appears in its original spectral location, whereas in modulation time cycle with field applied the line is shifted, sometimes even split. The radiation of the microwave oscillator precisely adjusted to the frequency of the absorption line can only interact with uninfluenced line, the shifted or split line cannot absorb at the frequency emitted by the source. Thus measuring signals I and incident signals I_0 alternately reach the detector. Fig. 6.19 shows the transmittance of a line at point ν_0 with the Stark field off and its transmittance of lines shifted or split at points ν_1 and ν_2 with the field on.

Moreover, by applying a suitable electric filter to this alternating signal and appropriate amplification, a stable, low-noise signal can be generated.

The diagram of a microwave spectrometer with sampling is shown in Fig. 6.20.

The microwaves generated by the microwave generator are led via an attenuator for power control into a wave guide system adapted to the measuring frequency. The wave guide system contains the measuring cell. A semiconductor diode serves as detector. Impedance matching is carried out with the help of a non-reflecting termination. The measuring cell part of the wave guide is hermetically sealed by windows (e.g. made of PTFE or quartz) transmitting microwaves. The Stark electrode is installed in the middle of the measuring cell. It has the shape of a flat sheet consisting of metal and is electrically insulated against the inner wall of the wave guide.

If a gas component of an exhaust gas is to be determined, the exhaust gas is sucked through the wave guide by a vacuum pump. At one end of the wave guide is a microwave emitter which radiates through the gas mixture in the experimentally required frequency specific to the measuring substance. At the other end, the detector measures the microwave radiation absorbed. As, in mean concentrations, only approx. one millionth of the radiated intensity is absorbed, direct measurement of power loss is not possible without amplification.

As a result of the slight magnitude of the measuring effect, measuring cell lengths of several meters are necessary when using microwave spectrometers. Nevertheless, wave guides of this length are manageable as they can be folded, i.e.,

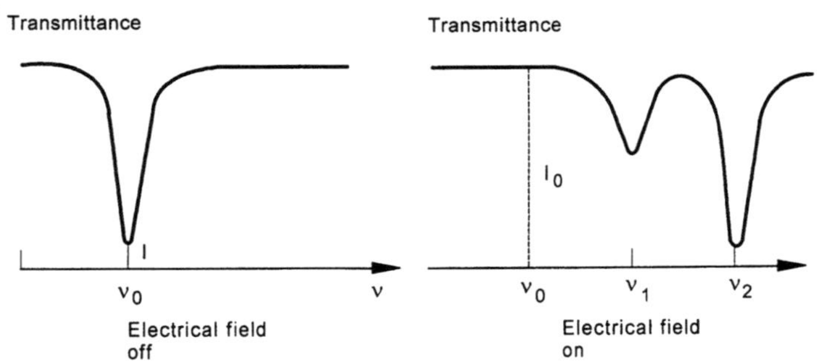

Fig. 6.19: Principle of *Stark* modulation (shifting and splitting of an absorption line, shown during transmittance)

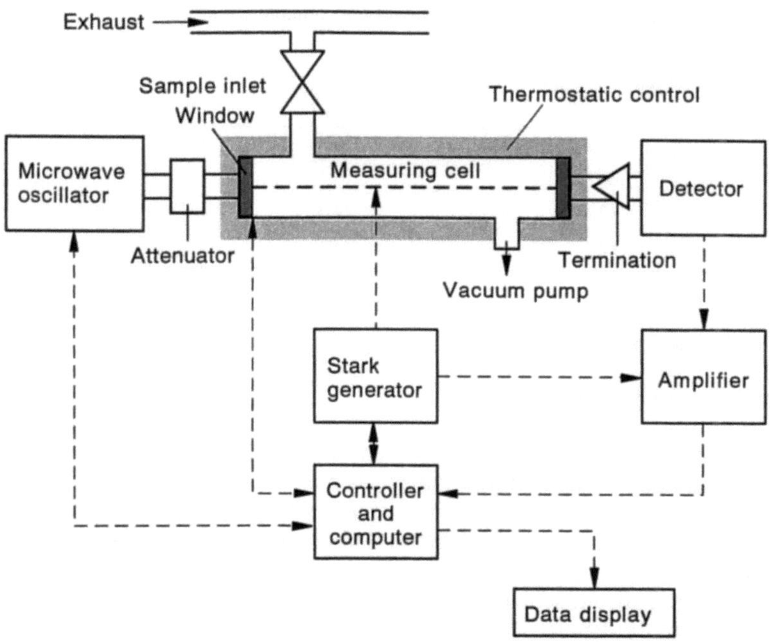

Fig. 6.20: Diagram of a microwave spectrometer with sampling

they can be connected with a 180° semicircular wave guide or they can be coiled in spiral shape. The wave guide measuring cell can be replaced by a resonator.

As described in Chap. 6.1 microwave spectrometry uses the molecule's dipole property of absorbing a specific wavelength by rotational excitation. High selectivity results from the fact that the slightest of changes in the molecule's mass or geometry lead to other excitation frequencies and that, at a pressure of 10 - 100 Pa in the wave guide as a measuring cell, refer to Fig. 6.20, the absorption lines have extremely narrow line widths. High selectivities can be achieved in a frequency range of 20 - 40 GHz and a half-value width of the absorption line of 0.5 - 1 MHz.

Pressure and temperature inside the wave guide are controlled very strictly as they have a great influence on measuring accuracy (temperature: ± 0.1 K; pressure: ± 2 Pa).

For some gas components wavelengths and frequencies of the rotation lines of the rotation quantum number $J = 0$ in the microwave range are shown in Table 6.3.

Assessment:
Microwave spectrometers have not become established as industrial measuring instruments. Frequency ranges of different components vary greatly and require different wave guide systems and wideband or frequency-specific oscillators. However, microwave spectrometers can be used for special purposes.

Table 6.3: Wavelengths and frequencies of some exhaust gas components in the microwave spectral range

Component	Wavelength mm	Frequency $10^9\ s^{-1}$
CO	2.6	114.8
NO	2.9	101.9
N_2O	12.3	24.3
NH_3	12.6	23.7

6.1.7 Fourier-Transform-Infrared Spectroscopy (FTIR Spectrometer)

6.1.7.1 Principle and Method

Fourier-transform-infrared (FTIR) spectroscopy has been recognized as a promising new method for automobile exhaust gas measuring technology. The VW and Nicolet companies jointly, as also Horiba and Pierburg, have developed a new analyzer for many exhaust gas components. Laboratory devices based on this principle have long been known in chemistry, but permit only low time resolutions.

In many absorption-spectroscopic methods a widebanded beam must be divided into sufficiently narrow-banded beams with the help of optical means in the radiation path (dispersive measuring instruments). The wave number band width of these beams must be narrow enough for the radiation to be absorbed only in the spectral range desired within one spectrum, e.g., in the rotation lines of a gas component. In dispersive spectroscopy this spectral dispersion is achieved by dispersive set-ups (e.g. monochromators). In Fourier-transform spectroscopy a double-beam interferometer is used for this purpose.

The double-beam interferometer is an optical device for splitting one into two beams whose relative difference of their optical path lengths can be varied. This creates a phase difference between the two beams. This leads to the phenomenon that after recombining the two beams, interference effects can be observed which are a function of the path length change Δs between the two beams in the interferometer.

In Fourier transform infrared spectroscopy - it must be kept in mind that only absorption spectroscopy in the medium infrared spectral range is of interest here - the most frequently used interferometer is the Michelson interferometer. The basic function of a Michelson interferometer is illustrated in Fig. 6.21.

The Michelson interferometer contains two plane mirrors perpendicular to each other. One mirror is fixed, the other can be moved along an axis placed at right angles to its surface. An optically semitransparent medium, the beam splitter, is placed between the two mirrors in diagonal position to their surfaces. The beam splitter is "ideal", when reflection and transparency are exactly identical; this is only possible at a certain frequency.

Just exactly how a Michelson interferometer provides spectral information is illustrated below by the simple, idealised case of a monochromatic radiation of the wave number $\tilde{\nu}$ and intensity $I_0(\tilde{\nu})$.

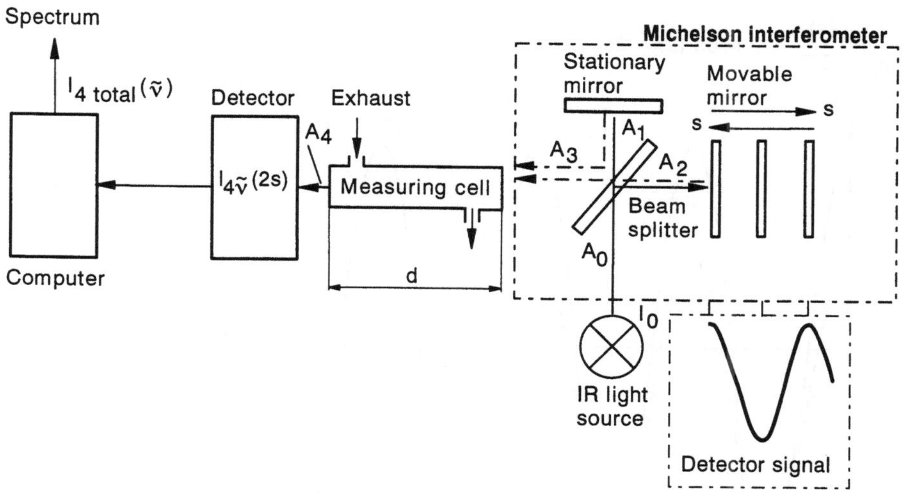

Fig. 6.21: Diagram of a Fourier-transform spectrometer

An ideal beam splitter splits beam A_0, emitted by the radiation source, of intensity I_0 into two beams A_1 and A_2 of intensities $I_{1\tilde{\nu}} = \frac{I_{0\tilde{\nu}}}{2}$ and $I_{2\tilde{\nu}} = \frac{I_{0\tilde{\nu}}}{2}$. One of these radiation paths has a fixed optical path length, the path length of the other can be changed by moving the movable mirror. When, after reflection, the beams are recombined at the beam splitter, they interfere due to their optical path length difference of $2s$ into beam A_3 of intensity $I_{3\tilde{\nu}}(2s)$, if losses at the mirrors are disregarded.

For the beam emerging from the interferometer (beam A_3), for the spectral intensity $I_{3\tilde{\nu}}(2s)$ in relation to $2s$, the expression of interfering waves known from spectroscopy results for the entire mirror movement $2s$ (forward and backward motion of the mirror)

$$I_{3\tilde{\nu}}(2s) = I_0(\tilde{\nu})\left[1 + \cos\left(2\pi\tilde{\nu}2s\right)\right] = 2I_0(\tilde{\nu})\cos^2\left(2\pi\tilde{\nu}s\right). \tag{6.44}$$

If the beam splitter and the mirrors are not ideal, a correction factor $A(\tilde{\nu})$ must be included which is attributed to the $I_0(\tilde{\nu})$, i.e., instead of $I_0(\tilde{\nu})$ the following holds good:

$$I_0^*(\tilde{\nu}) = A(\tilde{\nu})\, I_0(\tilde{\nu}) \,. \tag{6.45}$$

From this follows

$$\frac{I_{3\tilde{\nu}}(2s)}{I_0^*(\tilde{\nu})} = \left[1 + \cos\left(2\pi\tilde{\nu}2s\right)\right] = 2\cos^2\left(2\pi\tilde{\nu}s\right) \,. \tag{6.46}$$

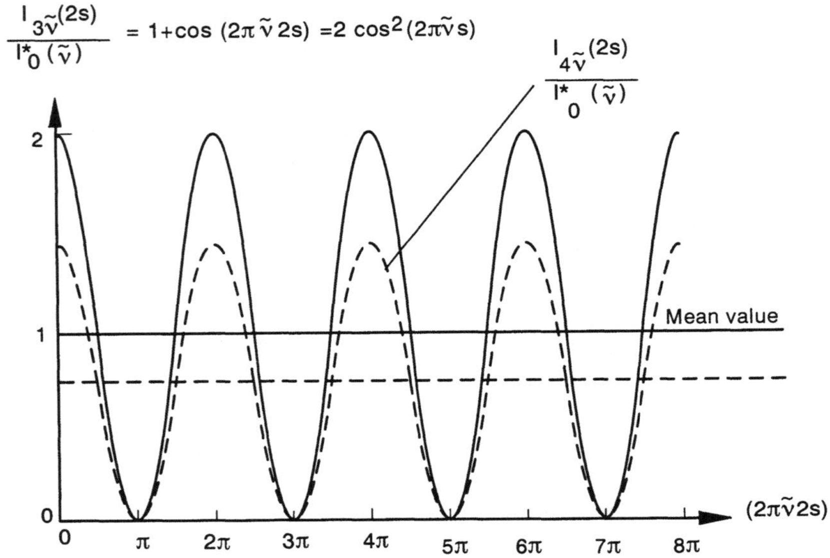

Fig. 6.22: Diagrammatic representation of the standardised spectral intensity density of the beam emerging from the FTIR spectrometer in relation to the mirror path with and without absorption

This standardized function is illustrated in Fig. 6.22 (fully drawn). Interference maxima or minima show for the emerging beam A_3, according to the cosine function, if the pathlength $2s$ of the mirror moved is for:

Maxima: $(2\pi\tilde{\nu}\,2s) = 0, 2\pi, 4\pi, \ldots,$

$$\text{i.e. } 2s = n \cdot \frac{1}{\tilde{\nu}} = n \cdot \lambda \qquad n = 0, 1, 2 \tag{6.47}$$

or:

Minima: $(2\pi\tilde{\nu}\,2s) = \pi, 3\pi, 5\pi, \ldots,$

$$\text{i.e. } 2s = \left(n + \frac{1}{2}\right)\frac{1}{\tilde{\nu}} = \left(n + \frac{1}{2}\right) \cdot \lambda \qquad n = 0, 1, 2, \ldots \tag{6.48}$$

Equation (6.46) contains two parts, one constant component and a modulated component.

The modulated component in (6.46)

$$\cos\left(2\pi\tilde{\nu}\,2s\right) \tag{6.49}$$

is the interference term of the intensity distribution, the so-called interferogram. The interferogram thus reflects a radiation spectrum as a function of a path (covered distance $2s$ of the mirror moved).

If the mirror is moved with a constant velocity v, then

$$s = \mathrm{v}\, t \tag{6.50}$$

For (6.49) the following is then obtained

$$\cos\left(2\pi\tilde{\nu}\, 2\,\mathrm{v}t\right) \tag{6.51}$$

$I_{3\tilde{\nu}}(2s)$ can thus be expressed as a function of time t. Acc. to (6.51) the corresponding frequency is

$$\nu \approx 2\,\mathrm{v}\,\tilde{\nu}\ s^{-1}\ . \tag{6.52}$$

The originally radiated monochromatic wave A_0 of intensity I_0 is thus modulated in cosine form and has the frequency $2\mathrm{v}\tilde{\nu}$.

If (as shown in Fig. 6.21 in a diagram) a measuring cell is installed in the optical path behind the interferometer, containing a medium which absorbs monochromatic waves - in our case gas or a gas component -, provided the Lambert-Beer law is applicable, (6.31) can be used and instead of $I_0^*(\tilde{\nu})$

$$I_0^*(\tilde{\nu})\ e^{-\varepsilon_i(\tilde{\nu})c_i d} \tag{6.53}$$

can be substituted for the transmitted beam A_4 of intensity I_4. Instead of (6.46) the following is then valid for the intensity of the transmitted beam A_4 behind the measuring cell

$$\frac{I_{4\tilde{\nu}}(2s)}{I_0^*(\tilde{\nu})} = e^{-\varepsilon_i(\tilde{\nu})c_i d}\left[1+\cos\left(2\pi\tilde{\nu}\, 2\, s\right)\right] = 2e^{-\varepsilon_i(\tilde{\nu})c_i d}\cos^2\left(2\pi\tilde{\nu}\, s\right)\ . \tag{6.54}$$

As shown qualitatively in Fig. 6.22, the amplitude would then be correspondingly smaller (dotted line).

In the idealised case of the monochromatic wave, the change of the maxima acc. to (6.54) yields the information on the concentration c_i of the gas component i, if the other quantities are known. When

$$\cos^2\left(2\pi\tilde{\nu}\, s\right) = 1 \tag{6.55}$$

then

$$c_i = \frac{1}{\varepsilon_i(\tilde{\nu})d}\ln\left[\frac{2\, I_0^*(\tilde{\nu})}{I_{4\tilde{\nu}}(2s)}\right] \tag{6.56}$$

is obtained. When two different wave numbers $\tilde{\nu}_1$ and $\tilde{\nu}_2$ are used, two $\cos^2$ functions overlap etc.

If the emitters are wideband, then based on (6.54), integration must be carried out including all wave numbers. Then

$$I_{4\tilde{v}\,total}(2s) = 2 \int\limits_{\tilde{v}=0}^{\infty} I_0^*(\tilde{v})\, e^{-d\,\varepsilon(\tilde{v})c(\tilde{v})} \cos^2(2\pi\tilde{v}s)\, d\tilde{v} \;, \tag{6.57}$$

is valid, as negative frequencies or wave numbers are not possible physically. ε and c become continuous functions of $\tilde{v}$ (refer to (6.37)).

However, $I(s)$, intensity as a function of path length, is not of interest, but spectral intensity density distribution $I(\tilde{v})$ as a function of the wave number. This inversion is obtained by the Fourier transformation, as is generally known. The complex Fourier transformation results in:

$$I_{4\,total}(\tilde{v}) = \int\limits_{s=0}^{\infty} I_{4\tilde{v}\,total}(2s)\, e^{-j(2\pi\tilde{v}\,2s)}\, ds \;, \tag{6.58}$$

where the fact was taken into account that negative distances s are not physically possible. This establishes the correlation between spectral intensity density distribution as a function of the wave number and thus of the frequency (the spectrum) and intensity distribution as a function of path length $2s$ (of the interfering beams in the interferometer), and the total spectrum is obtained as an interferogram. All the information on the spectra of the measurably absorbing gases contained in the measuring cell are obtained in the interferogram.

If there is no absorbing medium behind the interferometer in the measuring path of the spectrometer system, the interferogram is recorded by the detector as a likeness of the spectral intensity density distribution $I_0^*(\tilde{v})$ of the wideband light source. If an absorbing medium is present (gas mixture) in the measuring path, those proportions of the intensity distribution of spectral components which are absorbed by the different gases are missing at the detector during recording of the interferogram.

The recording reflects, as with all spectrophotometers, spectral absorption behavior of *all* components contained in the gas mixture as long as they absorb to a measurable degree, i.e., as long as they are present in sufficiently high concentrations. Named after Fellgett this is called the *Fellgett advantage*.

The equations shown last illustrate that an infinitely long path of the movable mirror would be necessary for complete resolution of the spectrum $I^*(\tilde{v})$. In practice, only a limited path is feasible.

$$\text{Path length:} \qquad 2s \le 2s_{max} \tag{6.59}$$

In this case a filter function must be introduced. If a step function is used

$$y(2s) = \begin{cases} 1 \text{ for } 2s \le 2s_{max} \\ 0 \text{ for } 2s > 2s_{max} \end{cases} , \tag{6.60}$$

then

$$I_{4\,total}(\tilde{\nu}) = \int_{s=0}^{2s\,max} I_{4\tilde{\nu}\,total}(2s) \cdot y(2s)\, e^{-j(2\pi\tilde{\nu}2s)} ds \qquad (6.61)$$

is obtained by multiplying the integrand of (6.58) with the step function $y\,(2s)$. Fourier transformation of the step function $y\,(2s)$ leads to

$$y(\tilde{\nu}) = 2s \text{ sinc } [2\pi\tilde{\nu}2s] \qquad (6.62)$$

with the sinc function

$$\text{sinc } x = \frac{\sin x}{x} \,. \qquad (6.63)$$

Hence, the real spectrum is obtained as a so-called convolution of the two functions having to undergo a Fourier transformation $I_{4\tilde{\nu}\,total}(2s)$ and $y(2s)$.

If a filter is disregarded a spectrometer's resolution, i.e., the band width of the sections into which the radiation spectrum is split, is dependent on the inverse maximum reflective path length:

$$\text{resolution:} \qquad \Delta\tilde{\nu} = \frac{1}{2\,s_{max}} \,. \qquad (6.64)$$

The left section of the example in Fig. 6.23 shows the interferogram of wide-band radiation with a narrow absorption line. The partial interferogram of the wide-band radiation (dashed curve) has a distinct structure around the point of $2s = 0$. The interferogram for the absorption line (dotted curve) has an even course over the path $2s$ with decreasing amplitude.

The resulting interferogram (solid line) shows that the middle section of the interferogram (short path length $2s$) is dominated by the interference image of the wide-band spectrum. Information on narrow-band spectral sections only have an effect on the smaller intensities of the resulting interferogram (longer path length $2s$). In the right-hand section of Fig. 6.23 the corresponding representation is shown after Fourier transformation. The absorption line (spectral line) is to be seen.

Usually, Fourier transformation is computed in a digital computer with the help of the FFT (Fast-Fourier Transformation) algorithm. Wider application of the FTIR analysis method has only been made possible by the enhanced capacity of computer systems.

The spectra of the gas components to be measured are determined individually beforehand as reference spectra with high-purity calibration gases and are recorded by the computer and stored. When measuring components of gas mixtures the

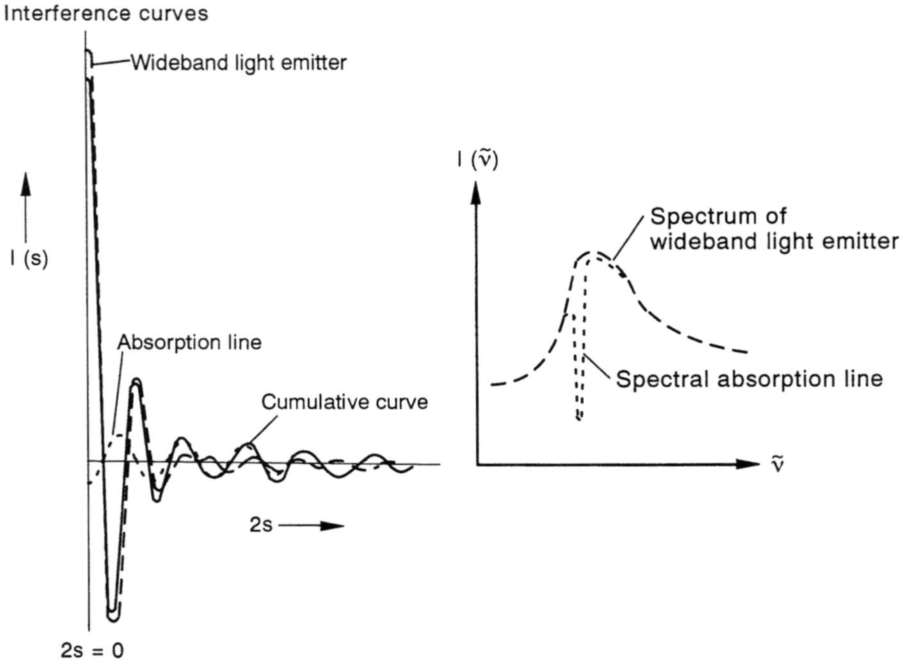

Fig. 6.23: Recording of spectral information in an interferogram with wide-band emitter and one absorption line

individual spectra are compared to the reference spectra stored and are thus identified and quantified.

Even a subsequent "calibration" for certain gas components with a calibration gas is possible. If, then, a gas mixture is examined to determine a gas component whose data have not been initially evaluated explicitly, later evaluation of the data stored is possible with subsequent calibration.
Fig. 6.24 shows the sequence of a measurement in a diagram.

Another advantage of the FTIR spectrometer is the possibility of finding a point without absorption between absorption lines and to thus make use of a reference interval where transmittance is uninfluenced by the gas. This can be used as a reference for all transmission lines. The "reference beam" with $I_0^*(\tilde{v})$ is provided automatically.

Fig. 6.25 shows the sampling and optical system. One part of the raw exhaust gas is conducted through a filter and a heated line through the measuring cell (gas cell) heated to a temperature of 185 °C ± 5 K and is then fed back into the exhaust line. Pressure is also kept constant during this. Another option is passing diluted exhaust gas through the measuring cell. Interferometer, measuring cell and detector are in a pressure-proof housing, which is rinsed with nitrogen (N_2), i.e., a protective gas which does not absorb infrared light.

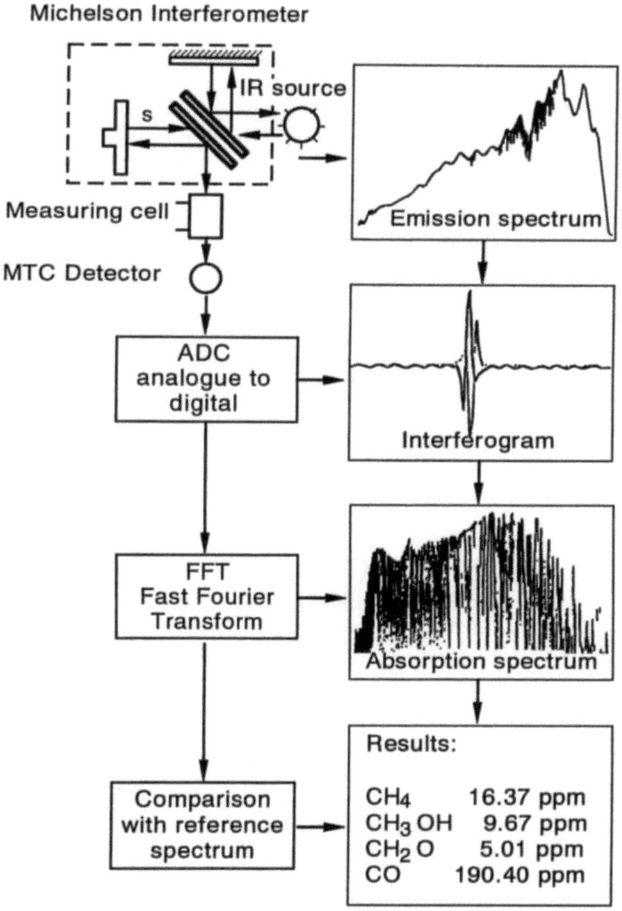

Fig. 6.24: Diagram of the measurement sequence

For a long period of time (≥ 6 months) high temperatures of measuring cell and measuring gas prevent soiling of the IR-transmitting windows which seal the cell.

The measuring cell is placed between interferometer and detector. A photograph of the device is shown in Fig. 6.26.

As an example Table 6.4 shows some measurable components of automobile exhaust gas with the corresponding sensitivity limit.

Instead of a simple measuring cell with single transmission one can also use a multiple reflection measuring cell (White cell), which has a longer radiation path and thus permits the recording of smaller concentrations. However, the multiple reflection measuring cell also involves some problems, e.g., concerning the required dimensions and the reduction of reflection intensity due to soiling, if the White cell is not heated.

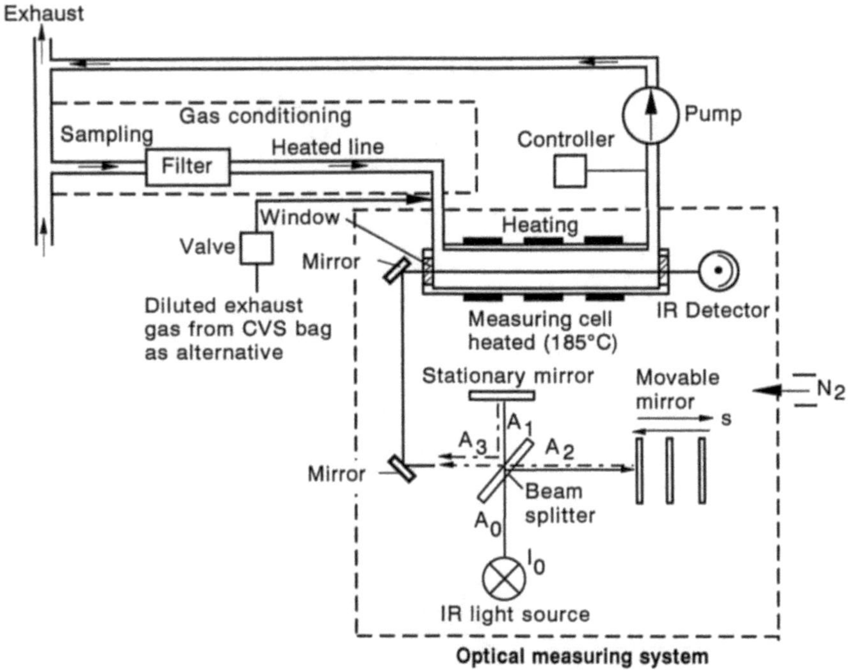

Fig. 6.25: Sampling with the FTIR device

Fig. 6.26: Photograph of a FTIR multicomponent exhaust gas analyser, Siemens AG

Table 6.4: Measurable components of an automobile exhaust gas

	Components	Sensitivity limit in ppm
NO	Nitrogen monoxide	0.8
NO_2	Nitrogen dioxide	0.2
N_2O	Dinitrogen monoxide	0.12
HNO_2	Nitrous acid	1.6
NO_x	Oxides of nitrogen	2.5
NH_3	Ammonia	1.0
CO_2	Carbon dioxide	2.0
CO	Carbon monoxide	0.2
CH_4	Methane	2.0
C_2H_2	Acetylene	4.5
C_2H_6	Ethane	1.5
C_2H_4	Ethylene	5.0
C_3H_6	Propylene	10.0
CH_3OH	Methanol	2.2
C_2H_5OH	Ethanol	2.0
CH_2O	Formaldehyde	0.8
CH_3CHO	Ethanal (acetaldehyde)	8.8
C_4H_6	1,3 Butadiene	8.0
C_4H_8	Isobutylene	7.3
THC	Total hydrocarbons	20.0
HCOOH	Formic acid	1.1
HCN	Hydrogen cyanide	0.9
SO_2	Sulfur dioxide	1.5
H_2O	Water	920.0
CF_4	Carbon tetrafluoride	0.012

Apart from the mirror set-up shown in Fig. 6.21 other optical arrangements are also used in the Michelson interferometer. Mirror movement is, e.g., effected by a rotary movement instead of rectilinear carriage. Two examples of these arrangements, one rotary and one pendulum arrangement are shown as diagrams in Figs. 6.27 and 6.28.

Two roof-shaped mirrors are installed on a bracket pivotable around its center. Two other mirrors are fixed in stationary positions. When the mirror carrier is at rest the beam splitter splits the beam emitted by the infrared source into two beams of approx. the same path length. If the mirror carrier executes an even, oscillating movement around its own middle axis, the path lengths of the two partial beams change reciprocally. Due to the phase difference effected by the path length differences between the two beams, an interference image is created, as is the case in the Michelson interferometer with rectilinear mirror movement, reflecting the intensity distribution of the spectrum of the incident beam as a function of the reflective path difference caused by the rotary motion.

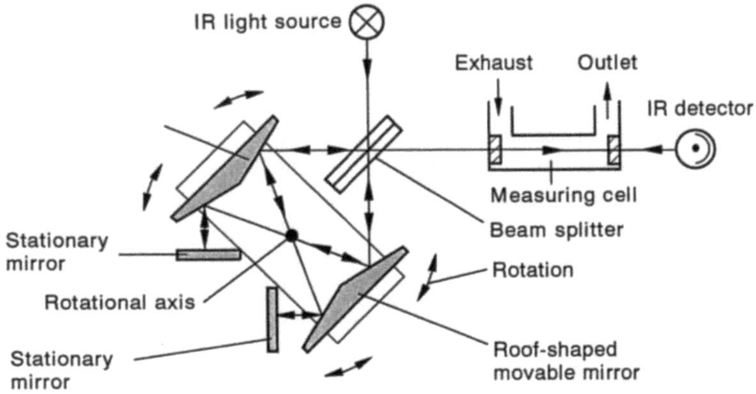

Fig. 6.27: Set-up of a Michelson interferometer with mirror movement by rotation

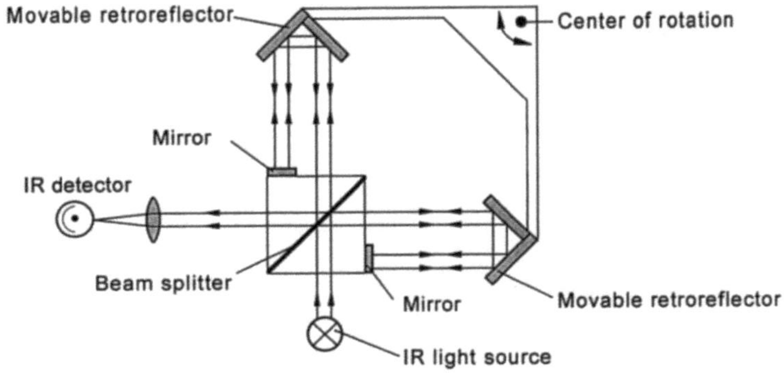

Fig. 6.28: Set-up of a Michelson interferometer with mirror movement effected by a pendular motion

In the set-up of Fig. 6.28 the path lengths of the two partial beams generated by the beam splitter are changed by the pendular counter-motions of the two rigidly connected retroreflectors around a joint axis. In addition, the retroreflectors guarantee that despite the mirror movement the reflected beams always strike the stationary mirrors fixed to the beam splitter. The path length counter-change of the two partial beams again is the cause for the emerging of the interferogram.

6.1.7.2 Assessment of Fourier Transform Spectroscopy for Measuring Automobile Exhaust Gases

- Simultaneous recording of several components (e.g. 25) is fundamentally possible (Fellgett advantage).

- Direct measurements at the exhaust pipe are possible. A distance of defined length (measuring cell) in the exhaust gas stream can serve as measuring distance over which infrared radiation can be sent.
- Selectivity (undisturbed recording of each component without spectral interference by adjacent components) is dependent on and given by the spectrometer's resolving power and thus by the possible path of the mirror moved inside the spectrometer. The mirror path must be long enough (several cm) to attain a resolving power reaching the half-value width of the spectral lines. However, the mirror must be moved plane-parallel with high precision. Over the whole distance deviations from plane-parallelism must be small, i.e. $< \lambda/10$, compared to light wavelength. This has been technically realized.
- Real time measurement requires a measuring cycle time (mirror path plus return to starting point) of one second. Also, data recording and processing (analogue-digital conversion and Fourier transformation) must be carried out within one second. Today's technology has reached this level.

As an example Fig. 6.29 shows the CO spectrum of a calibration gas measured with an FTIR spectrometer. This calibrates the device for this component, the value is stored in the computer and from then on will be compared with the values measured from gas mixtures. This is also illustrated by the Fellgett information which is even valid for individual rotation lines. If concentrations are high, the lines of lower absorption are evaluated; when CO concentrations are low lines of higher absorption. This requires only one cell length *d*, i.e., only one cell. Several lines can be evaluated for one component, thus raising the certainty of the measurement.

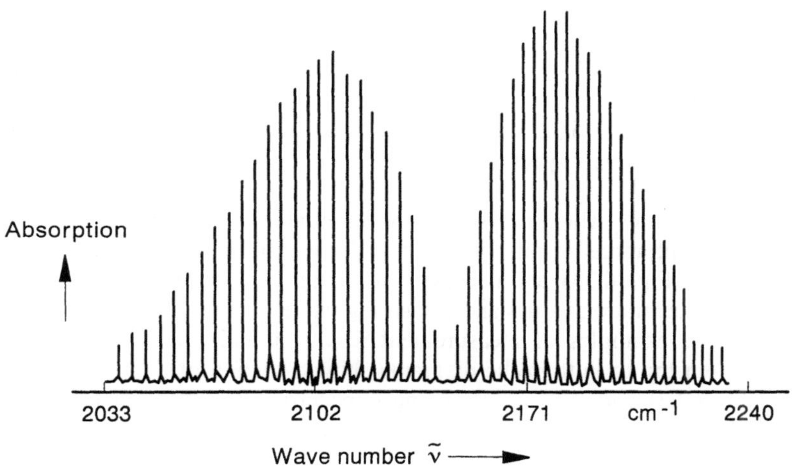

Fig. 6.29: CO spectrum, recorded with calibration gas of a known CO concentration, refer to also Figs. 6.9 and 6.10

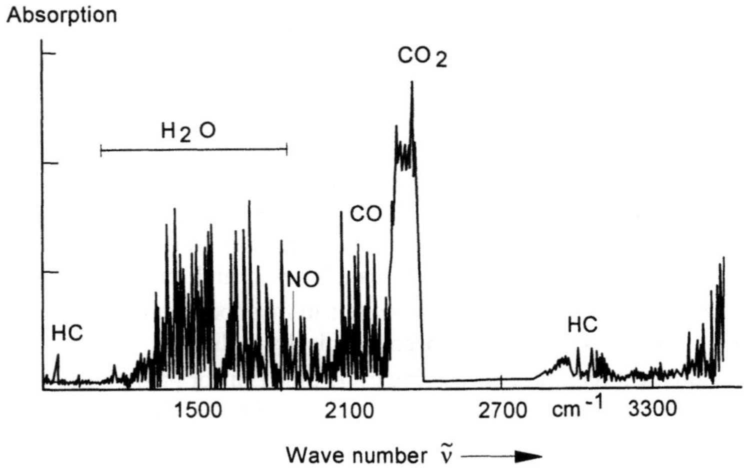

Fig. 6.30: Exhaust gas spectrum of gasoline engine exhaust gas of a vehicle without catalytic converter

As an example Fig. 6.30 shows an exhaust gas spectrum of the gasoline engine exhaust gas of a vehicle without catalytic converter measured by an FTIR device. The problem is water vapor, which is always present in engine exhaust gases and whose absorption is in the same spectral range as of the components of interest.

Of the IR-active main components of the exhaust gas, H_2O and CO_2, the H_2O bands overlap a wide IR range. Therefore accurate determination of the evaluation ranges with optimal resolution (wave number 0.5) and multipeak evaluation, e.g., with NO and CO, is necessary. Ranges must be chosen where the absorption of water vapor practically equals zero. Fig. 6.31 shows evaluation ranges roughly classified acc. to wave numbers. Overlapping with water-vapor absorption bands occurs in all ranges.

In contrast to this Fig. 6.32 shows exhaust gas spectra at high resolution. Sections where individual spectral lines have not been disturbed can be recognized.

The Lambert-Beer law which has been considered above, only applies roughly with the low concentrations of this case. For wider measuring ranges, calibration curve corrections (Fig. 6.33) are necessary to obtain linear relations between calibration gas concentration values and measuring values.

Fig. 6.34 shows examples of measurement results of individual exhaust gas components. From top to bottom the diagrams show a section of a driving curve (refer to Chap. 8.2.3) and the corresponding concentrations measured for some nitrogen compounds as a function of time for a vehicle with a gasoline engine and a three-way catalytic converter. The high time resolution of 1 s is apparent. The catalytic converter requires a minimum temperature of approx. 300 °C (light-off temperature) to be able to function. Ammonia (NH_3) forms only after the catalytic converter has lit off (warmed-up) and is produced by its catalytic effect. Laughing gas (N_2O) also forms only in the catalytic converter, but in the early phase of "lighting off".

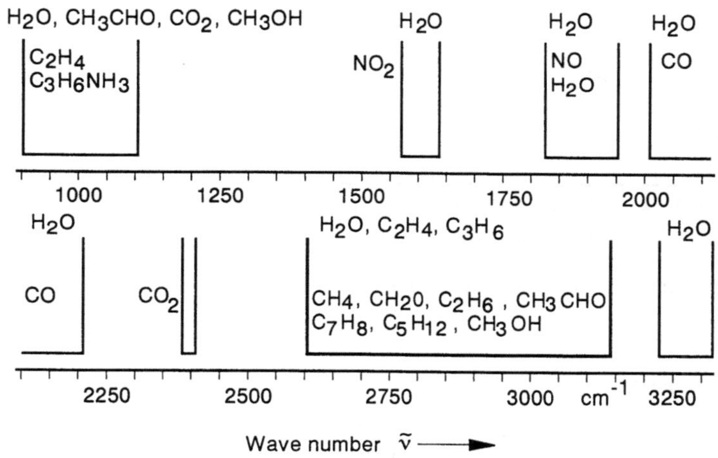

Fig. 6.31: Evaluation ranges

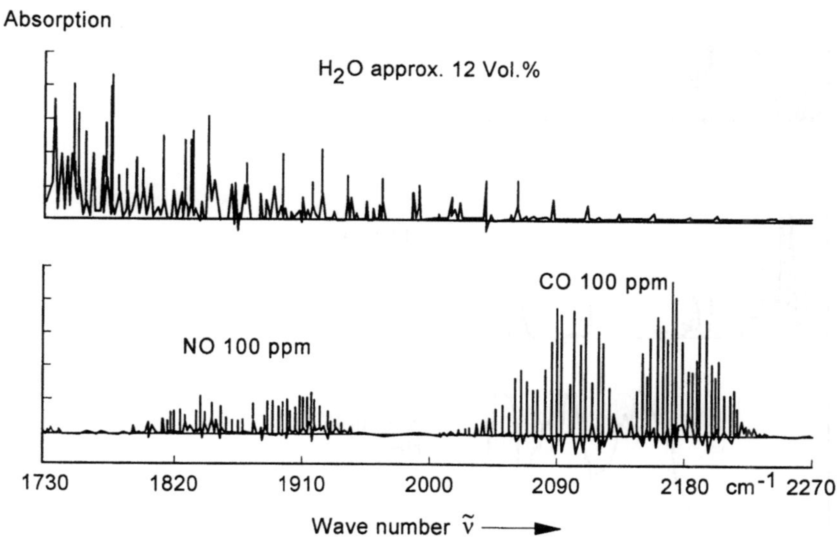

Fig. 6.32: Exhaust gas spectra at high resolution, upper diagram showing water vapor in high concentration, lower diagram showing NO and CO as examples of exhaust gas components, both in relatively small concentrations

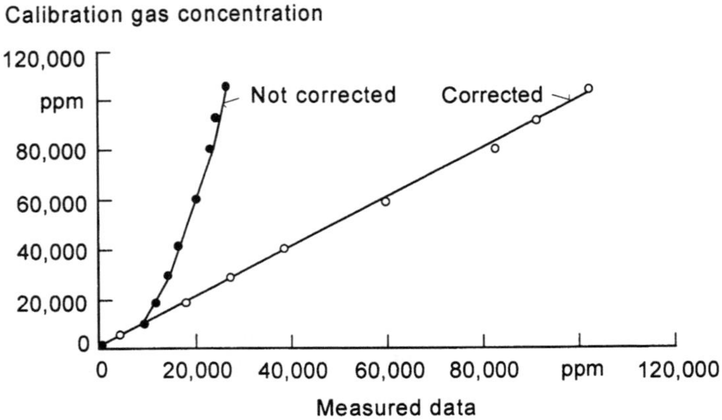

Fig. 6.33: CO_2 calibration curves with and without correction

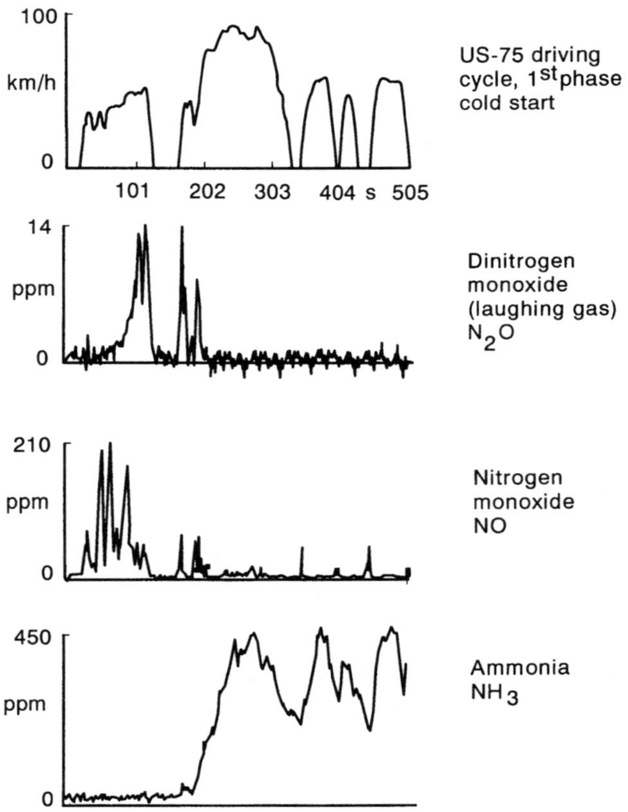

Fig. 6.34: US-75 driving cycle - 1st phase (505 s) at cold start and concentrations of N_2O, NO and NH_3 in relation to time

6.1.8 Semiconductor-Diode-Laser (SDL) Spectrometer as Selective Emitter

Unfortunately, there are no lasers which can be tuned in the medium infrared range over a wider wavelength range. In monomode operation per se the laser itself provides extremely narrow-banded, almost monochromatic radiation. Lasers achieve emission line widths of as narrow as 10^{-4} cm^{-1}, hence, the lines are very narrow compared to gas rotation lines (approx. $1 \cdot 10^{-1}$ cm^{-1} at atmospheric pressure). Naturally, the laser line must be in the range of an absorption line, and the laser must be tunable through at least one absorption line of the gas. This requires lasers in the medium infrared range. Semiconductor-diode lasers on a lead-chalcogenide basis are such lasers, they are tunable in a small wave number range.

Lead-chalcogenides are semiconducting compounds of lead with cadmium or selenium and tellurium (lead salts). Laser diodes are designed such that two differently doped (p- and n-doped) layers of the same material are adjacent to each other (Fig. 6.35).

Applying voltage to the metallic contact surfaces causes a current which drains charge carriers from the p- and n-layer and injects them into the other layer respectively. This creates an active zone on either side of the p-n transition. The crystal's front and reverse side are cleavage surfaces forming the laser resonator. Above a critical current value, the threshold current, the current-carrying part of the active zone emits laser radiation.

Diode lasers are manufactured in homostructures and heterostructures. In lasers with homostructure the p-n transition is diffused into the homogenous semiconductor material. Lasers with a dual heterostructure exhibit better properties where the active zone on either side is adjacent to a semiconductor material with electric and optical properties different from the primary material. In both homo- and heterostructures additional geometric structures, so-called banded structures

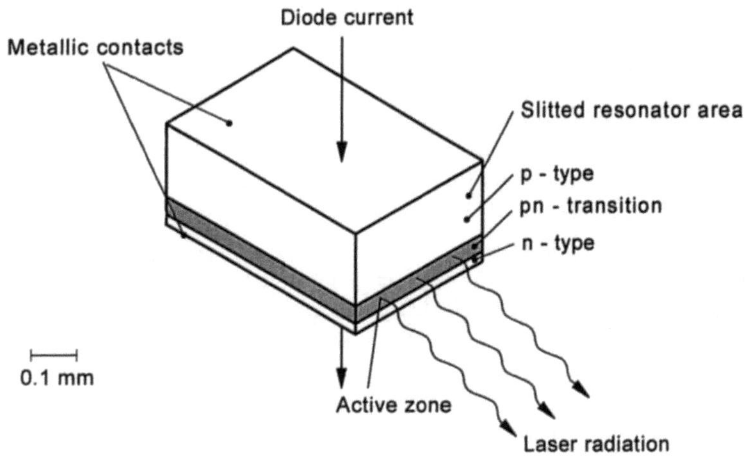

Fig. 6.35: Diagram of a laser diode

(between approx. 10 and 50 μm of width), provide improved lateral guidance for the laser light waves forming in the resonator. One disadvantage is the low operating temperature of 20 - 80 K necessary for physical reasons, which requires a cooling device.

The wavelength or wave number range covered by different laser diodes and their laser emission depends on the type and composition of the laser material used. Some diode lasers currently in common use and their corresponding operating temperatures are compiled in Fig. 6.36. In addition, the absorption ranges of different gases are indicated.

Within the emission wave number range determined by lead salt composition the operating temperature necessary for a certain absorption line is set by varying the current. How a diode laser is tuned by means of temperature is illustrated in Fig. 6.37.

Rough tuning is carried out continuously by means of temperature over a certain wavelength or wave number range (Fig. 6.37a). Using a finer measure, however, (Fig. 6.37b) it can be seen that at certain temperatures emission changes suddenly, the laser emitting in several modes.

A controlled cooler must be used for keeping the operating temperature in a range between 20 and 30 K. Fine tuning is, however, usually not carried out by the cooler but, at a fixed cooler temperature by selecting the appropriate electric current intensity of the operating current, i.e., by changing the electric power loss. Fig. 6.38a shows the fine tuning of a diode laser by continuously changing the applied direct electric current, and again the mode jumps are obvious. Fig. 6.38b

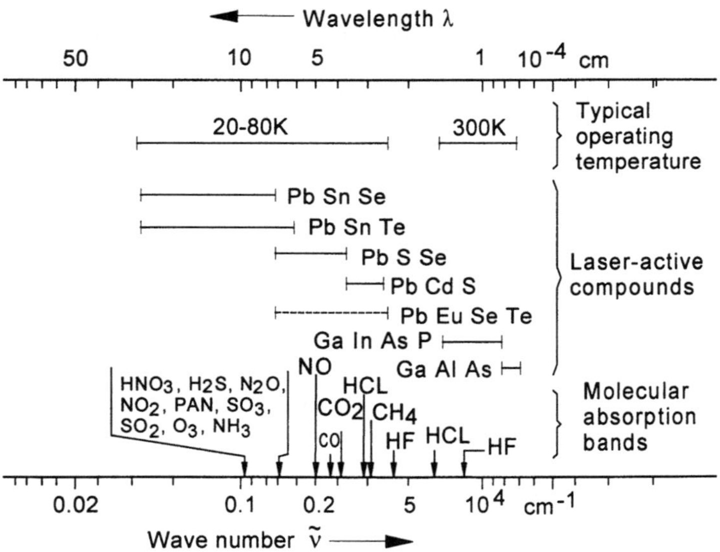

Fig. 6.36: Spectral emission ranges and operating temperatures of different diode lasers currently in use. Each range requires different laser diodes. In addition, the spectral positions of various molecular absorption bands are indicated

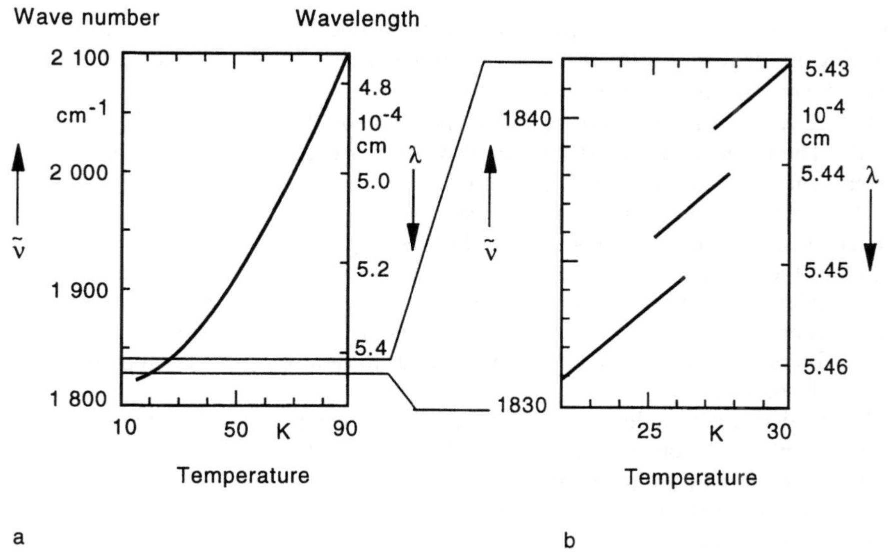

Fig. 6.37: Temperature tuning of an PbSSe diode laser showing (a) rough tuning and (b) fine tuning by varying electric current. Wave number or wavelength are drawn as a function of temperature in K

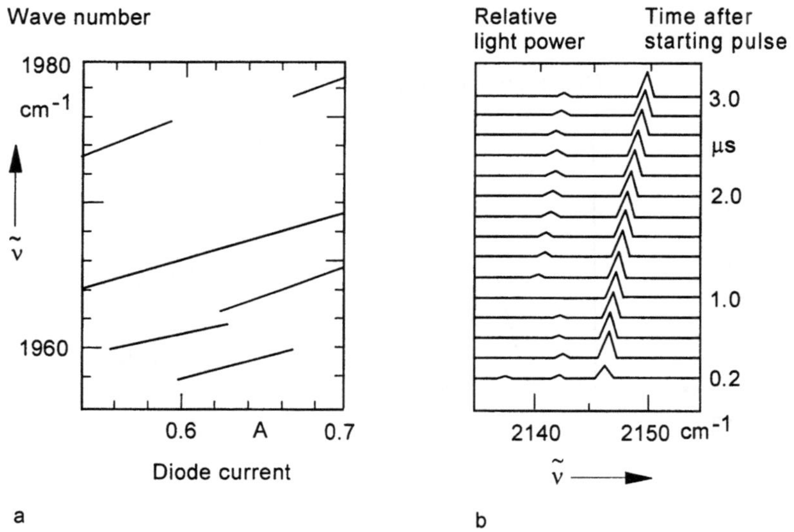

Fig. 6.38: (a) Electric current fine tuning of a PbSSe diode laser (several modes) and (b) tuning of a certain emission line during a 3 μs long electric current pulse

shows the tuning of a particular laser emission line by increasing the laser heat ensuing immediately after the electric current pulse has been switched on. At $\Delta\tilde{v} \approx 5\ cm^{-1}$ the tuning range is small.

Selective gas measurement requires operation of the laser diode in only one mode. The other modes must be excluded. This is done with the help of a mode filter, generally a grating monochromator.

Due to the diode systems' minute size which is in the mm range, housing included, several diodes can be accomodated jointly on one cooling finger which is a part of the cooling system. Temporal subsequent tuning of the diode lasers poses no problems, as tuning with an electric current pulse takes only several microseconds per diode. As each diode laser requires monomode operation several mode filters must be provided or the mode filter must also be quickly tunable, refer to Vol. B, optics.

In practice, each diode laser is installed in a housing and only a few (5 at the most) are placed on one cooling finger, which is cooled below the lowest operating temperature. Even a temperature change of 1 mK causes a shift of the central wave number of the emission line by approx. $10^{-3}\ cm^{-1}$, which is ten times the halfvalue width of the laser emission line of approx. $1 \cdot 10^{-4}\ cm^{-1}$.

It is not possible to tune a diode laser quickly from the absorption range of one gas to the absorption range of another gas.

This means that in multicomponent measurements a separate diode laser must be used for each measuring component, even for each spectral line of a gas component, thus rendering multicomponent measurement impossible.

For practical reasons so-called integrating spectrometry is used for absorption measurement, i.e., the laser emission line is tuned across an absorption line of the gas to be measured by applying an electric current pulse, as is shown in a diagram in Fig. 6.39.

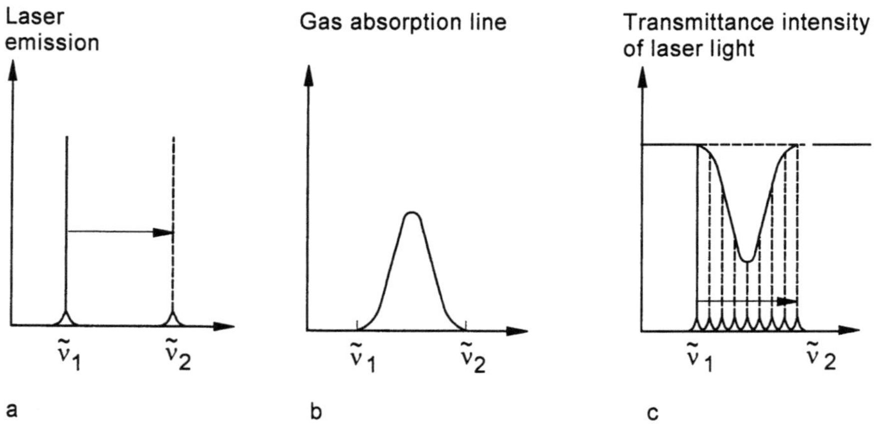

Fig. 6.39: Diagram showing the tuning of an SDL emission line across an absorption line of a gas:
(a) the laser emission line is tuned from $\tilde{v}_1$ to $\tilde{v}_2$,
(b) the absorption line of the gas is between $\tilde{v}_1$ and $\tilde{v}_2$,
(c) transmitted laser light in the wave number interval between $\tilde{v}_1$ and $\tilde{v}_2$.

The laser emission lines' extremely narrow halfvalue width permits integrating absorption measurement even at low gas pressures, i.e., when the halfvalue width of the gas absorption lines has a correspondingly smaller value. As is readily apparent, the possibility of measuring gas lines overlapping with neighboring interfering gas lines is limited when line width is reduced. As an example, Fig. 6.40 shows the transmission spectrum of 1000 ppm NO in the presence of H_2O (15000 ppm) and CO_2 (150000 ppm) at an absolute pressure of 10^5 Pa atmospheric pressure (wide absorption line) or $3 \cdot 10^3$ Pa (narrow absorption line) respectively.

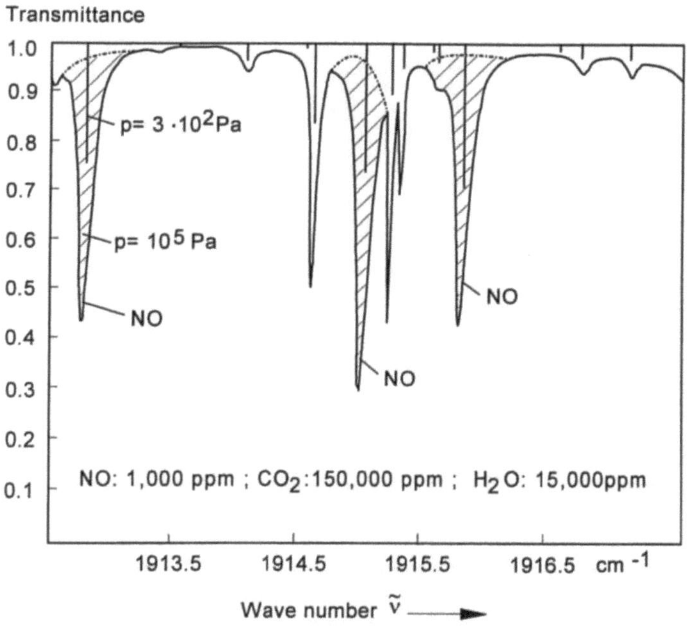

Fig. 6.40: Tansmittance spectrum of 1000 ppm NO in the presence of H_2O and CO_2 at 10^5 Pa or at $3 \cdot 10^3$ Pa respectively and a measuring cell length of 1 m. The gases' temperature lies at T = 470 K

Practical device concepts

Semiconductor diode laser exhaust gas measuring systems are offered for quasi-simultaneous measurement of a few selected exhaust gas components, for which, as a special feature and depending on the number of measuring components, rapid, high time resolution measurements at 100 Hz measuring frequency and higher are possible.

As an example Fig. 6.41 shows the diagram of a laser measuring system for dynamic exhaust gas measurements on engine test stands.

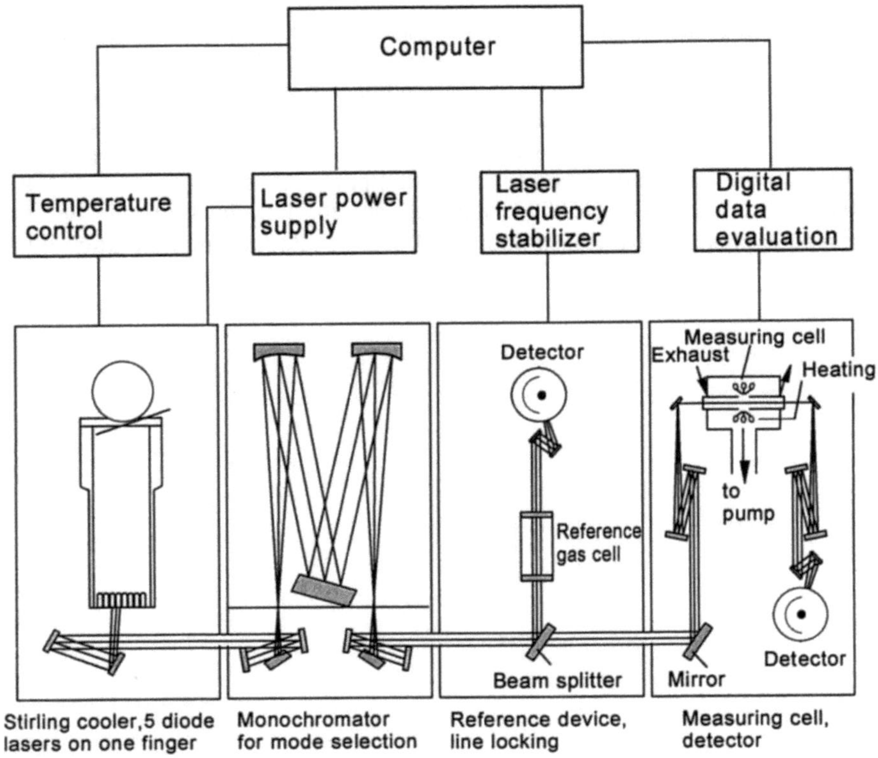

Fig. 6.41: Diagram showing the set-up of a laser measuring system for dynamic exhaust gas measurement on engine test stands (FhG-IPM). The triple-solid-lines demonstrate the optical beams

The measuring system has a modular structure. Several diode lasers, which can be tuned each to a different absorption line of a certain gas by selecting the operating temperature (first module), are mounted jointly on a cooling finger. The monochromator (second module) filters out the undesired modes of each laser diode. The laser emission wave number set by the selected temperature is stabilized by continuous comparison to a reference wave number, which can, e.g., be obtained by the central wave number of a pre-selected gas absorption line (line locking). For this purpose, a reference device which includes a calibration gas cell and a subsequent detector is installed in the path of the beam (third module). As a rule, the calibration gas cell contains the same type of gas as the component to be measured, e.g., CO. The fourth module contains the measuring cell and the detector. Due to the desired high frequency of repeated measurements the measuring cell is designed for a very fast gas flow rate. This is achieved by keeping the volume of the measuring cell small.

The development of semiconductor diode lasers is still in a state of flux, especially as far as operation at higher operating temperatures and mode purity is concerned.

6.2 Ionization Methods

6.2.1 General Observations

Automobile exhaust gas contains a high number of hydrocarbon compounds (refer to Chap. 7) consisting of non- or partially oxidated fuel compounds and newly formed compounds, refer to Fig. 6.42, which also absorb in the infrared range.

The lower diagram of Fig. 6.42 shows a gas chromatogram of the composition of a fuel (for gas chromatography refer to Chap. 6.5), and the upper diagram the composition of hydrocarbon exhaust gas emissions from gasoline-powered vehicles which, however, vary for different vehicles and different fuel qualities. This explains the difficulty of total hydrocarbon measurement, as higher hydrocarbons occur which can only be recorded very inaccurately.

Infrared analyzers (NDIR instruments) are unsuited for accurate measurement of total hydrocarbon emissions especially because of the low concentrations in the exhaust gas of modern automobiles, even though absorption bands in the infrared range do occur. The values established by these instruments do not reflect the sum of the actual hydrocarbon emission in automobile exhaust gas correctly as an analyzer calibrated with *n*-hexane merely records the paraffins of the total hydrocarbon content, as is shown in Fig. 6.43.

Thus the so-called flame ionization detector (FID) known from gas chromatography is used to obtain more realistic values of total hydrocarbon emissions. Measurement of total hydrocarbons (THC) in automobile exhaust gas is mandatory.

Sum measurement means that each individual hydrocarbon contributes to the total figure according to the number of carbon atoms (C atoms) contained in the molecules in question. For instance, hexane (C_6H_{14}) with six carbon atoms theoretically contributes the sixfold signal of methane (CH_4) which has only one carbon atom. The measuring result is indicated preferably in equivalents of C_1 (methane).

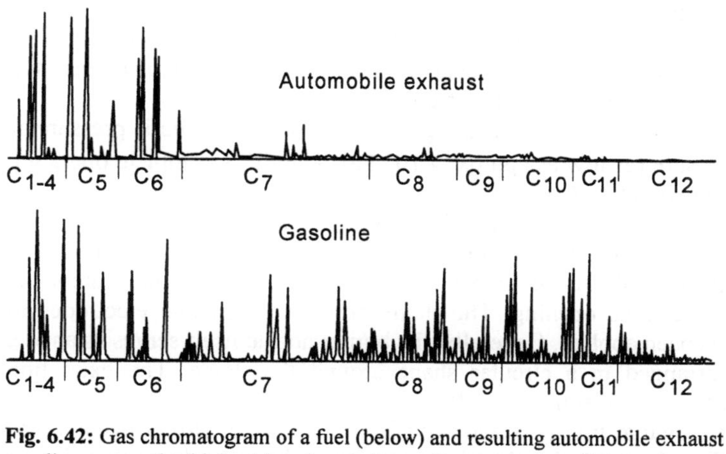

Fig. 6.42: Gas chromatogram of a fuel (below) and resulting automobile exhaust gas from gasoline-powered vehicles (above) scaled according to number of C atoms per molecule

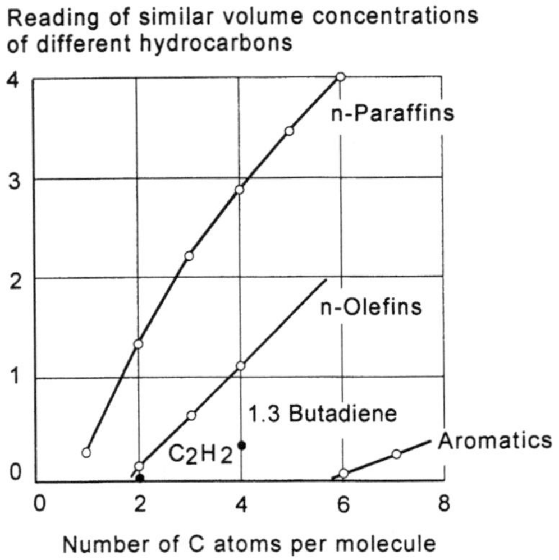

Fig. 6.43: Indication sensitivity of an NDIR instrument for different hydrocarbons

6.2.2 Flame Ionization Detector (FID)

6.2.2.1 Operating Principle

In gas chromatography (refer to Chap. 6.5) the individual components, isolated in separation columns, are successively recorded by the FID. During this, different sensitivities towards individual components can be compensated for by "calibration factors" (response factors). This is not possible when recording the sum of all hydrocarbons.

The FID measuring principle is based on the process of ionising hydrocarbon molecules in a hydrogen flame, with the number of ions corresponding to the number of carbon atoms in the molecules. Thus, the flame ionization detector essentially has an indication proportional to the number of carbon atoms (C). Fig. 6.44 shows the diagram of an FID set-up.

The FID can be operated in the measuring chamber with atmospheric pressure or with negative pressure (as low as approx. $4 \cdot 10^4$ Pa). The FID's complete operating principle is illustrated in Fig. 6.45.

Before it reaches the burner nozzle the gas to be analyzed is added to the combustion gas (hydrogen (H_2) or a mixture of hydrogen and helium (He)). The flame is operated as a diffusion flame; the air required for burning enters the burner through a separate opening. The flame burns between two electrodes to which voltage has been applied. Generally, the burner nozzle itself serves as anode, the cathode is arranged in a circular shape around the flame. The extraction potential applied to the electrodes pulls the ions formed in the flame to the cathode, so that a charge transport and with it a measurable current occurs. The current is to be directly proportional to the C content of the gas to be analyzed.

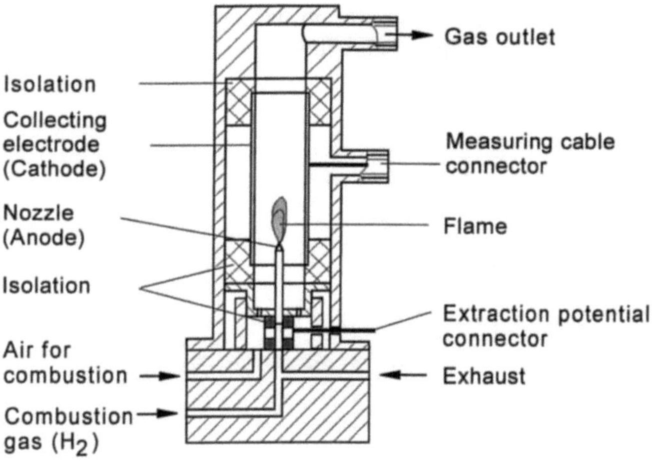

Fig. 6.44: Set-up of an FID (diagram)

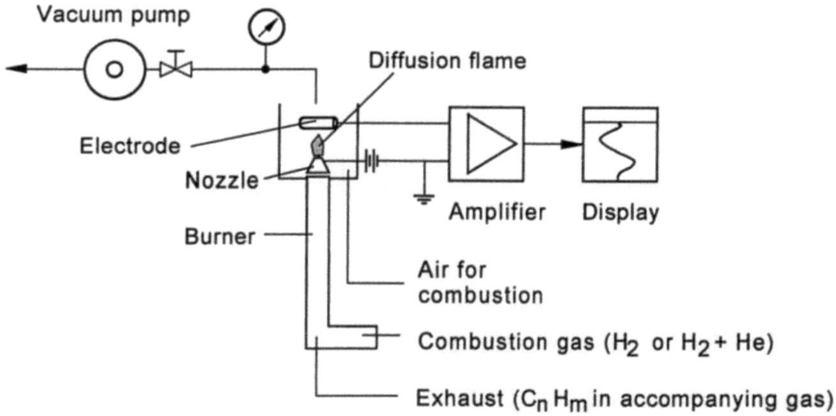

Fig. 6.45: Diagram of the operating principle of a flame ionization detector in negative pressure operation

The measuring signal of the flame ionization detector is determined by the number of hydrocarbon molecules reaching the flame per unit of time. Hence, the volume flow of the gas to be analyzed must be kept constant. The same is necessary for the combustion gas and the air for combustion.

Depending on the FID model voltage varies between 20 and 200 V. The ion current which is proportional to the carbon content lies in a range of 10^{-12} to 10^{-10} A.

Calibration with a reference gas is necessary. Regulations require calibration with propane.

The FID's pneumatic control which is similar in principle to that of the chemiluminescence analyzer (refer to Chap. 6.3) is shown in Fig. 6.46.

A photograph of an FID is shown in Fig. 6.47.

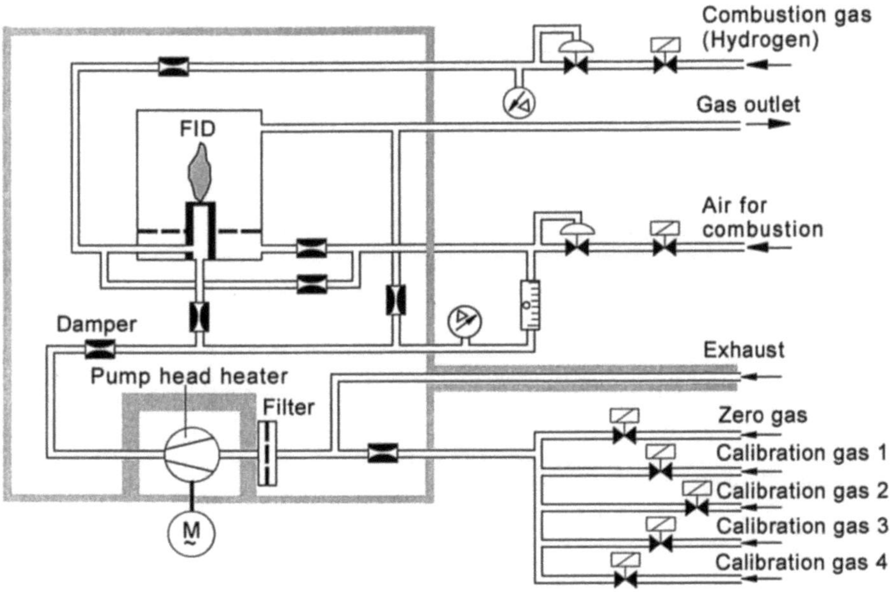

Fig. 6.46: Pneumatic circuit of an FID, with positive or negative pressure options

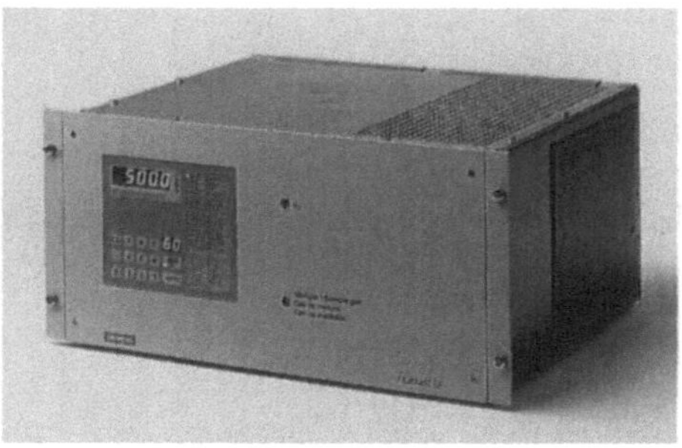

Fig. 6.47: Photograph of an FID (Siemens AG)

6.2.2.2 The Mechanism of Ion Formation

Coming back to ion formation, even today its mechanism has not been totally understood yet. A probable hypothesis will be looked into here. Before the actual ionization, hydrocarbon molecules containing several carbon atoms must be split into cracked products containing only one C atom each. The cracking process takes place in the so-called pyrolysis zone of the hydrogen flame. Basically, the mechanism proceeds via H atom reactions. Information in literature is sparse. One example given is:

ethane + hydrogen atom $\rightarrow$ ethyl radical + hydrogen

$$C_2H_6 + H^* \rightarrow C_2H_5{}^* + H_2 \qquad (6.65)$$

(* meaning excited)

and from this

$$C_2H_5{}^* + H^* \rightarrow 2\,CH_3{}^* \quad \text{(methyl radical).} \qquad (6.66)$$

Generally, a stripping of the crack products takes place:

$$OH^* + CH_3{}^* \rightarrow CH_2{}^* + H_2O \qquad (6.67)$$

and

$$OH^* + CH_2{}^* \rightarrow CH^* + H_2O\,. \qquad (6.68)$$

The cracking reactions are endothermic reactions.

A series of possible ion formation mechanisms via chemical ionization can be deduced from the above considerations. The most probable ones are:

$$CH^* + O^* \rightarrow CHO^+ \qquad (6.69)$$

and in the second reaction:

$$CH^*(2\,\Sigma^+) + O^* \rightarrow CHO^+ \qquad (6.70)$$

(CH^* excited in the electron term doublet Σ^+).

It can be deduced from the reaction mechanisms that apart from the hydrocarbon fragments the oxygen atom is the dominating co-reactant. The role of the CHO^+ ion as primary ion has been confirmed experimentally.

The hydronium ion H_3O^+ stemming from the water vapor around the reaction zone outside of the direct reaction zone of the flame was found to be the dominating ion. The following equation follows from a charge (proton) exchange mechanism:

$$CHO^+ + H_2O \rightleftarrows H_3O^+ + CO \qquad (6.71)$$

CHO^+ and H_3O^+ are balanced here, although the balance is shifted strongly towards the hydronium ion at a ratio of $1 : 3{\cdot}10^3$.

The hydronium ions occur at a balanced ratio with their hydratized forms:

$$H_3O^+ (H_2O)_N \qquad \text{(hydratized).} \tag{6.72}$$

The hydronium ions recombine with electrons or OH^- ions:

$$H_3O^+ + e^- \rightarrow H + H_2O\,; \tag{6.73}$$

$$H_3O^+ + OH^- \rightarrow 2\,H_2O\,. \tag{6.74}$$

Apart from these positive ions negative ions also form, e.g., C^-, C_2^-, CH_3^-, OH^- etc.

Under the conditions described above, the hydronium ions exist long enough outside the reaction zone for the ions to be recorded by a collecting electrode which is installed at a certain distance from the place of formation.

6.2.2.3 Assessment of FID Instruments

Among the measuring instruments for automobile exhaust gas analysis the flame ionization detector occupies a special position in that the measuring signal records the sum of all hydrocarbons present in the exhaust gas and not a particular gas component. Considering the very different structures of the great number of organic substances present in automobile exhaust gas, it is not surprising that measuring sensitivity is not determined exactly by the number of C atoms in one substance, but that molecular structure itself, too, has an influence on the measuring signal. Hence, there are varying degrees of deviations from the value corresponding to the C number in measuring practice. Even greater deviations are to be found in the measurement of hydrocarbon derivatives, particularly in oxygenic hydrocarbon compounds such as alcohols or aldehydes. The oxygen present in the molecules is released in the flame, so that the precondition of a pure diffusion flame is no longer given. The FID indication's deviations for the individual hydrocarbons or hydrocarbon derivatives, relative to a reference gas (e.g., methane, propane, n-butane) are expressed by so-called response factors. Using the example of three flame ionization detectors from different manufacturers, Table 6.4 shows the response factors for some compounds relative to n-butane as reference gas.

Response factors not only vary depending on the components, but also on the different manufacturers. They even differ in instruments from the same manufacturer.

When assessing FID instruments one must, on the whole, be aware of the limits of this method. One essential part of the method is the dissociation of the molecules and the ionization of the molecule fragments.

Apart from temperature and pressure, dissociation also depends on the molecule type. Ionization is also influenced by exhaust gas components other than the HC molecules to be measured. Varying concentrations of oxygen, carbon dioxide, carbon monoxide, water vapor, nitrogen oxide and nitrogen, such as those occuring

Table 6.5: Response factors for different hydrocarbons relative to n-butane using instruments from three different manufacturers

Components	Response factors		
	FID 1	FID 2	FID 3
Propane	1.03	-	1.00
n-butane	1.00	1.00	1.00
n-heptane	0.94	0.95	0.95
Cyclohexane	0.98	0.92	0.95
Isopropanol	0.71	0.78	0.81
Toluene	0.90	0.98	1.06
Acetone	0.69	0.76	0.74

in automobile exhaust gas, particularly during instationary operation, can lead to faulty measurement. Particularly variations in the exhaust gas sample's oxygen content show considerable influence on the hydrocarbon values measured. As a consequence, systematic errors in measurement are to be expected from the hundreds of chemical compounds of appreciable concentrations present in exhaust gas.

Fig. 6.48 shows the results of comparative measurements of the very same synthetic gas, which was mixed from different pure gases for a simulation of automobile exhaust gas. Its quantitative composition was known beforehand and verified for hydrocarbons by a gas chromatographic analysis (Fig. 6.48, upper section). This synthetic exhaust gas sample was analyzed by 44 FID industrial measuring instruments in different exhaust gas test laboratories (Fig. 6.48 lower

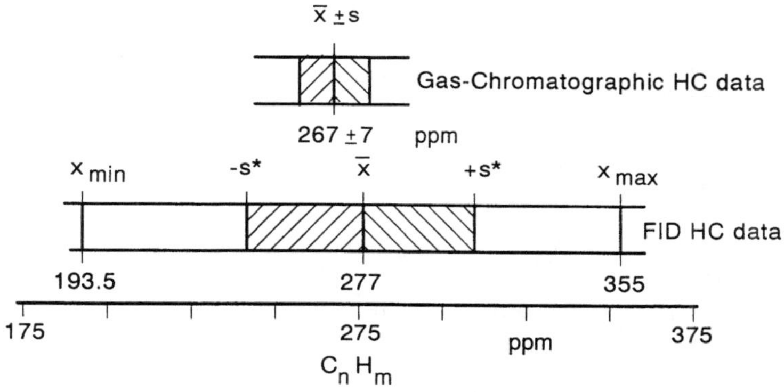

Fig. 6.48: Variation range of FID results, measured by 44 different industrial measuring instruments; $P = 95$ %, $\bar{x}$ = mean value, s = standard deviation, $2s_r = \pm 25$ %, $x_{max}/x_{min} = 1.8$; as compared to gas-chromatographic HC data (upper section)

section). The analysis error determined contains systematic instrument errors and incidental calibration and operational errors.

These considerable variations of up to ± 33% for extreme deviations can have a number of different causes. E.g., different FID instruments which were properly calibrated with propane, showed considerable deviations when other individual components were measured, with ethane a maximum of 5 % and with ethylbenzene a maximum of 6 %, but with acetylene + 73 to - 62 %.

Causes are, as mentioned, differences in the chemical structure and in the ionization behavior of hydrocarbon molecules.

In another study, exhaust gas from the same sample was measured simultaneously by 16 FID instruments from two manufacturers, namely relative to one of these measuring instruments (Fig. 6.49).

The results shown in Figs. 6.50 (gasoline engine exhaust gas) and 6.51 (diesel engine exhaust gas) show a considerable range of variance (of + 25 % to - 16 %) between the values measured by the individual instruments, although all of them were fed simultaneously with the same exhaust gas.

Systematic differences occur between the measuring values of both manufacturers' measuring instruments. Furthermore, it is interesting to note that when changing from gasoline to diesel engine exhaust gas (Figs. 6.50 and 6.51) deviations of instruments H5 and H9 change their signs relative to the measuring values of the reference instrument (0%).

This can be caused by the fact that these two instruments in particular have conspicuously different sensitivities for certain components in diesel exhaust gas than the other instruments.

Efforts to improve the FID method have not met with marked success. Despite its weaknesses it must be stated that although FID instruments have their

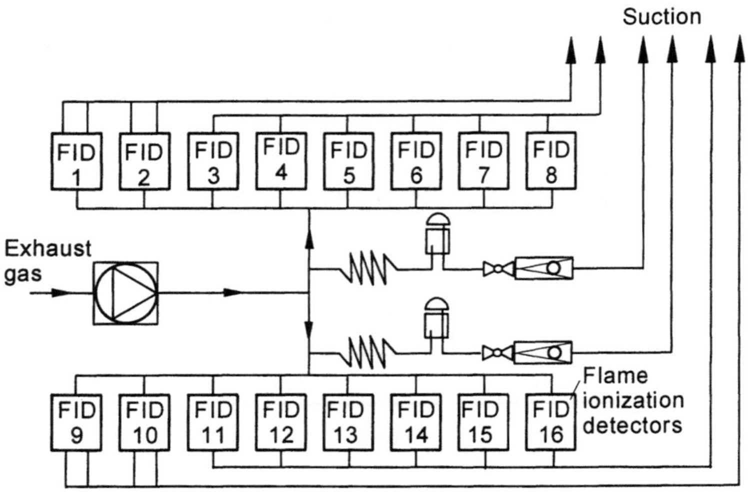

Fig. 6.49: Measuring set-up for investigating the comparability of values measured by different FID instruments when using the same exhaust gas for all instruments simultaneously

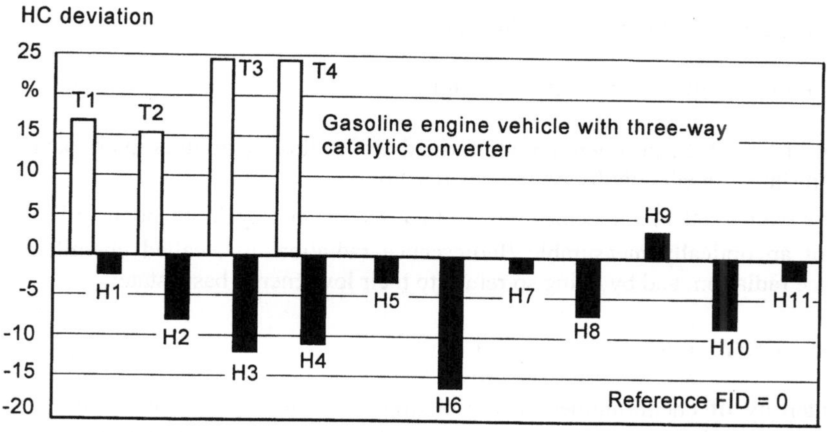

Fig. 6.50: Values of the FID comparative simultaneous measurement, based on one FID value set to zero. Exhaust gas of a gasoline vehicle with lambda sensor-controlled three-way catalytic converter (T or H meaning different instrument manufacturers)

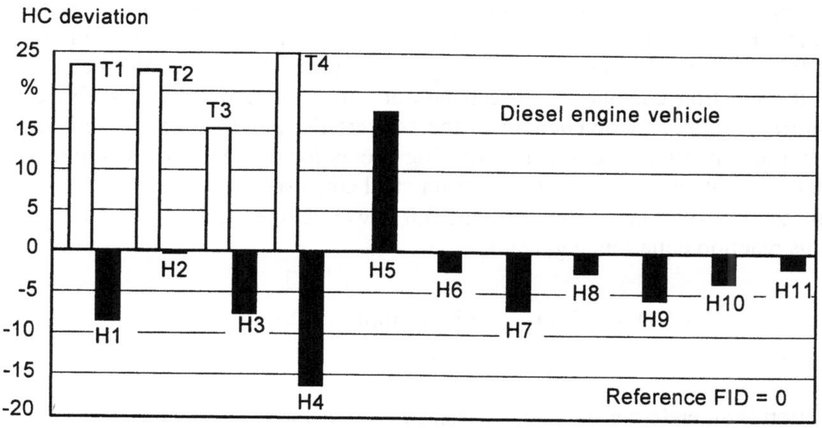

Fig. 6.51: Values of FID comparative simultaneous measurement, using exhaust gas of a diesel-powered vehicle, otherwise as in Fig. 6.50

limitations for determining the total hydrocarbon content of vehicle exhaust gas they are still mandatory.

6.3 Chemiluminescence Analyzer

6.3.1 Principle

The measuring principle of the chemiluminescence analyzer (CLA or CLD) is based on the spontaneous reaction between nitrogen monoxide and ozone in a reaction chamber acc. to the following equation:

$$NO + O_3 \rightarrow NO_2 + O_2 + 205\ \text{kJ / mol} . \tag{6.75}$$

Approx. 10 % of the nitrogen dioxide formed acc. to this equation is in an excited electron state immediately following this reaction.

These excited NO_2^* molecules release their excess energy spontaneously in the form of an optically measurable fluorescence radiation $h\nu$, called chemiluminescence radiation, and by doing so return to their low-energy basic state:

$$NO_2^* \rightarrow NO_2 + h\nu . \tag{6.76}$$

The intensity of chemiluminescence radiation $h\nu$ which can be observed in a spectral range of approx. 590 nm to approx. 3000 nm, is a direct measure of the nitrogen monoxide concentration.

This measuring principle was originally used for recording ozone (O_3), the NO serving as reagent. In a manner of speaking, the method was reversed, so that nitrogen oxides in the exhaust gas could be measured.

It is problematic, however, that only NO components are recorded with this measuring principle. The NO_2 contained in the exhaust gas sample which is formed from NO in the presence of oxygen, must be reduced to NO before it reaches the reaction chamber. This can be done in a thermal converter above a temperature of 925 K. Molybdenum in the converter's tube material or activated carbon in the converter function as catalyzing agents for the reaction. A minimum molybdenum content of 2.5 % is required for steel converters.

Optimized catalytic converters also permit lower temperatures (< 750 K). An analogous reaction equation, e.g, reads:

$$NO_2 \xrightarrow[\text{catalysis}]{925\text{K}} NO + \tfrac{1}{2} O_2 - 57\ \text{kJ / mol.} \tag{6.77}$$

The reaction is an endothermal one, i.e., heat energy must be added.

6.3.2 Method and Measuring Instrument

Fig. 6.52 shows the operating principle of the chemiluminescence analyzer in a diagram.

The exhaust gas sample's volume flow must be kept constant so that light generated is proportional to the concentration. The gas sample thus flows past a pressure controller and through particle filters as well as a sample metering capillary and then through the converter into the reaction chamber. Bypass circuits ahead of the metering capillary are necessary for achieving acceptable response times.The ozonizer, which generates ozone from oxygen and whose outlet also leads into the reaction chamber via a dosing capillary, is supplied from an oxygen tank or from ambient air. Immediately after leaving the reactor the exhaust gas of

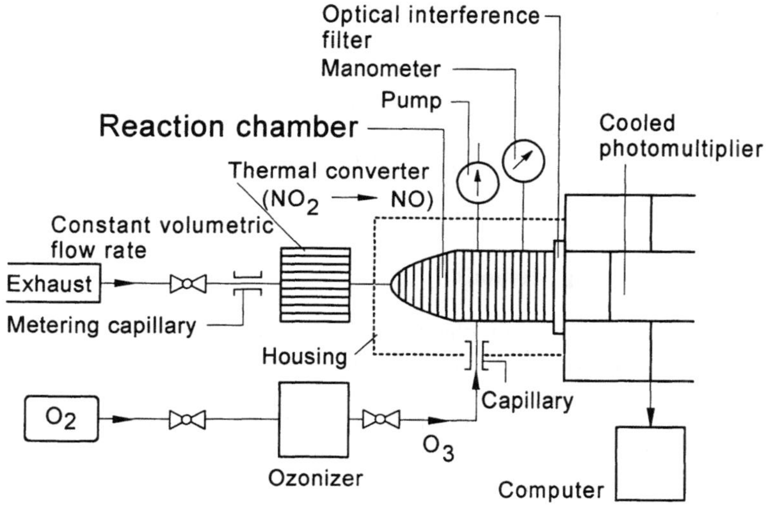

Fig. 6.52: Set-up of a chemiluminescence analyzer

the reaction chamber passes an "ozone trap" (not shown in Fig. 6.52) which consists of an activated carbon package for the removal of residual O_3, so that no ozone is released into the environment.

Through a narrowband optical interference filter the radiation produced in the reaction chamber reaches a photomultiplier, whose output signal can be recorded by plotters or be processed by computers.

When nitrogen oxide concentrations are very low the interference level can lead to a very uneven signal. A reduction of this error, i.e., improving the signal-noise ratio, is possible by lowering absolute pressure in the reacton chamber to approx. $1 \cdot 10^3$ Pa. Modern instruments operate at these low pressures.

Calibration with a test gas is necessary. It is mandatory that this operating principle has to be used. The converter is the critical item, as its efficiency decreases with increasing operation time. Regular converter check-ups are therefore both necessary and also mandatory.

6.4 Oxygen Measuring Methods

6.4.1 General Observations

Oxygen (O_2) as a gas with homoatomic molecules shows no radiation absorption in the range of rotational-vibrational spectra of gases. This means, that the oxygen content of automobile exhaust gas cannot be measured with the methods of absorption spectroscopy in the infrared range. However, oxygen can be recorded by mass spectrometry. A measuring method for oxygen must meet the following basic requirements:

- Smallest measuring range: ≤ 1.0 Vol. % O_2;
- Cross sensitivity (other components overlapping the oxygen value measured): ≤ 1% of the measuring range;
- Response time (90 % time) T_{90}: < 1 s;
- Measurement directly in the exhaust gas stream, alternatively in the bypass as a continuous measurement.

6.4.2 Paramagnetic Measuring Methods

Oxygen is paramagnetic, i.e., oxygen molecules exhibit a very high paramagnetic susceptibility. They are drawn into the magnetic field, whereas most other gases are diamagnetic and are repulsed by the magnetic field. Paramagnetic measuring methods for determining the oxygen content of gas mixtures are based on this particular oxygen characteristic. When examining exhaust gas, the fact that nitrogen monoxide (NO) and nitrogen dioxide (NO_2) are also paramagnetic must be taken into account.

Common measuring methods are the magneto-mechanical (torsion balance) and the magneto-pneumatic (pressure differential) measuring method.

6.4.2.1 Torsion Balance Method

The principle of the magnetic torsion balance is illustrated in Fig. 6.53.

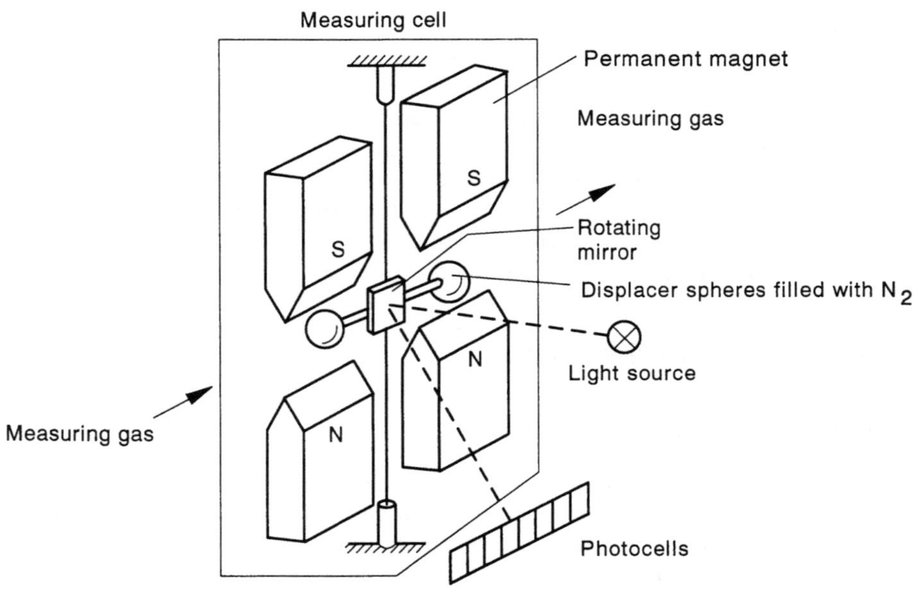

Fig. 6.53: Principle of the magnetic torsion balance

Two permanent magnets generate a non-uniform magnetic field in a measuring cell through which the measuring gas flows. In the magnetic field there is a two-part, dumbbell-shaped displacer made of diamagnetic material (e.g., a thin-walled quartz glass body with a nitrogen content; N_2 is diamagnetic) which is suspended on flexible bands and can be rotated. If the measuring gas contains paramagnetic oxygen it will be drawn into the magnetic field. The local gas compression taking place during this causes the two displacers to be pushed out of the magnetic field. As a result of this, torque *M* acts upon the "dumbbell" which is expressed by

$$M = (\kappa_1 - \kappa_2) V \mu_0 H \frac{\partial H}{\partial x} d, \tag{6.78}$$

where κ_1 and κ_2 are the susceptibilities of the measuring gas or of the gas in the displacer, *V* is the volume of the displacer, *d* their distance from center to center, $\frac{\partial H}{\partial x}$ the non-uniformity of the magnetic field and μ_0 the magnetic field constant.

The deflection of the torsion balance can be measured, e.g., with the help of a beam of light, which is reflected by a rotating mirror placed in the dumbbell's center of gyration and whose angle position, which is proportional to the O^2 partial pressure, is scanned by photocells.

In the versions in operation a balancing torque is created by current flow through wire loops placed around the dumbbell. The current required for a balance is proportional to the oxygen partial pressure in the measuring gas.

6.4.2.2 Differential Pressure Methods

A diagram of this measuring method is shown in Fig. 6.54. Paramagnetic oxygenous measuring gas and a diamagnetic reference gas (e.g., nitrogen) are simultaneously fed into a measuring cell.

The measuring gas flows directly through the measuring cell, the reference gas stream is split into two equal parts and is fed via channels into the measuring cell from opposite sides. One of the inlets is inside a non-uniform magnetic field (represented by a circle in Fig. 6.54). At the area of contact between the oxygenous measuring gas with the susceptibility κ_1 and the reference gas with the suceptibility κ_2 a pressure p' is built up where

$$p' = const. \mu_0 H^2 (\kappa_1 - \kappa_2) . \tag{6.79}$$

This pressure p' is higher than the pressure p in the measuring cell's space without magnetic field. If the magnetic field is pulsed, a pulsed differential pressure is created which continues into the transverse canal connecting the two reference channels. The pressure variations, which develop inside of this transverse canal and which are proportional to the oxygen content of the measuring gas can be measured, for instance, with the help of a flow microsensor.

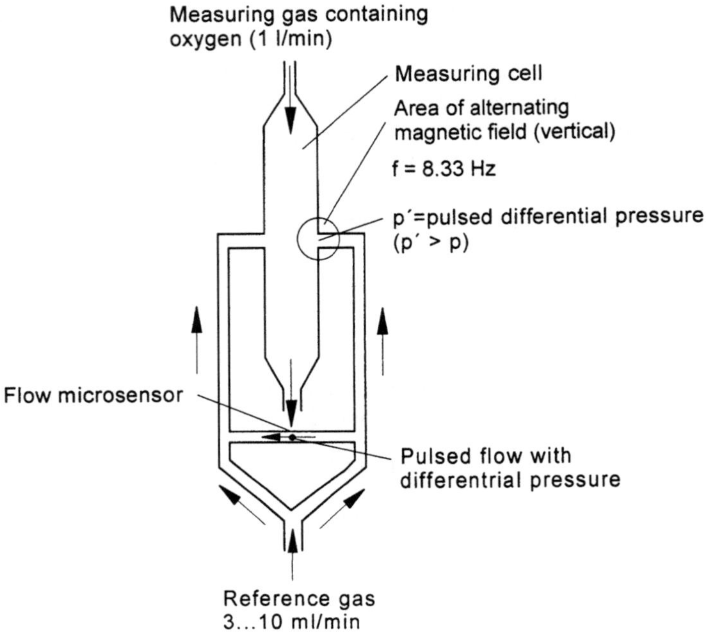

Fig. 6.54: Diagram of a differential pressure oxygen analyzer

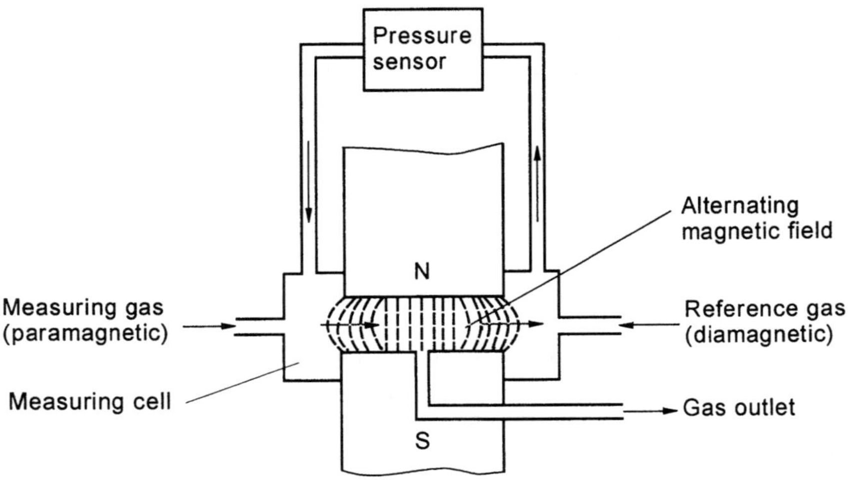

Fig. 6.55: Diagram of the operating principle of magneto-pneumatic oxygen measurement

As a further example, the operating principle of the magneto-pneumatic method is shown in a different way in Fig. 6.55.

In a measuring cell where an alternating magnetic field between magnetic poles is generated, the oxygenous measuring gas and a diamagnetic reference gas (e.g. nitrogen) are admitted through inlets from opposite sides. The paramagnetic oxygen is drawn into the field, the diamagnetic reference gas is pushed out. This causes pressure variations on both sides of the cell, which are, if the reference gas flow is kept constant, proportional to the oxygen content of the measuring gas and can be measured by a pressure sensor.

Assessment of magnetic methods of oxygen measurement:

- The smallest feasible measuring range is at 1 Vol.-% O_2.
- Paramagnetic gases such as NO and NO_2 cause unavoidable (as physically caused) cross sensitivities: e.g., in a measuring range of 1 vol.-% O_2 an NO concentration of 5000 ppm causes a corresponding cross sensitivity to 0.2 vol.-% O_2, i.e., 20% of the oxygen measuring range.
- Response times (90 % times) are longer than 2 s.
- A defined measuring gas flow of approx. 60 l/h must be fed into the measuring cell. This can only be achieved via a bypass and a special sampling device.

6.4.3 Electrochemical Oxygen Measurement

Electrochemical measuring methods are based on the electrochemical conversion of oxygen in an electrochemical measuring cell. An electrochemical oxygen measuring cell is illustrated in Fig. 6.56 in a diagram. The cells consist of an electrolyte between a cathode and an anode (both of noble metal). Some models contain an additional third electrode which keeps the potential applied constant. Good selectivity for oxygen is achieved by selecting suitable electrode materials and electrolytes. Before the cathode there is a microporous layer (in some models a capillary) through which the measuring gas reaches the cathode by diffusion. Diffusion rate depends on the oxygen's partial pressure in the measuring gas. The oxygen which is diffused to the cathode, is converted electrochemically. The conversion rate is measured in the cell as an ion flow and is a measure of the oxygen concentration in the exhaust gas.

Assessment of the electrochemical method:

- Measuring ranges smaller than 1.0 vol.-% O_2 can be achieved.
- Cross sensitivity is small, as satisfactory oxygen selectivity can be achieved by selecting the appropriate electrolyte and anode material.
- Response times depend on the cell's design. In special versions it is less than 1 s.
- Cells are unsuited for direct measurements in the exhaust gas flow at temperatures of several 100 °C, as the operating temperature is restricted to approx. 50 °C due to the aqueous electrolyte.

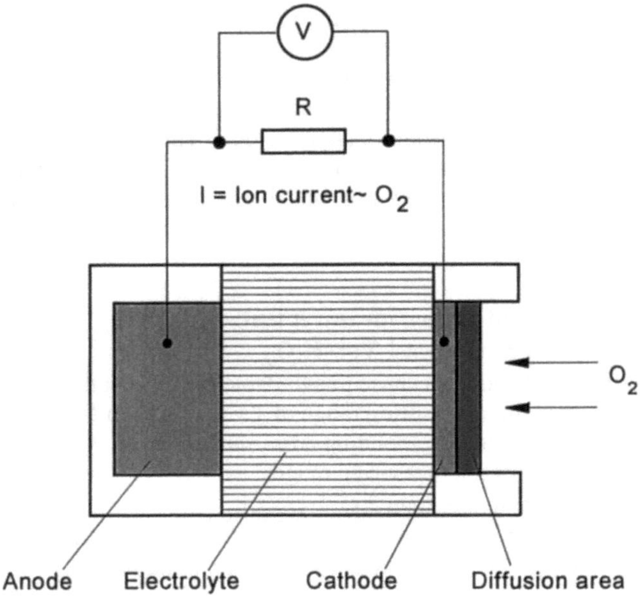

Fig. 6.56: Operating principle of an electrochemical oxygen measuring cell

6.5 Chromatographic Methods

6.5.1 Basics

Due to the general importance of chromatographic methods in trace analysis some basics of chromatography will be mentioned first.

The term "chromatography" comprises a series of microanalytic separation processes serving for the separation of individual compounds from a given substance mixture.

The principle of chromatographic methods - represented in a highly simplified form by the example of gas-liquid chromatography - is the following:

An inert carrier gas (mobile phase) is passed through a "chromatographic separation column" in a constant flow. Fig. 6.57 shows a section of such a column in a diagram.

It is filled with an inert granular support material. The support material's surface is coated with a higher-boiling liquid (stationary phase) where the compounds to be examined dissolve to a more or less high degree.

Then a predetermined small amount of the substance mixture to be examined is added to the carrier gas flow. A substance component with high solubility is then dissolved rapidly by the high-boiling liquid coat and is retained by it until it is slowly expelled by the carrier flow. At the end of the separation column it thus emerges later than a substance with low solubility, which dissolves poorly and is

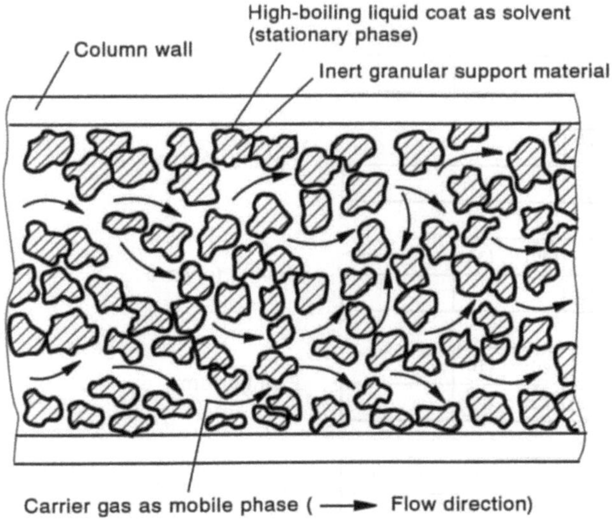

Fig. 6.57: Cross-section of a separation column for gas-liquid chromatography

expelled quickly. In this way, the individual components move through the column at different velocities and appear consecutively at the outlet of the separation column. One speaks of different distribution coefficients.

The alternation adsorption - desorption in gas-solid chromatography, refer to Chap. 6.5.2, proceeds in exactly the same way.

If one imagines the separating process in sections, permitting a balanced condition and continuing transport and again a balanced condition etc., the separation of two substances can be seen schematically in Fig. 6.58.

The two components symbolised by a square and a circle have different distribution coefficients (e.g., different solubilities), which is the precondition for a separation.

In the example shown, 90 parts of each component are present in the mobile phase in the initial condition. In the balanced state, the concentration of the "round component" is twice as high in the stationary phase as in the mobile phase, with the "square component" it is exactly the other way round.

Of the "square component" 60 parts remain in the mobile phase after the balance has been reached, 30 parts migrate into the stationary phase. Of the "circular component", 30 parts are in the mobile and 60 in the stationary phase.

Now the parts in the mobile phase are moved a small distance through the separation column. Again balance is achieved and a transport follows. Even after only these two balance/migration cycles the segment of the mobile phase, lower right in Fig. 6.61, contains the two components used in a concentration ratio of 40 : 10 as opposed to 90 : 90 in the initial concentration at starting position.

Over one thousand of these balancings take place during such a chromatographic separation and are accorded one "theoretical plate" each in the column. This model has led to the efficiency of a chromatographic system being defined by the number

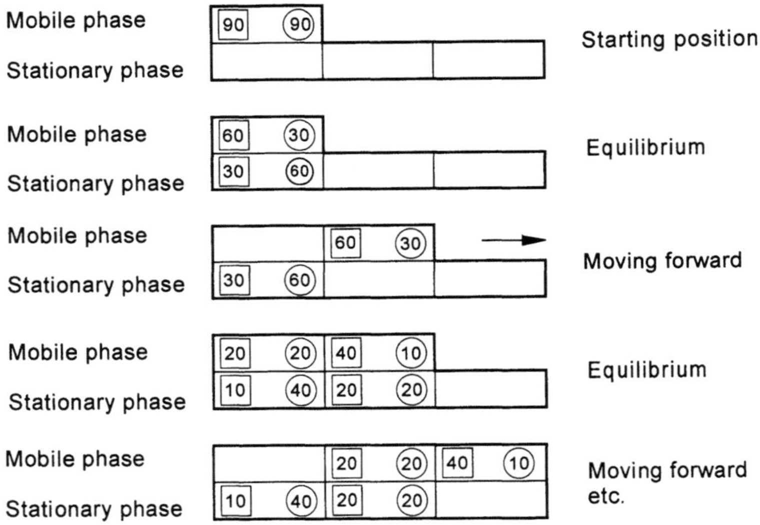

Fig. 6.58: Movement of two components with different distribution coefficients ($K_{\square}$ = 2:1 and $K_{\bigcirc}$ = 1:2) through the first separating stages of a chromatographic separation column (diagram)

of theoretical plates or the height of a theoretical plate, similar to the system of a distillation column.

The joint characteristic of all chromatographic methods is the multiple repeated (multiplicative) distribution of the components between the stationary phase and a (gaseous or liquid) mobile phase, which flows by the stationary phase.

In the course of the separation process the components are thus separated completely and are then moved through the mobile phase and out of the separation system in chronological order, or they remain fixed at one point in the stationary phase.

According to the type of the mobile phase, chromatographic methods can be divided into gas-chromatography with a carrier gas (N_2, He, H_2 or other gases) as mobile phase and into liquid chromatography with suitable organic solvents or alternatively water or aqueous solutions as mobile phase.

The stationary phase can be a solid or a liquid on a, if possible, inert carrier. Thus, there are four basic kinds of chromatographic separation methods which are listed in Table 6.6 with the following abbreviations

G = Gas; L = Liquid; S = Solid; C = Chromatography

The first letter of the abbreviation describes the mobile phase's physical state, the second one the stationary phase's physical state. All separation methods listed are used in exhaust gas analysis.

Table 6.6: Basic types of chromatographic separation methods

Stationary phase	Mobile phase	
	Gas	Liquid
Solids		
Adsorbent	GSC	LSC
Liquid	GLC	LLC

6.5.2 Gas-Solid Chromatography (GSC)

As an example, the set-up of a GSC instrument shown in a diagram will be explained below (Fig. 6.59).

A pressure cylinder with a pressure regulator and a manometer effects the flow of the mobile phase (gas) through the separation system. Ahead of the column is the sample application instrument which administers the sample to be separated into the carrier gas flow.

The sample should reach the beginning of the separation column as a short "plug", so that all individual components have the same starting point and starting time.

If the sample is liquid, it is necessary to provide for a rapid and complete vaporisation of all sample components in the sample inlet. Hence, the sample inlet has a temperature control. Dosing of the gaseous or liquid sample is carried out via milli- or microliter injection syringes (Fig. 6.60).

Depending on the physical state and volume of the sample to be injected there are different syringe models available.

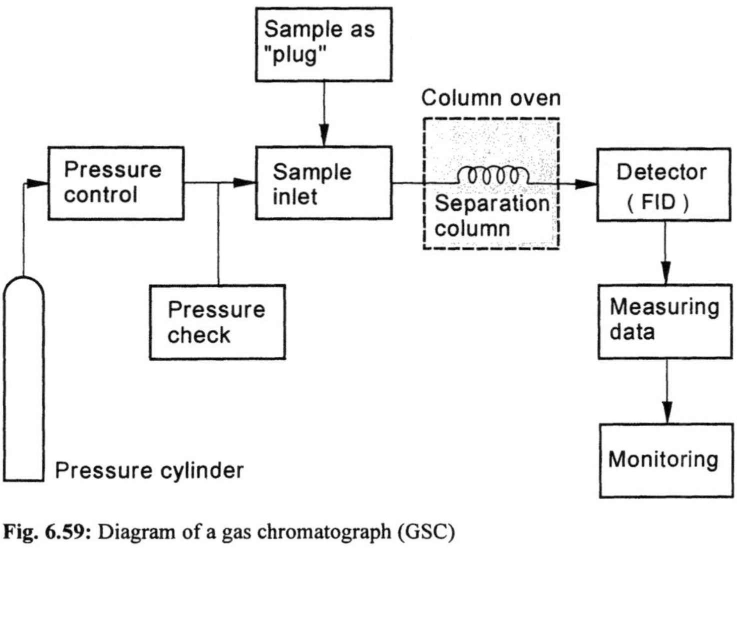

Fig. 6.59: Diagram of a gas chromatograph (GSC)

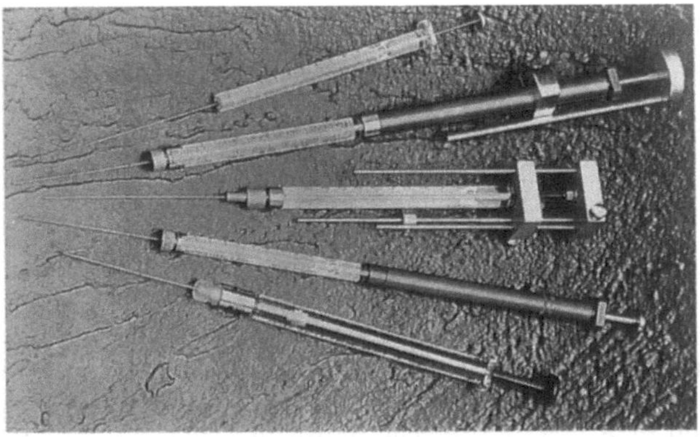

Fig. 6.60: Photograph of injection syringes for sample dosing

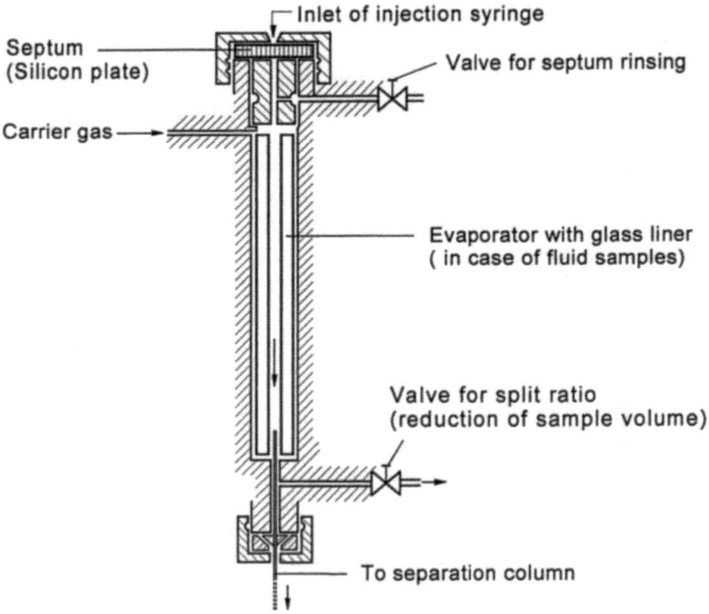

Fig. 6.61: Diagram of sample injection system with sample divider (splitter) and feeding for septum rinsing

The sample is injected into the carrier flow with the syringe (Fig. 6.61).

The carrier gas flow is sealed off by a silicon rubber disc (called "septum"). After taking out the injection syringe this septum seals itself again. Vaporizers with glass inserts permit vaporisation of the components to be examined when the sample is liquid.

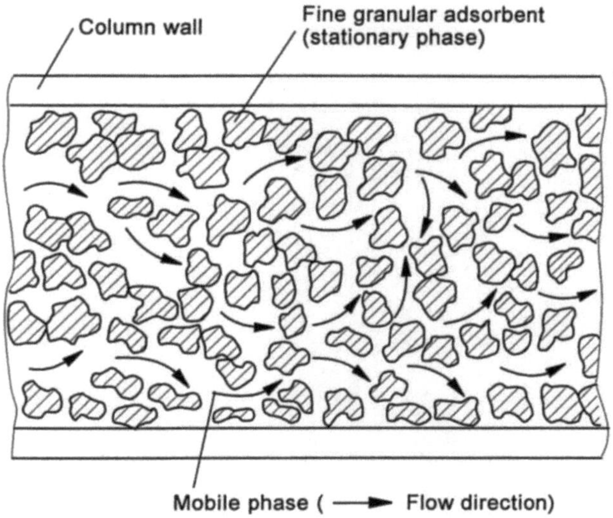

Fig. 6.62: Cross-section of a packed column of a GSC (diagram)

The sample inlet illustrated allows "septum rinsing" which prevents volatile components of the rubber seal from reaching the separation column and falsifying the analysis results. Moreover there is the possibility of injecting only a part of the sample into the separation column ("splitting"). Thus, - if necessary - excessive loading of the separation column with sample material and the consequent deterioration of separation efficiency are prevented.

A separation column with a stationary phase of fine-granular adsorbing material (adsorbent) in a cylindrical tube (Fig. 6.62) is called a "packed column".

The mobile phase, i.e., in GSC a gas, should flow around the stationary phase (the adsorbent) as evenly as possible. Hence, the ideal shape for the stationary phase is spherical, which has almost been achieved in practice, as can be seen from the following scanning electron micrograph (Fig. 6.63). In this case the spherical diameter is approx. 100 µm.

Mainly glass, but also metals such as steel or copper are used as materials for the separation column housing. For practical reasons, the columns are usually designed as spirals as in the following example of Fig. 6.64.

The separation column is installed in a separate oven (refer to Fig. 6.59).
The adsorption/desorption balance is temperature-dependent and can be influenced selectively by the temperature setting. For this, the oven temperature is either kept constant or it is changed according to pre-programmed profiles. Details on this will be given below.

Detector

Behind the column is a detector which converts the separated substances in the carrier gas arriving in chronological order into electric signals which are proportional to the components' concentration in the mobile phase (refer to Fig. 6.59).

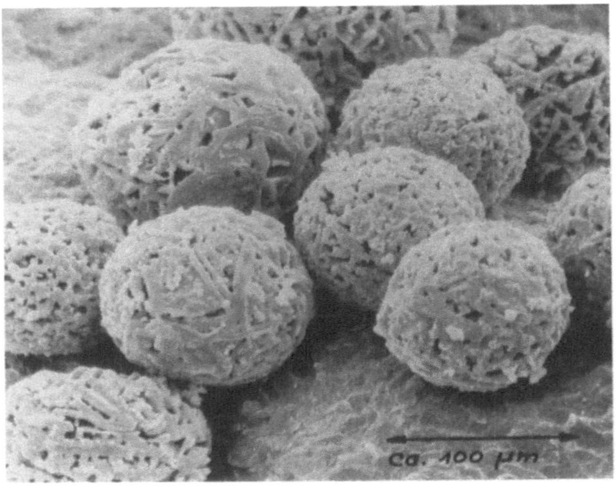

Fig. 6.63: Scanning electron micrograph (1000:1) of an adsorbent for GSC

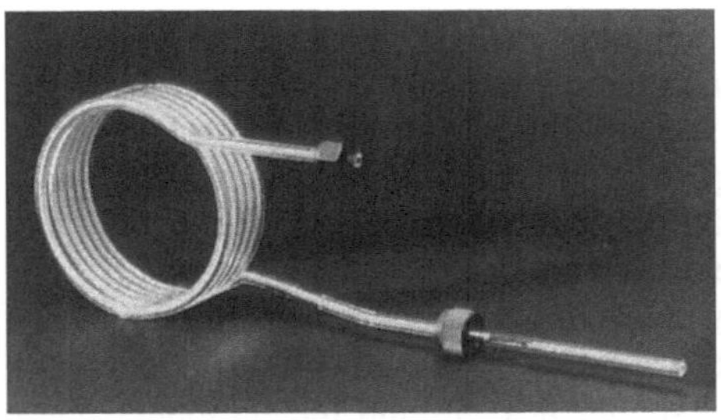

Fig. 6.64: Photograph of a packed glass column designed as a spiral, length: 100 to 1000 cm

The flame ionization detector (FID) has already been mentioned among the large number of detectors developed for all kinds of applications. The FID is a universal, highly sensitive detector with a wide dynamic range. It is used mainly for recording hydrocarbons and other organic compounds. By installing an alkali salt bead in the combustion chamber, the FID can be sensitized for nitrogenous compounds.

Calibration

In contrast to measuring the limited sum of *all* hydrocarbons in the exhaust gas, separate individual compounds are recorded here. Thus, the FID can be calibrated with samples of known compositions (previously weighed standard mixtures). Calibrations are carried out successively for each component to be measured, thus minimizing errors.

Chromatogram

The measuring signals recorded by the corresponding detector over the period of analysis provides the so-called "chromatogram" (Fig. 6.65). The upper section of Fig. 6.65 reflects the result of a gas-chromatographic analysis of C_6 - to C_{20} - alkanes. An FID has been used as detector.

Each chromatographic signal, called "peak", stands for a certain chemical compound which, under reproducible gas-chromatographic conditions, always appears at the same spot, i.e., it always has the same "retention time" (distance from zero). Retention time measurements are the basis of qualitative analysis (result of separation).

The areas below the individual peaks are used for the quantitative evaluation for determining concentration (in this case, e.g., for C_{20}). As is common practice today, area calculation can be carried out electronically with the help of "integrators", i.e., computer-aided. After calibration the area is a measure of the component's concentration (result of measurement).

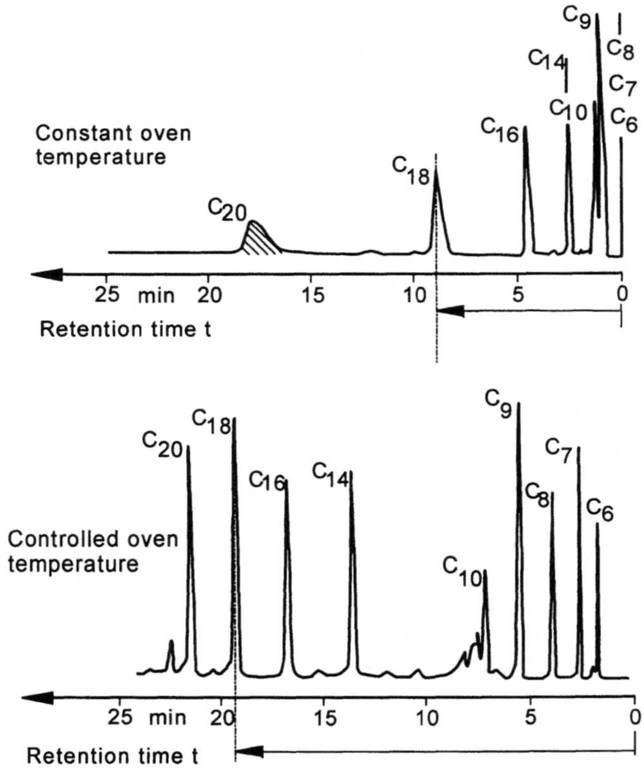

Fig. 6.65: Analysis of $C_6 ... C_{20}$ n-alkanes (paraffinic hydrocarbons) in isothermal operation at 200 °C (upper half) and with temperature control of 60 to 280 °C/minute (lower half)

The difference between the two chromatograms shown will also be pointed out:

The upper section of Fig. 6.65 shows the result of an analysis of substances (hydrocarbons, C contents are marked) with different distribution coefficients during isothermal (constant temperature) operation of the separation column oven. It can be clearly seen that in the upper half of the diagram peaks are very narrow and tightly spaced initially. This complicates quantitative evaluation. The components are not sufficiently separated. On the other hand, peaks become increasingly smaller and wider with increasing retention time, until they become completely indistinguishable in the noise of the base line.

The lower section of Fig. 6.65 shows the result of an analysis where the process was influenced selectively by changing the oven temperature. Analysis is started at a low column temperature. The migration speed of the components with the carrier gas is reduced, causing the highly volatile compounds in a GLC or the highly adsorbent compounds in a GSC to be separated sufficiently. The column temperature is then slowly raised. Migration velocities accelerate correspondingly. Skillful selection of the temperature program can accomplish that all components are recorded as peaks which can be easily evaluated. Moreover, analysis time is greatly reduced as low-volatile compounds in a GLC or low adsorbent compounds in a GSC emerge from the separating column faster at higher temperatures towards the conclusion of the temperature program.

The analyst's task then consists of selecting the separation conditions in such a way that the desired components appear as individual peaks. To this end, the following parameters can be varied:

1. material and dimensions (length, diameter) of the separation column;
2. stationary phase:
   a) carrier material and liquid phase or,
   b) adsorbent ;
3. mobile phase (type of gas and flow velocity);
4. temperature program (initial temperature, heating rate, final temperature);
5. detection method.

6.5.3 High-Performance-Liquid-Chromatography (HPLC)

Gas- and liquid-chromatographic methods differ in several ways as far as equipment and area of application are concerned (Fig. 6.66).

The liquid acting as mobile phase is pushed through the separation column either by its own weight or in most cases - as shown in Fig. 6.66 - by a high-pressure pump. In this case one speaks of high pressure liquid chromatography - HPLC or lately of high performance liquid chromatography - HPLC.

The sample, which must be liquid, is introduced into the liquid flow via a sample loop and a multiport valve. Fig. 6.67 shows the operating principle of such an instrument.

The left section shows how the sample solution fills the loop, which has a defined volume. After turning the 6-port valve, solvent flows into the loop (right section of diagram) and pushes the sample (whose volume is defined) into the separation column.

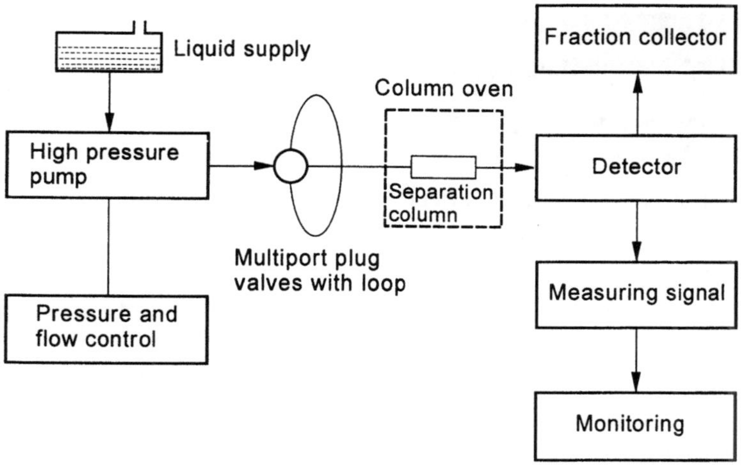

Fig. 6.66: Diagrammatic representation of a liquid chromatograph (HPLC)

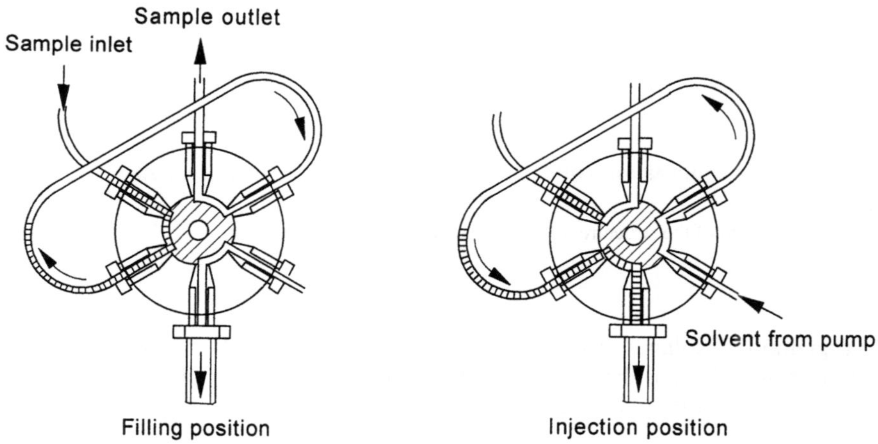

Fig. 6.67: Multiport plug valve as sample delivery system (diagram)

The separation columns used in HPLC consist mainly of stainless steel and are much more compact (Fig. 6.68) than those used in gas-chromatographic processes.

This limits the pressure drop in the separation column, which is considerably higher in liquids than in gases. Typical column lengths are between 5 and 25 cm as compared to 100 to 1000 cm in gas chromatography.

Physically LLC is mostly analogous to the described GLC. The stationary phase is a thin liquid film on the surface of the pellets packed into the column (Fig. 6.69).

The pellets as solid supports should have a surface as porous as possible, so that a large contact area is available to the liquid phase.

Fig. 6.68: Photograph of separation columns for HPLC

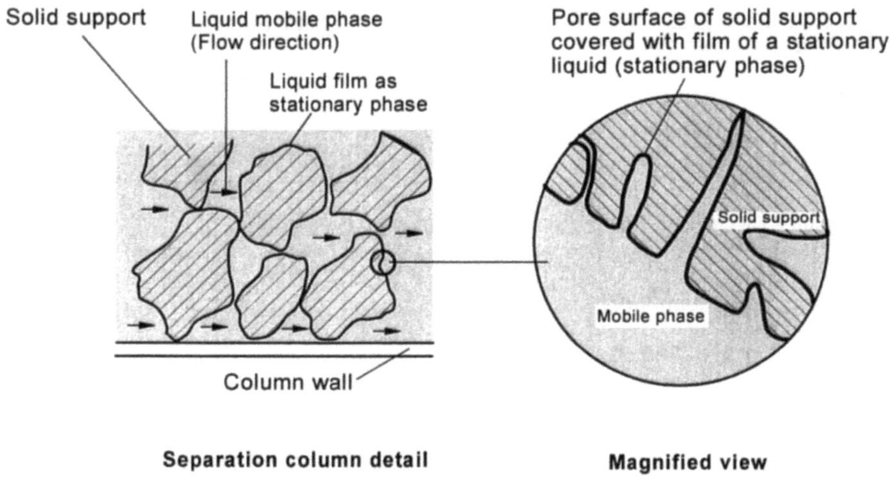

Fig. 6.69: Diagram of the structure of a packed column for HPLC

Programming column temperature - as described in GSC or GLC - is not customary in HPLC, as the great heating capacity of separation column and mobile phase does not permit reproducible control.

Gradient programming
However, comparable effects can be achieved by using two or more solvents whose ratios can be varied as mobile phase, instead of only one ("Gradient programming", Fig. 6.70).

Detectors
In contrast to gas chromatography, mainly optical and spectroscopic methods are used for the detectors. The great advantage of this method is that the sample components are not destroyed (not burnt) but can be retrieved unchanged, e.g., by a fraction collector.

Among others, the following instruments are used as detectors for HPLC:

- differential refractometer,
- UV/VIS- and IR detectors with fixed and tunable wavelengths,

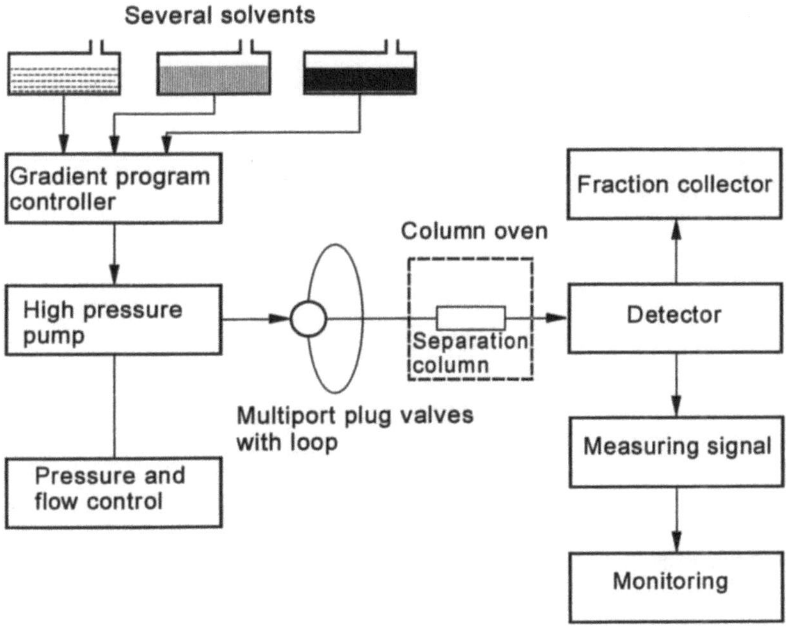

Fig. 6.70: HPLC with gradient programming (diagram)

- fluorescence detectors,
- conductivity detectors,
- and electrochemical and radiometric detectors.

The diode-array photometer, available on the market, will be introduced here as being representative of these detector concepts (Fig. 6.71).

The complete light spectrum of a deuterium lamp (wavelength range: 190 to 600 nm) is focussed through an achromatic lens onto the measuring cell, it is absorbed through the sample, dispersed into individual wavelengths through a holographic reflection grating and then focussed on a diode-array detector.

The array consists of a series of 200 photodiodes, arranged over a length of 10 mm. The light intensities of wavelengths of 190 to 600 nm are read simultaneously (10 ms read-out time) by these photodiodes. Resolution is approx. 4 nm. Recording, processing and transmission of data is done by microprocessors. A so-called "shutter" is the only movable part of the photometer. It is swivelled into position for dark current measurement and for wavelength calibration.

Complete spectra from UV light until far into the visible spectrum can be recorded time-resolved with this detector without interrupting the mobile phase's flow - something which is necessary when using an instrument with wavelength tuning through a grating. Due to its relatively low resolution, the diode-array photometer has only limited applicability for analyses requiring higher precision.

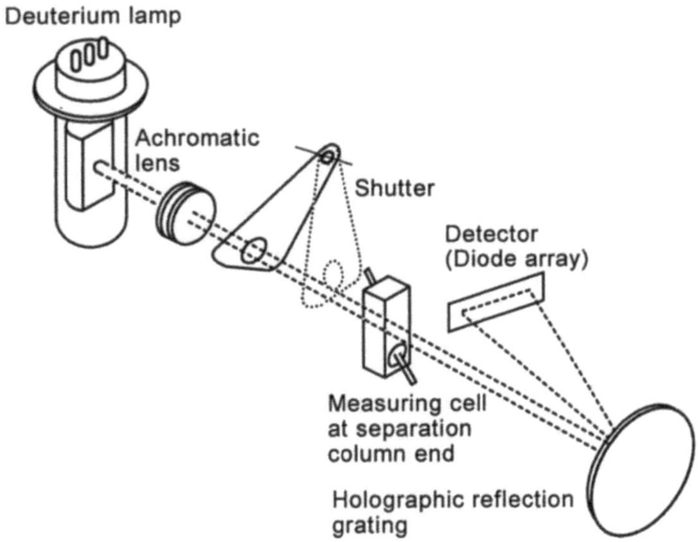

Fig. 6.71: Set-up of the diode-array photometer (diagram)

In practice, gas chromatography and liquid chromatography are less commonly used as alternatives; their possibilities rather complement each other. While gas chromatography is the method of choice when examining gases and highly volatile, undecomposably vaporizable substances, HPLC offers advantages when analysing dissolved high-molecular and strongly polar compounds.
Hence, HPLC is also used for the pre-separation of complex mixtures in fractions with few individual components.

6.6 Mass Spectrometry (MS)

6.6.1 Basic Observations

Not only through their interaction with electromagnetic radiation can molecules be analyzed but also separated and identified acc. to their mass, and their concentration thus be measured. Mass spectrometry serves this purpose; its basic principle is the generation of ions from molecules of inorganic or organic substances, to sort these ions in a separation system according to their ratio of mass and charge (m/q)[1] and to record them with a detection system selectively and according to their frequency. The mass spectrum is then plotted as a line spectrum on the mass scale. The "lines" are generally not sharply defined lines, but,

[1] m in atomic mass units ($1\ u = 1.66 \cdot 10^{-27}$ kg) and $q = z \cdot e$ in elementary charges e ($1\ e = 1.6 \cdot 10^{-19}$ C), $u/e = 1.04 \cdot 10^{-8}$ kg C^{-1} where z = number of elementary charges e

depending on the sharpness of the image of the instrument (ion scattering), they are more or less widened.

In contrast to absorption-spectrometric methods where the measuring gas flows through a measuring cell with a defined volume, generally under atmospheric pressure conditions, the mass separation system must, for physical reasons (avoiding impact between the ions generated and neutral molecules) be operated under high-vacuum conditions (pressure smaller than approx. 10^{-3} Pa).

Fig. 6.72 shows the diagram of a mass spectrometer system. Its individual components will be discussed in more detail below.

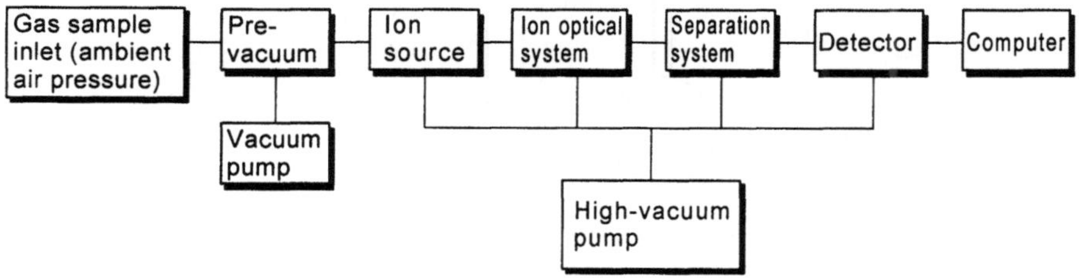

Fig. 6.72: Diagram of a mass spectrometer system

6.6.2 Inlet System

For continuous mass spectrometric measurement of gases, the gases which are generally at hand at atmospheric pressure must be fed into the ion source, which is kept in a vacuum, via a special inlet system. For one, this inlet system must reduce gas pressure from atmospheric pressure to the ion source's vacuum without changing gas composition; for another it must keep the gas's mass flow over capillaries and critical nozzles constant. It is necessary to keep the mass flow constant as, in quantitative measurements, the amount of compounds to be detected enters into the measurement result and not, as in absorption-spectrometric methods, their density. Fig. 6.73 shows a diagram of a continuous inlet system.

From the measuring gas flow under atmospheric pressure a measuring gas probe is passed via a capillary into a holder with an intermediate volume, which will be evacuated by a vacuum pump through a restriction. Via a critical nozzle part of the gas flows into the ion source, which in turn is evacuated by the high-vacuum pump of the mass spectrometer system.

6.6.3 Electron Impact Sources

Ionization of gas molecules takes place mainly by electron impact (electron impact: EI). In electron-impact ion sources, which are generally operated in a high vacuum at approx. 10^{-3} Pa, ions are generated by impacting sample molecules with electrons emitted by a heated cathode and accelerated by a voltage in the direction

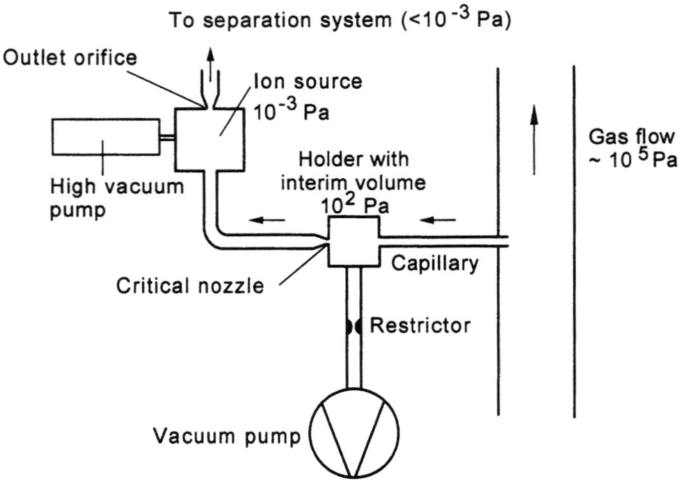

Fig. 6.73: Diagram of a continuous mass spectrometer inlet system (mass flow must be constant)

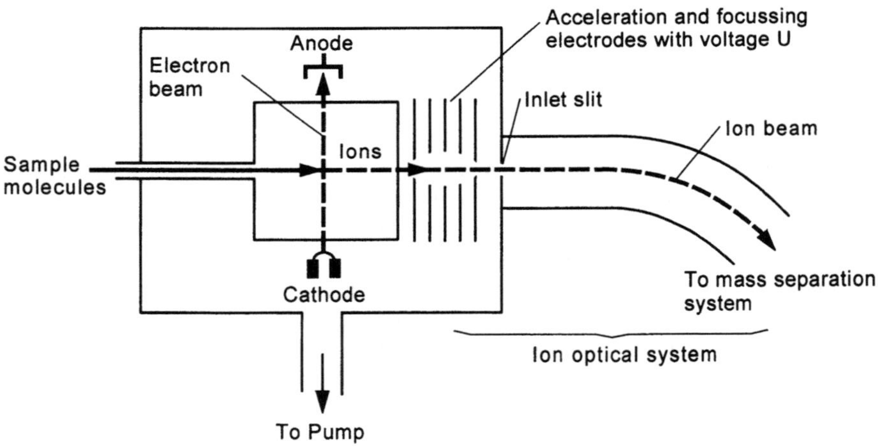

Fig. 6.74: Diagram of an electron-impact ion source

of the electron trap (anode). Fig. 6.74 shows the diagram of an electron-impact source.

The electrons emitted by the cathode are accelerated towards the anode by a potential of between approx. 50 to 150 volts. A magnetic fields (not included in Fig. 6.74) provides the most tightly focussed electron beam possible. The sample gas molecules introduced into the ionization region start to interact with the electrons. Some molecules will lose an electron during this (in individual cases two electrons are possible), they are ionised positively and entered into the mass separation system with the help of accelerating and focusing electrodes. Apart from molecule ions, fragment ions are also formed in most cases, these are ionized parts of the original molecule.

The ion flow I^+ generated in the ion source is given by

$$I^+ = N \cdot Q \cdot l \cdot i, \tag{6.80}$$

where i is the intensity of the electron flow, l the length of the ionization region, N the number of sample gas molecules per cm^3 and Q their effective ionization cross-section. The effective ionization cross-section Q of some molecules is listed in Table 6.7 for an ionization potential of 70 eV.

Table 6.7: Effective ionization cross-sections Q of some gases at 70 eV electronic energy

Type of gas	Q in cm^2 10^{-16}	Type of gas	Q in cm^2 10^{-16}
Ar	3.5	CH_4	4.6
H_2	1.2	C_2H_4	6.7
O_2	2.5	C_2H_6	8.3
N_2	2.9	C_3H_6	9.7
H_2O	3.0	C_3H_8	11.1
CO	3.0	C_6H_6	16.9
CO_2	4.3		

Ion yield in the electron impact ion sources is poor. Only approx. one out of 10^4 molecules present in the ionization region will be ionized. At energy values of the impacting electrons of 70 to 100 eV ionization probability reaches a maximum. When impact energies are higher the collision partners' interaction time becomes so brief that ionization probability decreases rapidly.

During the ionization process different ion types of the molecules, their fragments and atoms are formed:

- singly charged ions;
- multiple charged ions;
- fragment ions.

As an example Table 6.8 lists the percentage distribution of molecule ions and fragment ions related to the individual mass lines for the gases carbon monoxide (CO) and carbon dioxide (CO_2).

The "atomic mass unit (u)" is commonly used in mass spectrometry. As ion indication is based on the quotient mass by charge (m/q) doubly positively charged ions (e.g. CO^{++}) are thus recorded with half their mass according to $m/2e$.

Table 6.8: Percentage distribution of molecule ions and fragment ions for the gases CO_2 and CO

Type of gas	Mass line in u	Percentage %	Ion type
CO_2 (mass 44)	45	1.0	Isotope (Molecule ion)
	44	72.70	CO_2^+
	28	8.3	CO^+
	16	11.70	O^+
	12	6.2	C^+
CO (mass 28)	29	1.9	Isotope (Molecule ion)
	28	91.30	CO^+
	16	1.1	O^+
	14	1.7	CO^{++}
	12	3.5	C^+

6.6.4 Ion Optics

The ions created in the ion source must be accelerated out of the source and focussed into the mass separation system. This task is taken over by ion optics consisting of electrostatic lenses as a diaphragm or a system of diaphragms. Ion optics is structured thus that all ions occurring in the specific mass range are introduced into the separation system without a concentration shift as to the composition of the sample gas, refer to Fig. 6.74.

6.6.5 Mass Separation Systems

In the separation system of a mass spectrometer the ions leaving the ion source are sorted according to their mass/charge ratio and are fed to the ion detection system in spatially or temporally separated condition (similar to the process in chromatographs). The sequence of the recorded signals provides the mass spectrum. The separation systems can be divided into two groups: static and dynamic systems.

6.6.5.1 Static Separation Systems

A set-up is called a static separation system when its separating fields (magnetic field, electrical field) remain temporally constant during the ions' travel time through the system.

The common factor of all static separation systems is a magnetic sector field, homogenous magnetic fields being those most commonly used. The functional principle of the magnetic sector field is shown in Fig. 6.75.

In the homogenous magnetic field H the ion (accelerated by the voltage U, refer to Fig. 6.74) with its mass m, its charge $q = ze$ and its velocity v moves on a

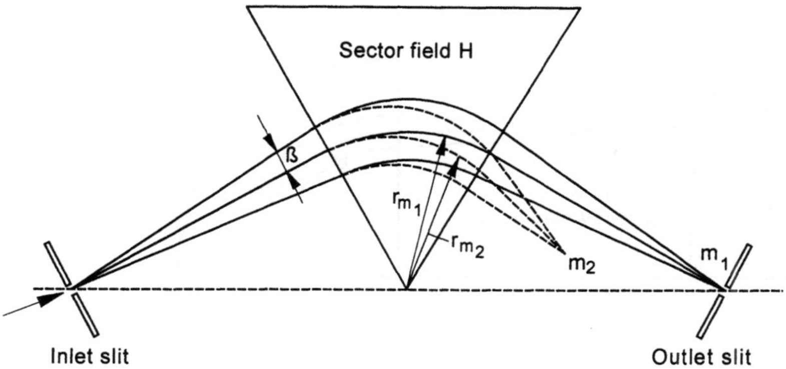

Fig. 6.75: Functional principle of the magnetic sector field

circular path having the radius r_m, with an equilibrium between the so-called Lorentz force $q\text{v}\mu_0\mu_r H = q\text{v}B$ and the centrifugal force $m\text{v}^2/r_m$:

$$q\,\text{v}\,B = \frac{m\,\text{v}^2}{r_m} \tag{6.81}$$

or

$$r_m = \frac{1}{B}\left(\frac{m}{q}\right)\text{v} \tag{6.82}$$

with

$$\text{v} = \sqrt{2\frac{q}{m}U}\,. \tag{6.83}$$

From this, the following relationship ensues with the accelerating voltage U for the path radius r_m

$$r_m = \frac{1}{B}\sqrt{\frac{2mU}{q}}\,. \tag{6.84}$$

With a given magnetic field H and a given accelerating voltage U only ions with a certain mass/charge ratio m/q will pass through the circle sector with the radius r_m and will reach the ion detection system behind the exit slit (e.g., molecule ions with the mass m_1 in Fig. 6.75). Other molecule ions (ions with the mass m_2) miss the slit. The magnetic field has an additional focusing effect so that identical ions with the same energy but entering the field at a small angle β relative to the ideal

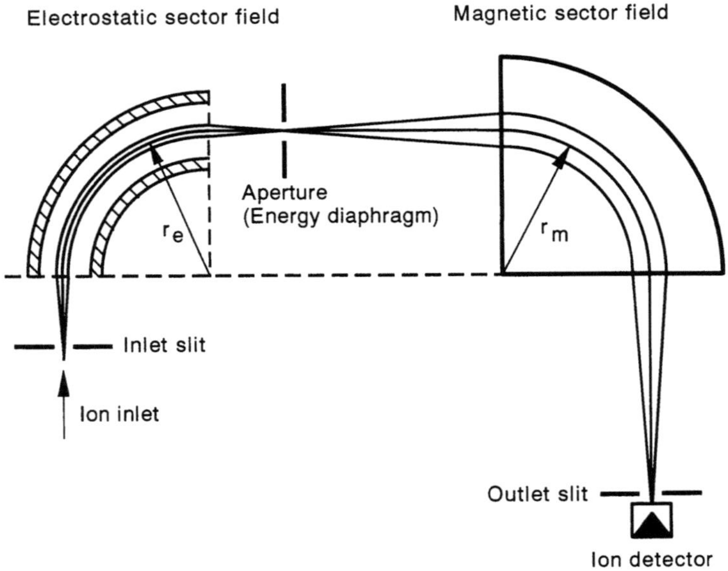

Fig. 6.76: Double focusing separation system in a mass spectrometer detection system

radius also find their way through the exit slit. Naturally, the wider the angle of the ion beam the poorer focusing becomes. This so-called energy dispersion of the magnetic sector field can be compensated for by combining it with an electric sector field. It is then a double-focusing separation system. Fig. 6.76 shows a diagram of the set-up of a double-focusing mass separation system.

In the electric deflector field the ions coming from the ion source are directed on a circular path with the radius r_e. The radius r_e depends on the ions' mass and velocity (m and v). The ions are focussed onto the energy diaphragm (aperture) and reach the magnetic deflector field in a pre-separated state. By tuning the magnetic field strength of the magnetic field ions of different masses are directed consecutively to their deflector radius r_m and recorded selectively and successively by an ion detector behind the exit slit. The deflector fields are tunable with frequencies < 1 Hz, this permits multi-component measurement with a corresponding cycle time.

6.6.5.2 Dynamic Separation Systems

A set-up is called a dynamic separation system if it utilizes, e.g., the influence of alternate magnetic fields or the ions' mass-dependent time of flight from the ion source to the ion detector for mass separation. Well-known systems are quadrupole mass filters and time-of-flight mass analyzers.

Quadrupole separation system

Quadrupole separation systems belong to the group of stable-path spectrometers, where only ions of a certain mass/charge ratio m/q can pass through the system on stable paths, whereas all others are filtered out: the quadrupole acts as a mass filter. The electric separating field is generated by four rod-shaped electrodes where a direct voltage U is superimposed by a high-frequency voltage V (frequency in the magnitude of 1 MHz).

$$\text{Total voltage:} \qquad U_1 + V = U_1 + U_2 \cos(\omega t) \tag{6.85}$$

Fig. 6.77 shows the principle of a quadrupole mass filter.

The ions enter in a direction parallel to the axis (z direction) with the same velocity with which they were introduced into the system. Under the influence of the total electric field the ions oscillate (in x and y direction) perpendicularly to the velocity direction. The ions' paths are described by the solutions of the Mathieu differential equations:

$$\frac{d^2x}{dt^2} + (a + 2\,b\cos(\omega\,t))\,x = 0 \ , \tag{6.86}$$

$$\frac{d^2y}{dt^2} - (a + 2\,b\cos(\omega\,t))\,y = 0 \tag{6.87}$$

with the parameters

$$\frac{a}{b} = 2\frac{U_1}{U_2}. \tag{6.88}$$

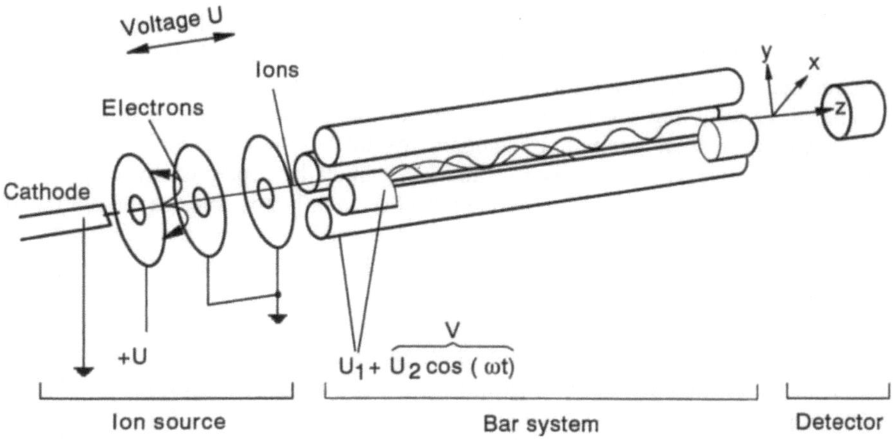

Fig. 6.77: Principle of a quadrupole mass filter

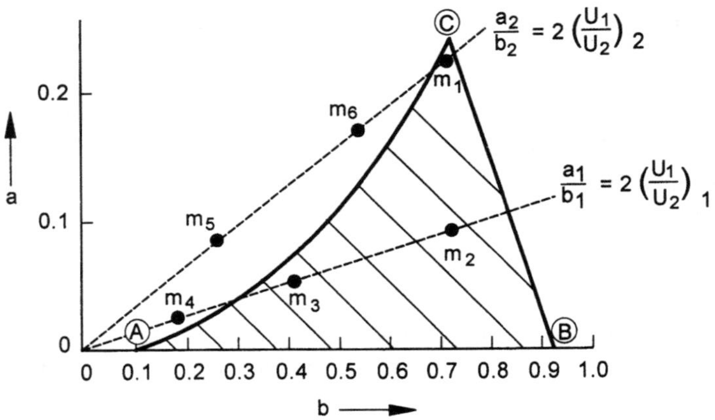

Fig. 6.78: Stability diagram of a quadrupole mass filter (with numerical values assumed as examples)

Only for certain values of the parameters *a* and *b* do the ions pass through the whole separation system on stable paths without building up perpendicularly to their paraxial movement to such an extent that they collide with one of the four electrodes and are then filtered out. The values of *a* und *b* important for stable paths are illustrated in a so-called stability diagram (Fig. 6.78).

To each ion (example m_1, m_2, m_3 in Fig. 6.78) flying on a stable path through the separation system there is a corresponding point (a, b) within the approx. triangular area A,B,C (area with accentuated lines in Fig. 6.78). Ions on unstable paths correspond to points outside of this area (example m_4, m_5, m_6). By appropriate selection of quotient $a/b = 2\,U_1/U_2$ the separation system can be adjusted in such a way that only one ion of a certain mass is in the area A,B,C, i.e., in the transmission band of the mass filter (e.g., the one with the mass m_1 in the tip of the triangular area). The system is then a selective one. The mass filter can be tuned by synchronously varying the direct voltage U_1 and the high-frequency voltage V, so that the ions with different masses selectively and successively pass the transmission band of the filter, i.e., reach the tip of the stability "triangle".

The mass resolving power of a quadrupole mass spectrometer is limited. The usual resolution is at $\Delta m/m = 1$ (standard resolution). With it, molecules of the same mass and charge, such as CO and N_2, cannot be separated by way of the molecule ions (CO^+ and N_2^+), separate detection is only possible via the matrix of the fragment ions (C^+, O^+ and N^+). Due to the small fragment ion percentage of total line intensity this is difficult, particularly if the concentrations of both components vary to a great degree.

Measuring time in quadrupole systems is very small (< 1 ms). In combination with the dynamic gas time constant of the gas's inflow velocity, which has the same magnitude, measuring frequencies of 1 kHz can be achieved. However, this only applies for the measurement of one component, in multi-component measurement the mass analyzer must be tuned. Then, depending on the number of the mass fragments to be detected, cycle time is between 10 ms and 1 s.

Other features of the quadrupole system are:

Measurable components:	all gases (problems arise in the measurement of molecules of identical mass);
Detection limits:	highly dependent on the composition (type of gas and concentration) of the measuring gas: 1 ppm to several hundred ppm;
Selectivity:	dependent on mass resolving power;
Measuring range dynamics:	10^5 to 10^6;
Handling:	quadrupole mass spectrometers are compact, have a stable construction and are easy to operate. Commercial instruments are in use for exhaust gas analysis, but their time resolution is not very high.
Advantage:	The instruments are calibrated with test gases prior to measurements, mass spectra of different molecules of different test gases are stored in the data storage of a computer and compared with the measuring data. This is how identification and concentration computation is carried out (comparable to the FTIR, refer to Chap. 6.1.7).

6.6.6 Ion Cyclotron Resonance (ICR) Spectrometer

One model of an ion cyclotron resonance (ICR) spectrometer is shown schematically in Fig. 6.79.

There is a high-frequency field between two plate electrodes, perpendicular to this there is a homogenous magnetic field. The ions are generated by electron impact in the separation system itself and are directed by the magnetic field onto circular paths in dependence of their mass. Only ions of certain masses can reach the ion trap on a stable cyclotron path in resonance with the high-frequency field. The system can be tuned by varying the high-frequency field or the magnetic field. The formula of the cyclotron angular frequency ω_c or frequency f_c reads as follows

$$\omega_c = \frac{q\,B_z}{2\pi\,m} \quad \text{or} \quad f_c = \frac{q\,B_z}{m}, \tag{6.89}$$

where q/m is the ion charge/ion mass ratio and B_z the magnetic flux density of the homogenous magnetic field.

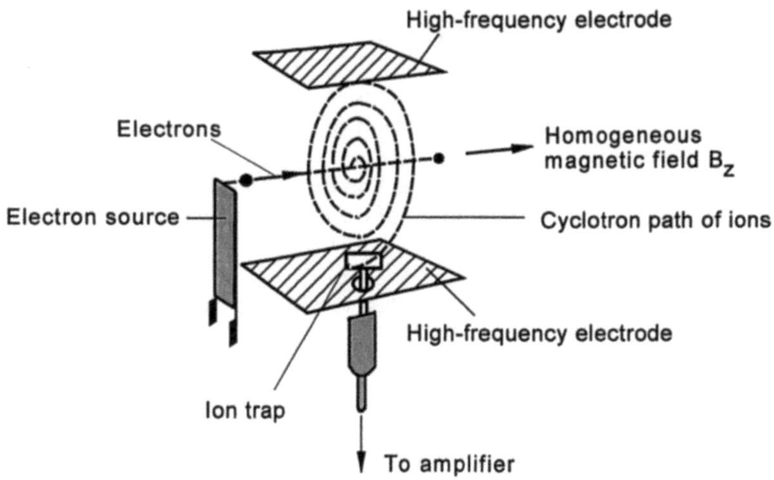

Fig. 6.79: Diagram of an ion cyclotron mass spectrometer

This method has two advantages:

- Ion formation and mass analysis take place in the same space.
- Ions are recorded according to the mass-dependent frequency of their orbit when tuning the system.

This method requires elaborate equipment, though.

In a more recently developed version of the ion cyclotron resonance method the ions' cyclotron frequency is measured directly as a time-dependent signal via two additional electrodes. The mass spectrum is obtained by Fourier transformation as in FTIR (refer to Chap. 6.1.7) from these time-dependent signals.

6.6.7 Mass Resolving Power

Resolution R which must be attained for two ions of equal charge and mass m and $m + \Delta m$ to be separated is expressed by the equation

$$R = m / \Delta m \,. \tag{6.90}$$

Δm is the smallest mass distance where two neighboring mass lines of equal intensity can still be considered separate. The term "separate" requires definition. Preferably, the so-called %-dip-definition is used, where the depth of the "dip" between two mass lines is expressed as percentage of the line height h. As an example Fig. 6.80 shows a 10% (0.1 h) dip between two mass lines.

One criterion of the resolving power of a mass separation system is its ability to separate ions of equal charge and equal nominal mass (doublet spectra such as for CO and N_2, whose molecules have a mass of 28 each). In this case separation is

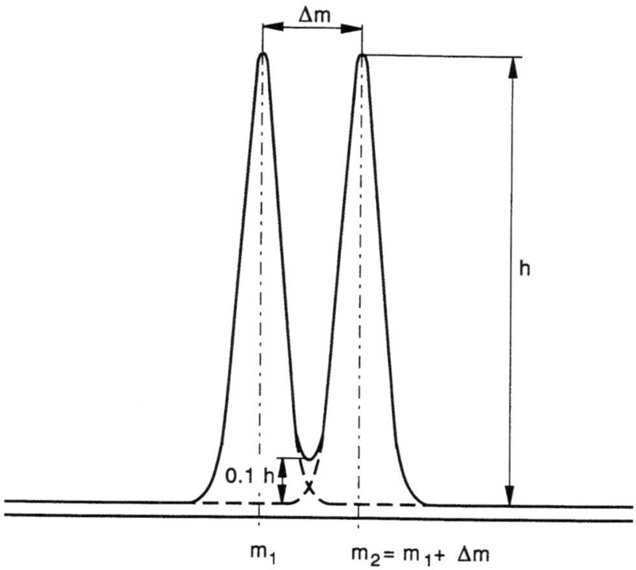

Fig. 6.80: Definition of the resolving power of a mass separation system using the example of a 10% dip

only possible as the precise masses differ by small amounts due to the different nuclear bonding energies of the atoms in the two molecules.

Example: For the double spectrum of N_2 / CO the following holds good:

$m\,(^{14}N\,^{14}N)$	=	28.00615
$m\,(^{12}C\,^{16}O)$	=	27.99491
Δm	=	0.01124

The theoretically required resolving power is then

$$R = m/\Delta m = 28 / 0.01124 = 2491.$$

Practical measurements, however, require a much greater resolution, particularly when there is a great difference in the components' concentration.

The resolving power R of the different mass separation systems varies greatly. It spans a range from $R = m$ (unit resolution) up to $R > 100{,}000$ in double-focusing sector field systems. A certain independence from the resolving power of the mass separation systems can be achieved with a selective ion source as this permits pre-sorting the sample gas's components.

6.6.8 Exemplary Result

In mass spectrometric gas analysis methods where the gas components to be analyzed are ionized by electron impact, a variety of dissociation products (fragment ions) usually form, as is shown in Fig. 6.81 with the example of benzene.

Thus, when analyzing a gas mixture with many components, there are numerous overlappings of different dissociation products on the same mass lines of the spectrogram. As a result of this, quantitative analysis frequently requires a lot of computing effort considering different dissociation products.

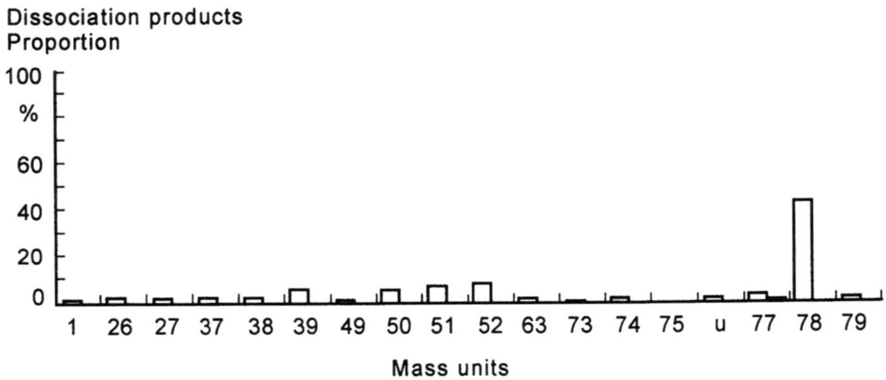

Fig. 6.81: Dissociation products of benzene during ionization by electron impact

6.6.9 GASP - Gas Analysis by Sampling Plasma

6.6.9.1 Basic Observations

In certain cases of application, particularly when determining hydrocarbons, is it advantageous to suppress the formation of fragment ions in favor of molecule ion formation as much as possible. The novel ionization technique of the GASP method is based on ion-molecule reactions. Dissociation of the molecule ions is avoided or reduced. This means, that for each neutral gas component to be analyzed there is, in many cases, only one molecule ion or few ionized molecule fragments, which, compared to standard mass spectrometry with electron impact ionization, simplifies allocating the ions detected to the neutral gas components at hand.

This is achieved by ionizing the individual gas components not directly by electrons, but predominantly by noble gas ions. Krypton or xenon are, e.g., used as working gases. The recombination energy of these noble gas ions is so low (12.1 and 14 eV) that, in many cases, the charge between the noble gas ions and the gas molecules is merely interchanged without dissociation taking place.

This is achieved by a "gentle" ionization method, namely chemical ionization (CI), where ionization of the sample gas molecules takes place via ion-molecule reactions. Reactive primary ions are formed by electron impact, though. Formation of a sufficiently large number of reactive ions requires relatively high pressure (approx. 10^4 Pa) in the ion source. Formation of the sample gas ions can be influenced by the selection of the reaction gas (of the reactive ions).

Chemical ionization in ion sources under atmospheric pressure is a special case. There, primary ions of the working gas are not formed by electron impact but in a corona discharge.

6.6.9.2 Principle

Low-pressure plasma ionization corresponds to chemical ionization in a general sense. Atoms A of a noble gas are ionized by electron impact in a primary source and form primary ions A^+. These primary ions reach a reaction chamber and selectively react with the molecules B_i of the measuring component i according to the reaction equation

$$A^+ + B_i \rightarrow C_i^+ + D_i. \tag{6.91}$$

This causes a weakly ionized low-pressure plasma in the reaction chamber. The ions C_i^+ formed in the reaction acc. to (6.91) characteristically represent the measuring component B_i; D_i are neutral components. Only the ions C_i^+ characteristic of the measuring component reach the detector system.

Primary ions A^+ can be selected such that they do not react or react only very slowly with those components of the sample gas which are not to be measured, but very rapidly with the measuring component. For instance, Kr^+ ions do not react with N_2, only slowly with O_2 (reaction rate approx. 10^{-11} cm^3 s^{-1}) and rapidly with CO_2 (reaction rate approx. 10^{-9} cm^3 s^{-1}). Preferably only one reaction gas should be in the reaction chamber.

A diagram of the basic set-up of a measuring system is shown in Fig. 6.82.

The basic gas is ionized in the ion source and reaches the reaction chamber via an electric lens. There, the molecules of the gas to be examined or analyzed are ionized by ion-ion reactions.

The ionized gas molecules and any possible ionized dissociation products are allocated to the mass lines of the spectrogram with the help of the mass spectrometer-detector system. Depending on the exhaust gas component to be analyzed, different noble gases, e.g. krypton or xenon, are to be successively used. This is one disadvantage of this method when considering simultaneous measurements. According to more recent findings, however, several reaction gases can be admitted, so that a kind of simultaneous measurement of several components seems possible.

As can be clearly seen in the example of benzene in Fig. 6.83 (accentuated column, extreme right) only one product is obtained by this method of ion-molecule reaction, in contrast to the ionization by electron impact which leads to many products of different mass units, refer to Fig. 6.81.

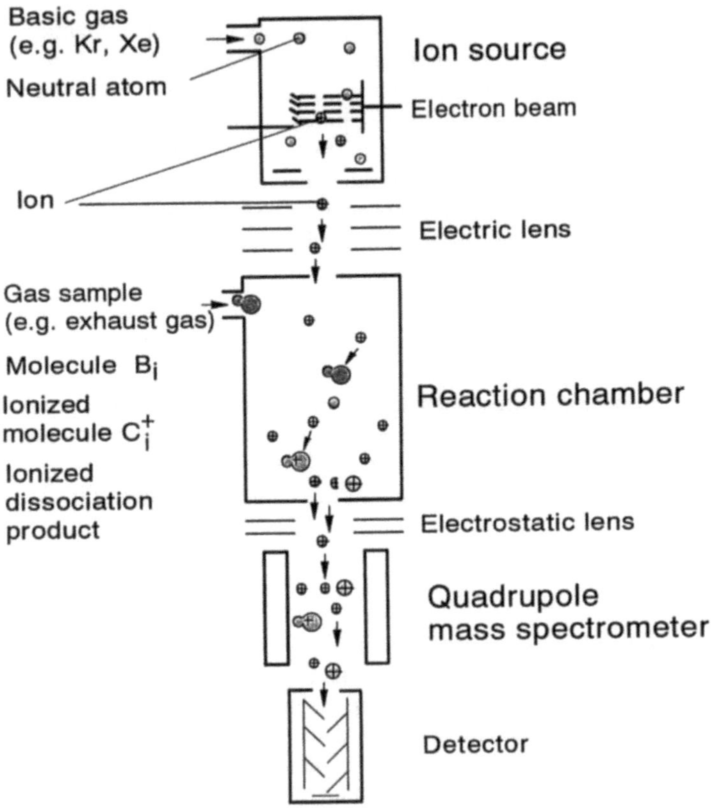

Fig. 6.82: Principle of a GASP measuring system

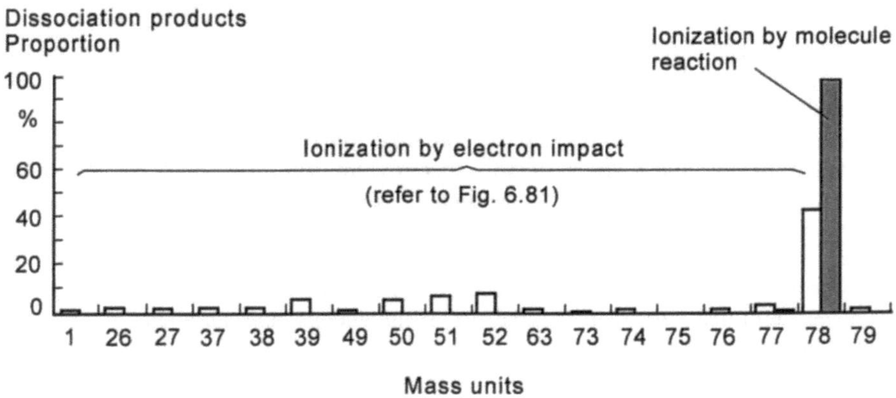

Fig. 6.83: Dissociation products of benzene during ionization by electron impact, refer to Fig. 6.81, or by ion-molecule reaction (Xe^+ ions), dashed column, extreme right

6.6.10 Time-of-Flight Mass Spectrometer

Time-of-flight (TOF) mass spectrometers have the operating principle illustrated in Fig. 6.84.

At a certain time the ions are generated by pulsing, e.g. by a laser pulse, and directly after that accelerated in an also pulsed electrical field with a potential U. Subsequently, the ions fly through a field-free section, where, depending on their mass m_1, m_2 or m_3 , they acquire different velocities and thus impact the detector at the end of this section at subsequent times. Time of flight t along the field-free drift section with the length d is given by

$$t = d\sqrt{\frac{m}{2qU}}, \tag{6.92}$$

refer to (6.83).

Time-of-flight difference Δt for, e.g., two ions with masses m_1 and m_2 and with the same charge q is thus

$$\Delta t = \frac{d}{\sqrt{2qU}}\left(\sqrt{m_1} - \sqrt{m_2}\right). \tag{6.93}$$

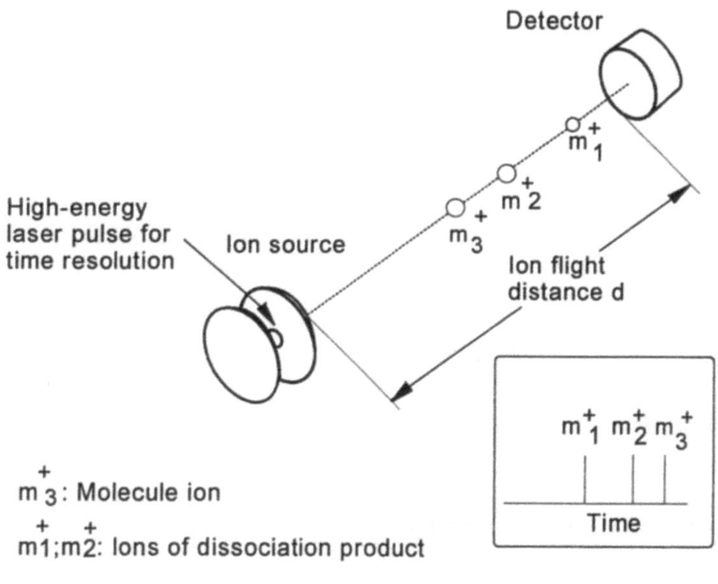

Fig. 6.84: Operating principle of the TOF mass spectrometer with results displayed (lower right)

In time-of-flight mass spectrometers mass resolving power is low, as ions of equal mass and charge impact simultaneously. More recently, an electrostatic ion reflector has been installed into the ions' flight path which influences the ions' flight time corresponding to their kinetic energy. In this way, resolving power could be enhanced.

6.6.11 Tunable Laser Ion Source for TOF-MS

A tunable laser ion source is a newly developed ion source in combination with the TOF-MS. Selective ionization is possible with the help of a tunable laser system.

In "resonant multiphoton ionization" gas molecules are ionized via multiphoton absorption by pulsed laser light in the visible or near UV range. As each type of gas has a specific absorption spectrum, refer to Chap. 6.1, certain molecules or molecule groups can be excited individually by selecting the appropriate laser emission wavelength. This, however, requires a tunable laser, for instance a dye laser (in combination with a YAG solid laser). This also permits separate recording of ions of equal mass.

As can be seen from Fig. 6.85, these instruments are quite complex in their structure.

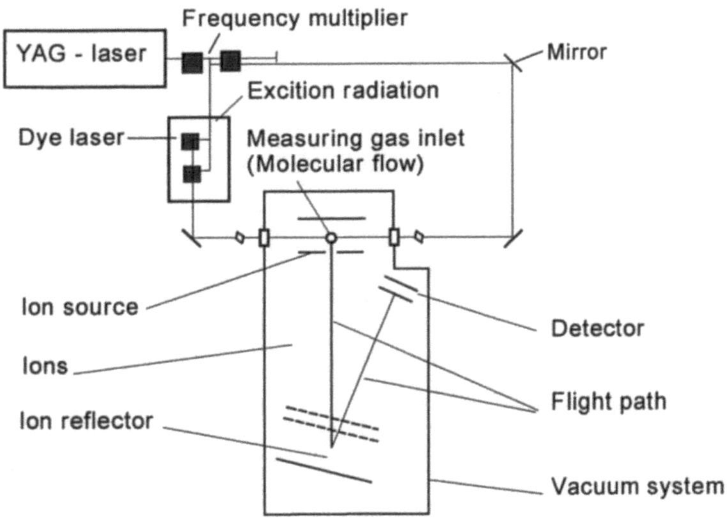

Fig. 6.85: Diagram of the set-up of a TOF mass spectrometer with laser ion source

6.6.12 Assessment of Mass Spectrometry

The possibilities offered by mass spectrometry for multicomponent measurement of automobile exhaust gases are the following:

- Mass separation systems are tunable, thus permitting multicomponent measurement in cycle times of 1 s.
- Homoatomic gases such as O_2 can be recorded with it.
- Mass indication is linear, providing a wide dynamic range.
- Measurement of molecules of equal mass number requires a high resolving power which can only be achieved by static separation systems.
- Direct measurement in the raw exhaust is not possible as special gas sampling systems are required.

6.6.13 Gas Chromatography/Mass Spectrometry (GC/MS)

The certainty of analytical results of chromatographic examinations can be further improved, if a mass spectrometer is coupled to the gas chromatograph as a kind of detector (direct connection of the separation column end with the ion source). Apart from retention times and peak areas, mass spectra of the substances examined are measured as an important aid to interpretation in such a GC/MS system. Mass spectra are not instrument-specific as far as their m/q values are concerned and, as to their ion intensities, only to a small degree. Hence they can be tabulated or stored in a computer. For identification it is thus not necessary that in every case each component to be measured is available in its pure state for calibration as a comparison of the mass spectrum measured with a stored "library spectrum" usually suffices.

Coupling high-resolution GC with double-focusing MS is currently one of the most efficient analysis methods both in qualitative and in quantitative examinations of samples with complex compositions. Not only does it provide mass spectra which can be used as identification aids, it also simultaneously contributes a chromatogram for quantitative evaluation, which is practically identical to a regular GC/FID chromatogram. This so-called total ion chromatogram is computed by determining for all mass cycles the sum of all ion intensities measured in each mass cycle, and these values are then plotted as a function of analysis time.

As an example Fig. 6.86 shows the block diagram of a gas-chromatograph-mass-spectrometer system with helium as carrier gas. Helium can be removed from the gas mixture easily and effectively by a molecule separator placed behind the GC column. After molecule separation the gas sample is ionized in the ion source by electron impact and the ions formed are deflected in the mass spectrometer's magnetic field according to their m/q values. Selected ion beams reach the detector through the collector slit. After reinforcing the measuring signal recording and plotting of gas chromatogram and mass spectrum are carried out.

Fig. 6.87 shows a photograph of a GC/MS laboratory

Fig. 6.88 shows the gas chromatogram of a methyl ester mixture and the mass spectrum of the component n C_{16} (marked in the chromatogram by an arrow).

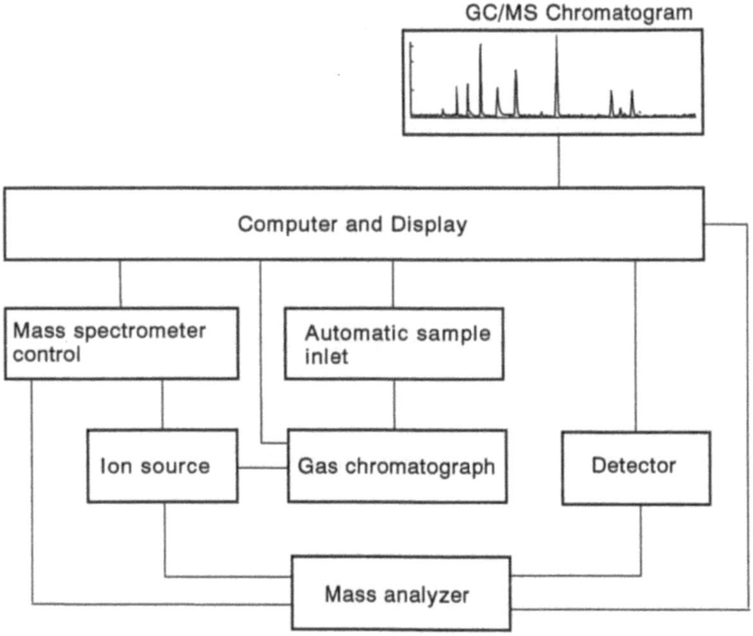

Fig. 6.86: Block diagram of a gas-chromatograph-mass-spectrometer system

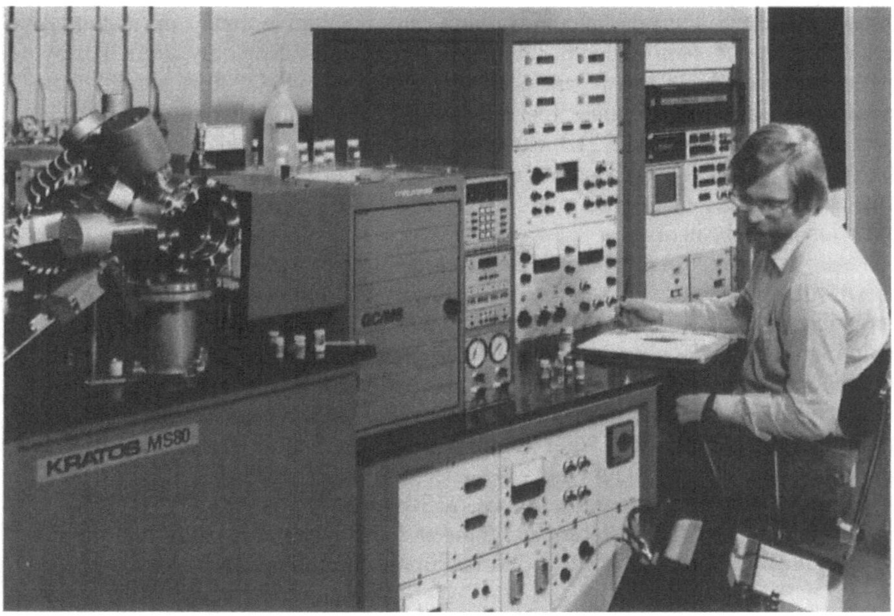

Fig. 6.87: Photograph of a GC/MS laboratory

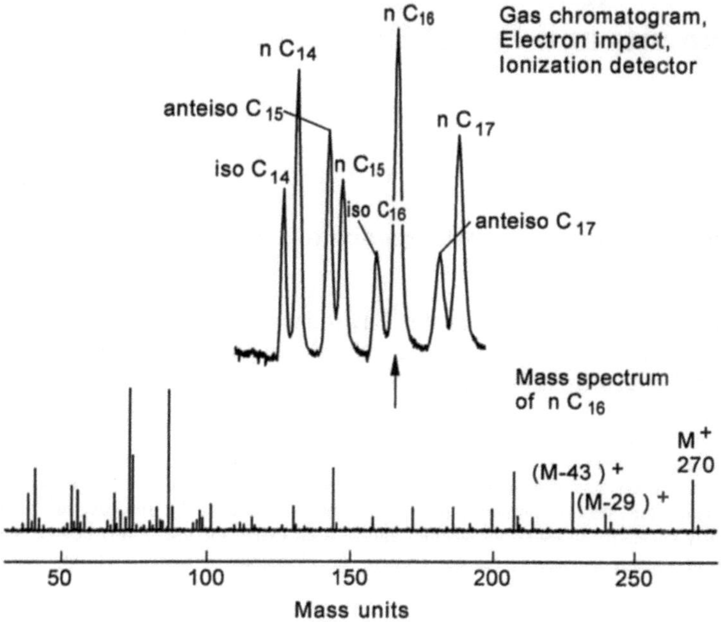

Fig. 6.88: Gas chromatogram of a methyl ester mixture and mass spectrum of the component n C_{16} (arrow)

7 Measurement of Unregulated Exhaust Gas Components and Diesel Exhaust Gas Particles

7.1 Legislative Situation with regard to Unregulated Components

As already described in Chap. 2, the Environmental Protection Agency (EPA) summarized its position in 1977 on unregulated components, requiring the following additional efforts from the automotive industry:

> **Directive of the EPA for limiting the unregulated components:**
> Vehicle manufacturers must prove in accordance to the "polluter pays principle" that exhaust gas purification systems or design elements, which are built into vehicles or engines to decrease exhaust emissions, do not threaten public health, welfare, and safety. In order to do this, it is particularly necessary to ascertain whether and to what extent the use of any device, system or element of design causes, increases, reduces or eliminates emissions of any unregulated pollutants.

This means for instance, that the systems etc. (e.g. the three-way catalytic converter), built in to meet emission standards must not produce additional components which cause risks (an example being the production of cyanide with rhodium-plated catalytic converters). Expressly considered are exhaust emission components for which no standards have been set yet. The term *unregulated exhaust emission components* therefore does not mean that there are no regulations at all.

With this, the EPA adopted a prophylactic method. Based on the idea of prevention, they defined a list of substances and groups of substances which should be investigated, independently of whether or not any proven knowledge about effects with regard to the concentrations of the components involved existed.

The most important substances on the EPA list are shown in Table 7.1.

In California the effects of automobile exhaust emissions are judged according to their "reactivity", as in earlier years. Basis of the judgment is the potential to produce ozone, which is determined by experiments for each hydrocarbon compound in question.

From the point of view of measurement technology this measure, accompanied by a general reduction of the emission standards, means the use of differentiated hydrocarbon analysis, i.e. the quantitative determination of the concentrations of individual hydrocarbon components.

In accordance with a proposal of the California Air Resources Board (CARB) this very expensive procedure can be carried out in two parallel steps:

> **Step 1:** Determination of the hydrocarbon components from C_1 to C_5. To do this a diluted exhaust sample is taken directly from the dilution tunnel and collected in a plastic bag, refer to

Table 7.1: Unregulated exhaust emission components from EPA list of substances (R = Hydrocarbon-fraction)

No.	Substance
1	Special nitrogen oxides (N_2O, N_2O_3, N_2O_5)
2	Organic amines ($R\text{-}NH_2$)
3	Nitrosamines ($R_2N\text{-}NO$)
4	Ammonia (NH_3)
5	Hydrogen cyanide (HCN)
6	Different hydrocarbons
7	Polycylic aromatic hydrocarbons (PAH)
8	Aldehydes and Ketones (R-CHO, R-CO-R)
9	Phenols (Phenol, Cresol)
10	Odorous substances
11	Sulfur dioxides (SO_2)
12	Sulfate ($SO_4{}^{2-}$)
13	Hydrogen sulfide (H_2S)
14	Carbonyl sulfide (COS)
15	Organic sulfur compounds
16	Particles
17	Particle bound organic compounds
18	Halogen compounds
19	Metallic compounds
20	Compounds with precious metals

Figures 7.4 and 7.5. For the subsequent analysis the sample can be directly injected into a gas chromatograph.

Step 2: For hydrocarbons > C_5, the samples are also taken from the diluted exhaust but the sample flows through a solid adsorbent (e.g. Tenax in small glass tubes, refer to Figure 7.6). A further treatment of the sample is not required since it is transported via a thermo-desorption cold trap injector into a gas chromatograph/mass spectrometer system.

Table 7.2 shows the results of such measurements. From this data and for this example, the relative reactivity of the exhaust of a vehicle equipped with a three-way catalytic converter operated with M-85 fuel (85 % of the fuel is methanol) was determined in relation to that of a vehicle operated with normal fuel.

The application of the reactivity factors (defined by CARB as preliminary) on the individual components results in a total reactivity of approx. 0.7 compared to a factor of 1.0 for normal fuel.

Besides the complex composition of the exhaust gas, the very low concentrations of the unregulated components as compared to the regulated components makes analysis very difficult. The low concentrations require very sensitive specific analytical methods, which are generally called "trace analysis". Furthermore, sampling methods have to be adapted to these conditions.

Table 7.2: Results of an analysis of different hydrocarbons in the exhaust gas of a mid-sized vehicle with M-85 and normal fuel

Group	Component	Emission values for an US-75-Test in mg/mi M-85 $\bar{x}$	M-85 ±v	Normal fuel $\bar{x}$	Normal fuel ±v
ALKANES	Methane	23.7	3.3	34.2	4.3
	Ethane	0.1	0.1	3.5	1.7
	Propane	n.d.	-	0.1	0.1
	Butane	2.6	2.3	4.1	3.0
	Isobutane	1.0	1.0	2.2	2.1
	Pentane	2.2	0.5	5.5	2.4
	Isopentane	3.0	1.1	8.3	2.7
	Hexane	1.0	1.0	2.8	0.8
	2-Methylpentane	0.8	0.4	5.5	0.8
	Heptane	0.5	0.3	0.8	0.4
	Methylhexane/ Dimethylpentane	1.2	0.4	2.9	0.4
	Isooctane	1.9	0.6	12.1	0.6
CYCLO ALKANES	Cyclopentane	0.6	0.4	0.4	0.3
	Cyclohexane	0.1	0.1	3.1	0.4
	Methylcyclopentane	0.6	0.4	1.0	1.0
ALKENES	Ethene	1.4	0.2	6.5	1.3
	Propene	0.4	0.3	3.4	0.3
	But-1-ene	n.d.	-	0.5	0.3
	But-2-ene	n.d.	-	0.2	0.2
	Methylpropene	0.3	0.3	3.0	0.4
	Buta-1,3-diene	n.d.	-	1.5	0.3
	3-Methyl-1-butene	n.d.	-	0.1	0.1
	2-Methyl-2-butene	0.4	0.4	4.2	2.8
ALKYNES	Acetylene	0.8	0.1	3.5	0.6
AROMATIC HC	Benzene	1.9	0.8	7.2	0.3
	Toluene	2.6	1.1	11.9	2.1
ALDEHYDES	Formaldehyde	16.9	3.5	1.8	0.5
	Acetaldehyde	0.2	0.1	0.8	0.3
	Benzaldehyde	0.3	0.3	0.6	0.3
ALCOHOLS	Methanol	167	32	-	-
	Ethanol	n.d.	-	-	-
Total HC	HC-FID	152	25	245	9
	Nonoxigenated HC	95	8	245	9

$\bar{x}$ = Mean, $v = \frac{t \cdot s}{\sqrt{n}}$ = confidence range (P = 95%, n = 3), n.d. = not detectable

Care has to be taken not to change the compositions of the exhaust gas probe by the sampling method, not to produce artifacts, and to keep the sampling method "realistic", so that the conclusions are valid for the emissions of vehicles in use. Similar to the test determining regulated components, the vehicle is driven on a chassis dynamometer following a driving cycle, refer to Chap. 8.2.3.

7.2 Sampling

7.2.1 Sampling from Undiluted Exhaust Gas

It seems to be obvious that the undiluted (raw) exhaust gas as it leaves the exhaust pipe should be taken as an initial sample. In undiluted exhaust gas, the substances to be studied are still present in higher concentrations. An example of this procedure is the sampling method used to measure polycyclic aromatic hydrocarbons (PAH) in motor vehicle exhaust gas.

7.2.1.1 Total Flow Method

Using this technique all the exhaust gas produced during a test (refer to Chap. 8) is fed through a device in which the condensable components are removed and the remaining gas is filtered. This method is therefore called "total flow method". Figure 7.1 shows a schematic presentation of the sampling device consisting essentially of a glass cooler, a condensate separator and a paraffin-impregnated glass fiber filter.

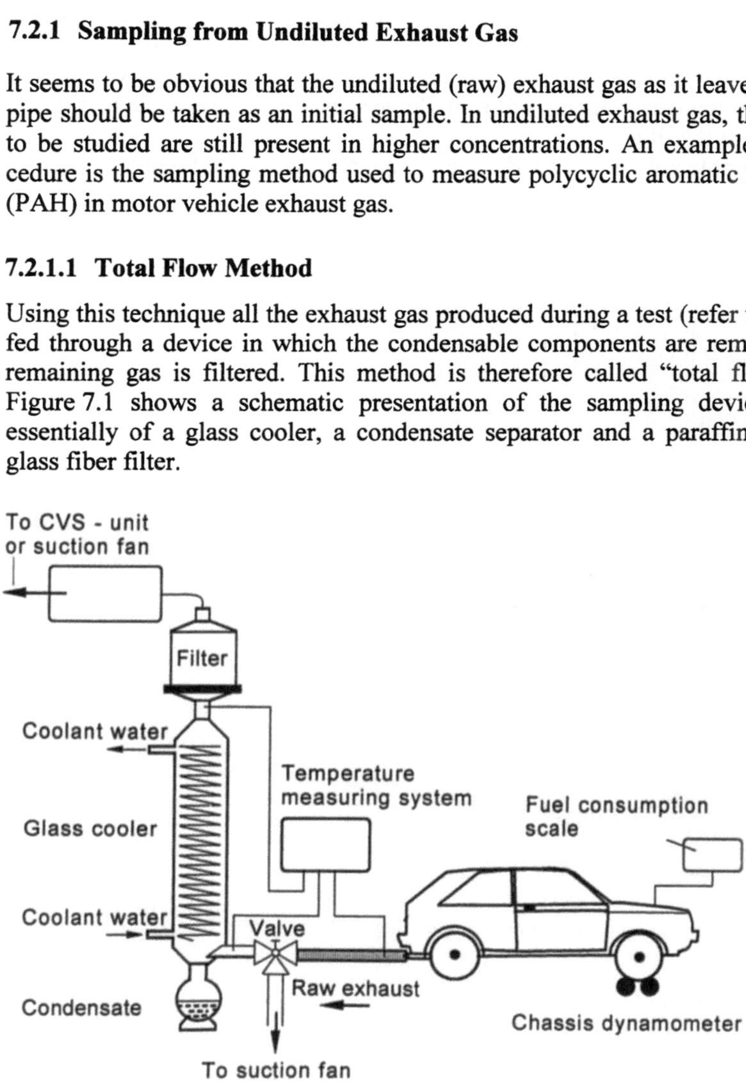

Fig. 7.1: Sampling according to the total flow method (schematic)

The cooler is fed with cold tap water so that all of the components condensing at the corresponding temperature can be obtained and collected as condensate in the glass container placed at the bottom of the cooler. The filter placed on the top on the cooler removes any uncondensed liquid or solid exhaust gas components which may pass through the cooler coil. The substances remaining in the glass cooler must be flushed out and the deposits on the filter must be extracted in order to obtain the total amount of the components to be determined.

A view of such a sampling device according to “Grimmer“ is shown in Figure 7.2.

The disadvantage of this procedure is that the sample sizes obtained are large. For one thing, there is the condensate which predominantly consists of water produced during combustion in the engine. For another, large amounts of solvents are used for subsequent cleaning of the cooler system and for the extraction of the substances from the filter.

The total sample volume for a US-75 test is approximately 3.5 liters.

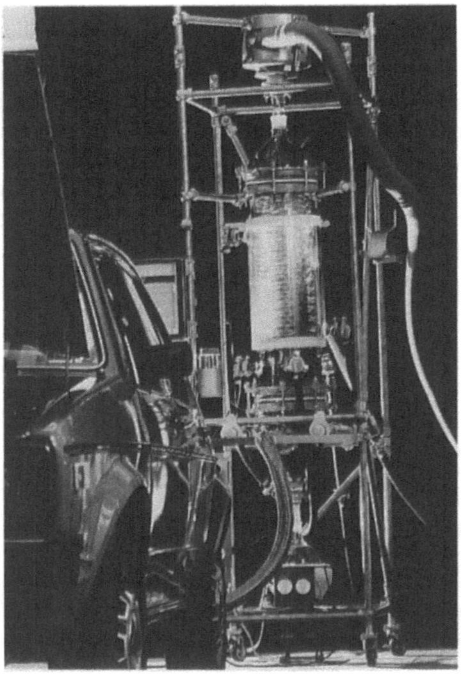

Fig. 7.2: View of a total flow sampling device according to “Grimmer“

7.2.1.2 Partial Flow Method

It is obvious that the separation of trace components from such large volumes involves a great deal of expense. Therefore, it is advantageous to develop procedures designed for smaller sample volumes.

One possibility is to divert a partial flow of the undiluted exhaust gas rather than using the entire flow of exhaust gas for sampling. However, taking a *representative* partial flow can prove to be problematic, as the amount of exhaust gas produced during a test (e.g. US-75 test) varies with the driving cycle (refer to Chap. 8). Controlling the partial flow via the amount of intake combustion air, e.g., in the carburetor, is relatively complicated. The partial flow taken per time-interval must be proportional to the total flow in this same time interval.

In spite of this, sampling according to the partial flow method for undiluted exhaust gas is used in exceptional cases.

Especially the analysis of highly volatile exhaust gas components which do not condense when using the total flow method discussed above is performed in this way.

In this case, agents of decreasing temperature are used successively (Figure 7.3).

By choosing different numbers and temperatures of the cold traps a virtually arbitrary fractionation of the exhaust gas at different boiling ranges is possible.

However, the technical complexity of the apparatus justifies its use only in exceptional cases.

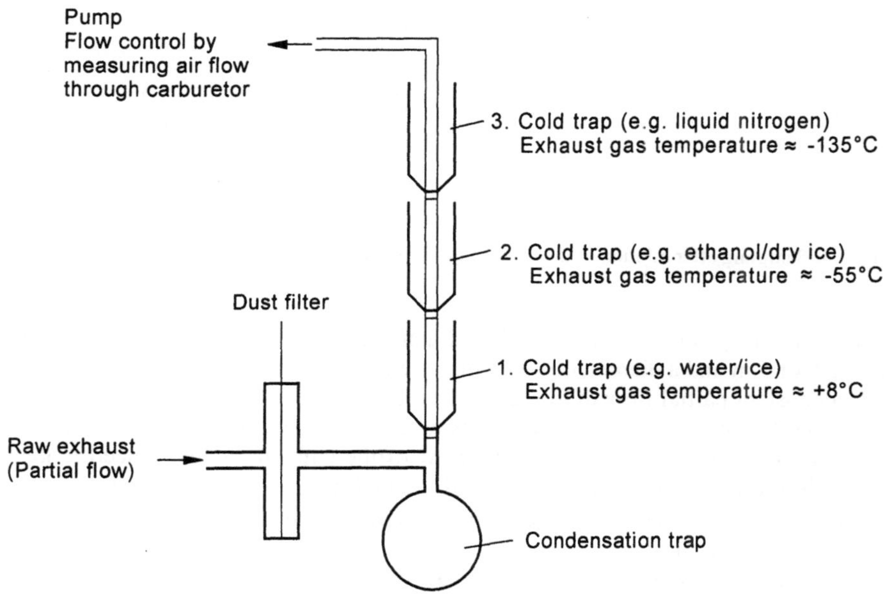

Fig. 7.3: Fractional condensation of a partial flow of undiluted exhaust gas.

7.2.2 Sampling from Diluted Exhaust Gas

7.2.2.1 Isokinetic Sampling from the Dilution Tunnel

There are several reasons not to sample from raw exhaust gas but from exhaust gas diluted with ambient air.

On the one hand, the sampling is more realistic because the exhaust gas from vehicles on the road is immediately diluted when it leaves the tail pipe. By diluting the exhaust gas, possible chemical reactions caused by contact between the exhaust gas components and the ambient air are thus taken into account. On the other hand, the dew point of the water formed during combustion in the engine can be surpassed so that condensation of this component (water) can be avoided during sampling. Furthermore, this sampling is representative if the dilution follows the scheme shown in Figure 7.4.

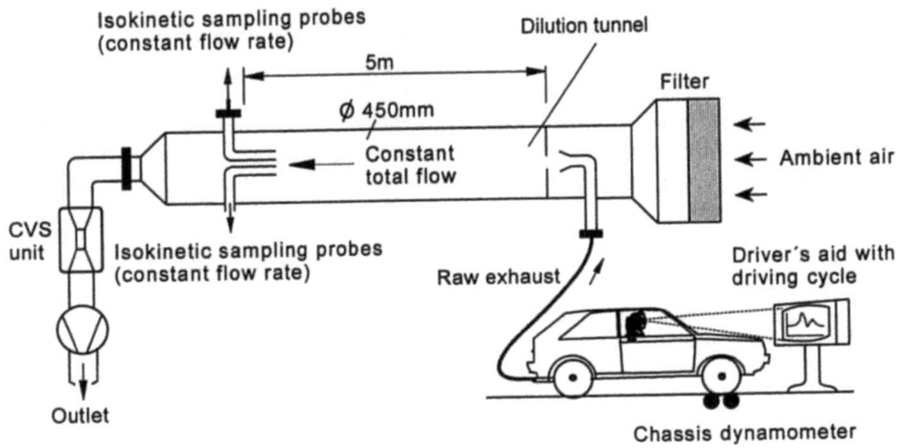

Fig. 7.4: Sampling after dilution of the exhaust gas in a "dilution tunnel"

The exhaust gas is diluted with filtered ambient air in a big dilution tunnel with a length of 5 m and a diameter of 0.45 m, for example.

If a constant partial flow of this diluted exhaust gas is removed isokinetically for sampling, the amount of the exhaust gas as present in the partial flow of air plus exhaust gas is kept constant. Thus the flow of the ambient air introduced is kept inversely proportional to the total flow of the undiluted exhaust gas. In this way a relatively simple representative sampling is possible (principle of the CVS method, refer to Chap. 8.2.6).

However, this method has the great disadvantage that the already low concentrations of the unregulated exhaust gas components are reduced corresponding to the degree of dilution. Therefore the detection sensitivity required of the analytic devices is correspondingly greater. For certain components the detection limit can be reached so that the total flow method must be used instead.

7.2.2.2 Collection in Sample Bags

For analysis of gaseous substances present at sufficiently high concentrations, the diverted partial flow can be collected in a plastic bag (Figure 7.5).

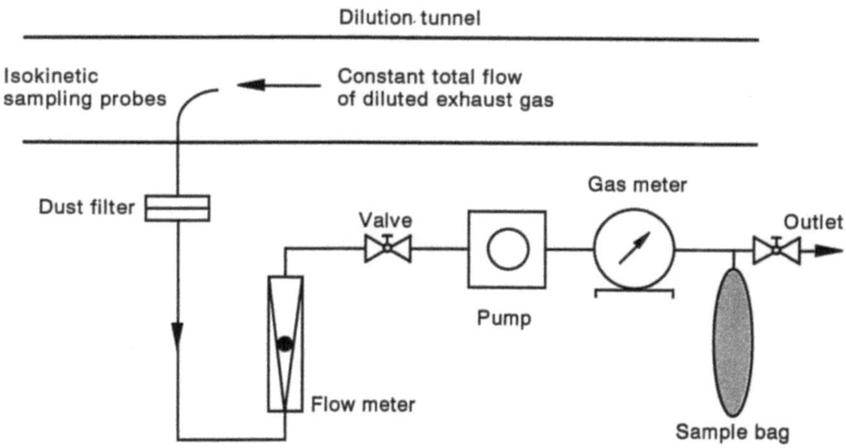

Fig. 7.5: Principle of collecting gaseous exhaust components in a sample bag

The partial flow is taken from the dilution tunnel by means of a pump through an interposed filter to remove solid exhaust gas components, refer to Chap. 7.6. The flow rate is set and controlled with a flow meter. The exact gas volume of the partial flow is measured with a gas meter. The partial flow is sampled in a plastic bag until the end of a test phase or the end of the total test is reached (refer to Chap. 8). Then the bag is closed and its contents are available for subsequent analysis of the gaseous components.

With a microsyringe aliquot partial volumes (i.e. the partial volume) can be taken from the bag and subsequently directly injected into a gas chromatograph for analysis (refer to Chap. 6.5). In this way, individual hydrocarbons are analyzed.

This method is more suitable than the total flow method for analyzing polycyclic aromatic hydrocarbons as well.

7.2.2.3 Adsorption and Absorption

Often the concentration of components of interest in the diluted exhaust gas is too low for direct analysis. Besides the above-discussed method of sampling undiluted exhaust gas probes, there is a method of sampling the compounds to be analyzed *selectively* from the partial flow of the diluted exhaust gas, thus concentrating these compounds (similar to the cooling trap system but here from diluted exhaust gas). The test must be repeated in this case as often as necessary to obtain enough material.

To sample gaseous exhaust components such as volatile hydrocarbons for example, adsorbing materials can be used, e.g. Tenax. These materials consist of tiny plastic porous pellets with a diameter of 0.2 to 2 mm and a specific surface area of 20 to 50 m^2/g. The partial flow is transported through this adsorbing material as shown schematically in Figure 7.6.

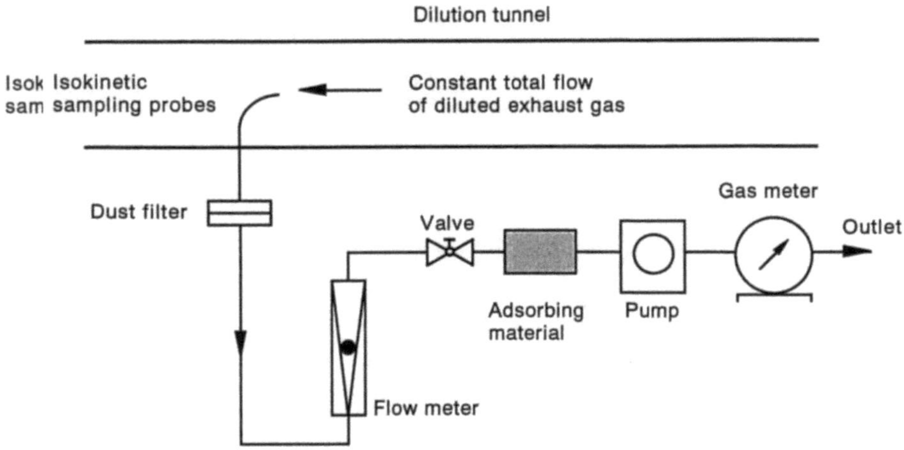

Fig. 7.6: Principle of collecting by absorption on solids (schematic)

The adsorption substances can either be thermally desorbed or extracted by using proper solvents.

The selectivity of this porous plastic is not very large and the spectrum of adsorbed components is correspondingly wide. Moreover, desorption or extraction of the adsorbed substances is not always complete, which can result in determination errors.

In addition to adsorption on solid materials, various substance classes may be isolated from the exhaust gas by absorption in liquids. For this purpose, the filtered partial flow which is now free of particles is fed through consecutive gas wash bottles containing liquids with corresponding absorbing properties (refer to Figure 7.7).

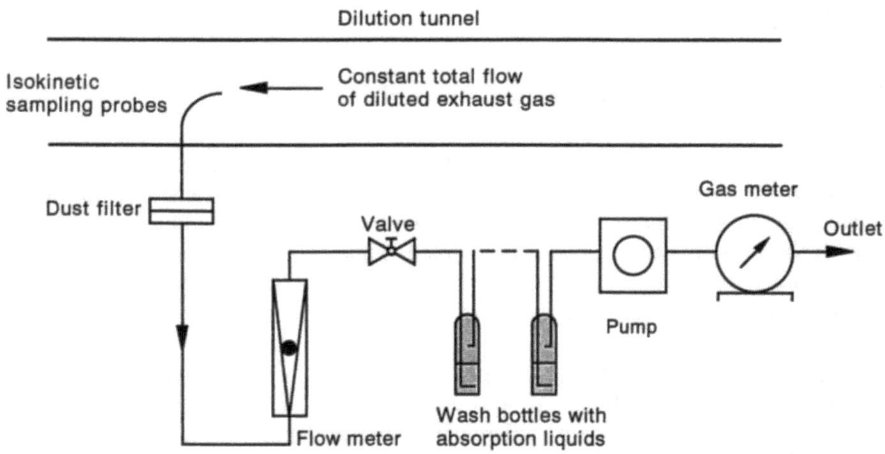

Fig. 7.7: Principle of collecting gaseous exhaust gas components by absorption in liquids

For example, if the wash bottles are filled with an acid, the alkaline components of the exhaust gas are absorbed and vice versa.

But again, it is not generally possible to separate individual compounds. However the achieved classification of the numerous exhaust gas components into groups with certain chemical or physical properties is a very helpful step if analysis of individual compounds is required.

7.2.2.4 Automatic Sampling System

An automatic system can be used in order to make sampling from diluted exhaust gas by collection in bags, by adsorption, absorption and filtration easier (refer to Figure 7.8). Computer-controlled samples of 8 gaseous components (in gas wash bottles or via adsorber tubes) and 5 particle samples (via filters) can be taken simultaneously per test or test phase.

The device essentially consists of a metering cabinet which houses the computer, flow meters, gas meters and pumps, and a thermostated unit to accommodate the gas wash bottles (see Figure 7.8 on the right of the test vehicle). The plastic sample bags are located in the top part of the collection device, whereas the sampling of particulate components takes place directly at the dilution tunnel, refer to Chap. 7.6.

Fig. 7.8: Chassis dynamometer with an automatic sampling system to collect unregulated exhaust gas components (VW and VEREWA)

7.3 Sample Treatment and Analysis

7.3.1 General

The analytical and treatment methods described in the following can obviously only be a selection. The great number of methods available can not be fully presented. It is not the intention to describe the complete analyzing program for each component listed in the EPA substances table but rather to describe different measuring principles for typical examples.

Despite the measures described which lead to a certain fractionation of the exhaust gas components, the result is still a mixture of many components of up to several hundred or even a thousand components. These mixtures must be treated additionally according to certain criteria before they can be analyzed.

The treatment of samples consists of:

- separating the compounds to be analyzed from those disturbing the analysis,
- enrichment of the compounds to be analyzed above the respective sensitivity of the measuring procedure,
- and if necessary, derivatization of the compounds to be analyzed by chemical reactions, where the resulting products can be more easily analyzed than the original substances.

A sample treatment is therefore only then unnecessary when the analyzing method in question allows one

- to differentiate between the compounds to be measured and other disturbing compounds, i.e., if there is no cross sensitivity
- to be able to analyze the compounds without enrichment and derivatization.

One such procedure is the gas chromatography described in Chap. 6.5 if used to analyze individual hydrocarbons of automobile exhaust gas.

Sample treatment taking place between the sampling and the analysis stage is often a complicated and expensive procedure. Illustrated by the example of the total flow method to determine polycyclic aromatic hydrocarbons (PAH) the expense for a sample treatment process will be shown. In this case, the sample to be investigated is split into three fractions: the water condensate, the cooler covering, and the filter housing, although generally the filter contains the largest amount of PAH.

In a multistage process (Figure 7.9), the PAH are first extracted from the three fractions, reunited and the accompanying substances are then removed by a double liquid separation step. After further prepurification on a silicon gel column, the PAH are divided into two fractions using column chromatography on Sephadex LH 20. The first fraction contains components with 2 or 3 rings, the second those with 4 to 7 rings. The solution obtained in this manner is then used for final analysis.

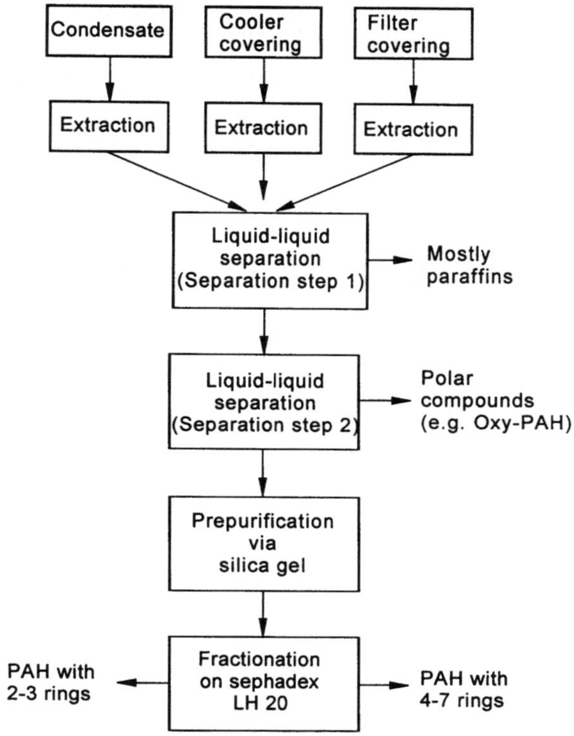

Fig. 7.9: Outline of the sample treatment method for polycyclic aromatic hydrocarbons (PAH) collected by using the total flow method

7.3.2 Analysis of Gaseous Exhaust Components without Preceding Treatment

Three different sampling methods for collecting different groups of substances have already been described:

Sampling from diluted exhaust

1. Collection of gaseous exhaust components *in bags*
2. Collection of gaseous exhaust components *by adsorption and absorption*
3. Collection of particulate exhaust gas components *by filtration*

Let us start with the first case, namely the gaseous exhaust emission components collected in a bag.

An elegant method of determining the gaseous exhaust components without preceding treatment of the probe is gas chromatography (GC) in its two variants, a solid (GSC) or a liquid (GLC) as stationary phase (refer to Chap. 6.5).

By using GC the separation of components from mixtures and their quantitative determination is possible when detection sensitivity is adequate.

By appropriate combinations of the separation parameters a lot of substances and groups of substances can be determined directly from the gaseous exhaust probe (Table 7.3).

Table 7.3: Gas chromatographic determination of gaseous exhaust components without preceding treatment.

Component	Stationary Phase	Detector
Differentiated Nitrogen Oxides	liquid solid	ECD
Organic Amines	liquid solid	NFID
Ammonia	liquid	NFID
Differentiated Hydrocarbons	liquid solid	FID
Organic Sulfides	liquid	FPD
Sulfur Dioxide	liquid solid	ECD

FID - Flame Ionization Detector
NFID - Nitrogen specific FID
ECD - Electron Capture Detector
FPD - Flame Photometer Detector

As an example the analysis of differentiated hydrocarbons is discussed here in more detail. The most important representations of this group of hydrocarbons are listed in Table 7.4. According to a definition from the EPA this group is called: "Individual Hydrocarbons".

Table 7.4: Most important representatives of "Individual Hydrocarbons" (EPA definition)

Component	Empirical Formula
Methane	CH_4
Ethane	C_2H_6
Ethene	C_2H_4
Ethyne	C_2H_2
Propane	C_3H_8
Propene	C_3H_6
Benzene	C_6H_6
Toluene	C_7H_8

After collection of the gaseous exhaust components in a plastic bag, an aliquot partial volume is taken from the bag with a microsyringe and directly analyzed chromatographically without any further interim measures.

Being restricted to the determination of eight individual components in no way simplifies the analytical problem because the physical-chemical properties of these eight substances (e.g. boiling point, polarity) are too varied, as shown in Table 7.5.

Table 7.5: Physical-Chemical Properties of the "Individual Hydrocarbons"

Component	Empirical Formula	Molecular Weight kg/kmol	Melting Point °C	Boiling Point °C
Methane	CH_4	16.04	-182.48	-164.00
Ethane	C_2H_6	30.07	-183.30	- 88.63
Ethene	C_2H_4	28.05	-169.15	-103.71
Ethyne	C_2H_2	26.04	- 80.30	- 75.00
Propane	C_3H_8	44.11	-189.69	- 42.07
Propene	C_3H_6	42.08	-185.25	47.40
Benzene	C_6H_6	78.12	5.50	80.10
Toluene	C_7H_8	92.15	- 93.00	110.60

At present there is no separation column enabling the separation of all components from each other and from the accompanying substances (e.g. atmospheric nitrogen or atmospheric oxygen). Therefore, following a proposal made by the EPA, several separation columns are used (Figure 7.10), each switched into the carrier gas flow (mobile phase) via a relatively complicated magnetic valve system when it is needed.

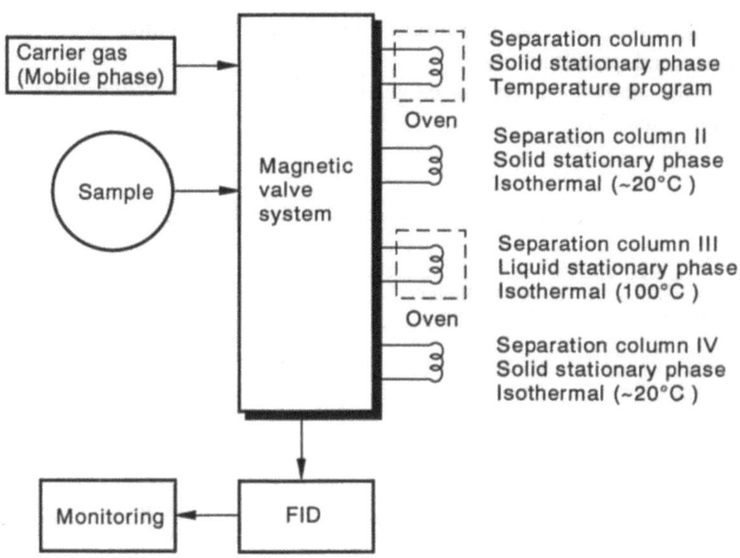

Fig. 7.10: Multidimensional Gas Chromatograph to determine "Individual Hydrocarbons"

All together, 4 columns with different stationary phases and different temperature gradients are used. Such a system is often called "Multidimensional Gas Chromatography". One gas chromatogram of a test mixture determined in this rather elaborate way is shown in Figure 7.11.

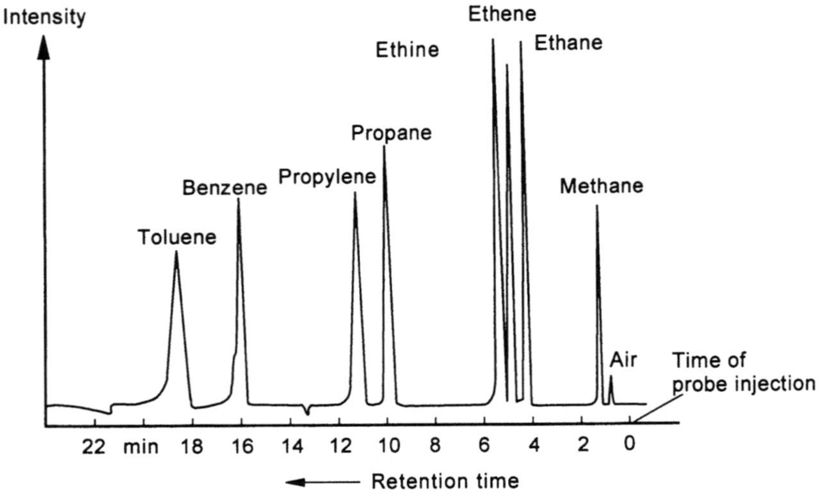

Fig. 7.11: Gas chromatogram of a test mixture of "Individual Hydrocarbons" (8 peaks of components and one of air)

All 8 components of an injected test mixture give distinct individual peaks and can be thus determined quantitatively.

The results of the analysis of typical exhaust from a gasoline engine-powered vehicle is given in Table 7.6.

Table 7.6: Concentrations of the "Individual Hydrocarbons" in the exhaust of a gasoline engine powered vehicle

Component	Empirical Formula	Concentration in exhaust in ppm	C Number	Concentration as C_1 in ppm
Methane	CH_4	20.24	1	20.24
Ethane	C_2H_6	0.61	2	1.22
Ethene	C_2H_4	11.16	2	22.32
Ethyne	C_2H_2	10.00	2	20.00
Propane	C_3H_8	0.05	3	0.15
Propene	C_3H_6	5.50	3	16.50
Benzene	C_6H_6	4.70	6	28.20
Toluene	C_7H_8	7.16	7	50.12
Sum		59.42		158.75

The concentration values determined vary in a relatively wide range, from 0.05 ppm for propane to approx. 20 ppm for methane. The sum of the concentrations of all 8 components in the test is approximately 60 ppm.

In the last column of table 7.6 the concentrations are calculated again in C_1-carbon units. In other words, the concentration values were multiplied with the C number of the respective substance. The sum of the "individual hydrocarbons" is then approximately 160 ppm C_1.

These values can be compared to the total hydrocarbon concentrations measured given in ppm C_1, determined by measuring the diluted exhaust gas collected in a bag by means of an FID. The total HC value is approximately 4000 ppm C_1 as compared to the values of 160 ppm C_1 for the "individual hydrocarbons". The 8 "individual hydrocarbons" have only a proportion of approximately 4 % of the total HC content if given in ppm C_1.

The rest is spread over a great number of other hydrocarbons from which up to 200 compounds can be determined using other gas chromatographic conditions.

The concentrations in this case are already in the ppb (10^{-9}) and ppt (10^{-12}) order. The detection limits of these analytical procedures are therefore reached. For further investigations the components must be enriched and disturbing components must be separated.

7.3.3 Combined Gas Chromatography/Mass Spectrometry (GC/MS)

The accuracy of the analysis result of a chromatographic determination can be improved considerably if a mass spectrometer is linked up to the gas chromatograph as a detector (the end of the analytical column is directly connected to the ion source). Such a GC/MS system measures retention times and peak areas, and in addition the mass spectra of the substances examined. This helps as an important interpretation aid. Mass spectra are not instrument-specific with regard to their m/q values, and are only slightly instrument-specific with regard to the ion intensities. Therefore, they may be tabulated or stored in a computer. Thus it is not necessary in every case to have a reference standard available for identification of a component, as often a comparison of the measured spectrum with a stored reference spectrum suffices, refer to Chap. 6.6.13.

Currently, the combination of high-resolution GC and double focusing MS is one of the most efficient analysis methods with regard to both the quantitative and qualitative investigations of samples with complex compositions. It provides not only mass spectra which may be used as identification aids, but also provides a gas chromatogram at the same time, which can be quantitatively evaluated and is practically identical to a normal GC/FID chromatogram. This total ion chromatogram is determined by a computer by consecutively calculating the sum of all ion intensities measured in a mass scan, and plotting these values over the analysis time.

7.3.4 Further Analytical Procedures

Since each detail of the applicable analytical methods cannot be described to its full extent, the procedure for determining derivatives of polycyclic aromatic hydrocarbons, the nitro- and oxy-PAH is briefly outlined as an example.

In addition to the technically complicated combined gas chromatography/mass spectrometry, one mainly uses investigations in series for the determinations of nitro-PAH:

- capillary gas chromatography with a nitrogen-specific thermoionization detector (GC/TID)
- and the High Performance Liquid Chromatography (HPLC) with on-line reduction of the nitro-PAH to their amino derivatives and final fluorescence detection.

For qualitative analysis of oxy-PAH, combined gas chromatography/mass spectrometry (GC/MS) is imperative, whereas after identification the quantitative analysis is usually carried out using a gas chromatograph/flame ionization detector (GC/FID).

Figure 7.12 shows a GC/FID chromatogram with numbered peaks for a fraction of a diesel particle extract. In Table 7.7 the names of the components are listed corresponding to their number on the GC peak sequence in Figure 7.12. Peaks for both the oxy-PAH and the nitro-PAH are indicated in the chromatogram, as the FID detection principle does not differentiate between them. Similar signals of underivated PAH which could not be fully removed during the sample treatment are also recognizable. The identification or characterization of PAH and PAH derivatives is mainly carried out by interpreting the mass spectra produced in a GC/MS experiment.

However, the concentration of the substances to be investigated is not always large enough to produce a completely interpretable mass spectrum. For example, only 9-nitro-Anthracen (Peak No. 20) and 1-nitro-Pyrene (Peak No. 42) of the

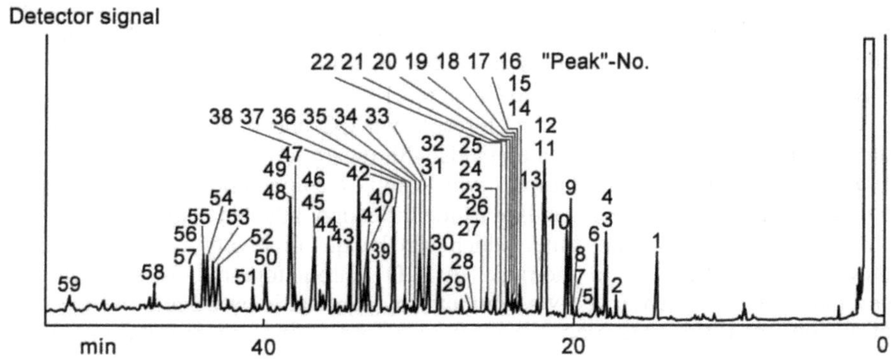

Fig. 7.12: GC/FID chromatogram of the oxy-PAH fraction of a diesel particle extract. (The names of the components for the numbered peaks are listed in Table 7.7)

nitro-PAH class are present in sufficient quantities. Often, the expected fragmentation pattern has numerous background peaks superimposed on it, making quick interpretation difficult. Some peaks could not be identified and they are therefore not listed in Table 7.7.

Table 7.7: Identified characteristic PAH and PAH derivatives in the oxy-PAH fraction of a diesel particle extract (x, y, z = unknown positions of isometric compounds of the group), refer to Figure 7.12

Peak No.	Component	Peak No.	Component
1	9-Fluorenone	29	PAH cumarin
2	x-Methyl-9-Fluorenone	30	Benzo(x)fluorene-y-one
3	x-Methyl-9-Fluorenone	31	Benzo(x)fluorene-y-one
4	Acenaphthenequinone	32	4,4'-Dinitro-Biphenyl (IS)
5	x-Methyl-9-Fluorenone	33	Benzo(x)fluorene-y-one
6	1-H-Phenalene-1-one	34	Benzo(x)fluorene-y-one
7	x-Methyl-9-Fluorenone	35	Benzo(x)fluorene-y-one
8	x-Methyl-9-Fluorenone	36	Benzo(x)fluorene-y-one
9	Anthraquinone	37	Chrysene
10	Phthalate	38	7-H-Benz(de)anthracene-7-one
11	1,8-Naphthalenedicarboxylic acid anhydride	40	Phthalate
12	4-H-Cyclopenta(def)-phenanthrene-4-one	41	Benz(a)anthracene-7,12-dione
13	Fluoranthene	42	1-Nitro-Pyrene
14	2-Nitro-Fluorene	43	ß-ß'-Binaphthyl
15	x-Methyl-Anthraquinone	45	Benzo(b)fluoranthene
16	Pyrene	46	Benzo(xy)pyrene-z-one
17	Phenanthrene-x-aldehyde	47	Benzo(e)pyrene
18	Anthracene-x-aldehyde	48	6-H-Benzo(cd)pyrene-6-one
19	x-Methyl-4-H-Cyclopenta(def)-phenanthrene-4-one	49	Benzo(a)pyrene
20	9-Nitro-Anthracene	50	Benzo(xy)pyrene-z-one
22	x-Methyl-4-H-Cyclopenta(def)-phenanthrene-4-one	52	1,12-Pyrenedicarboxylic acid anhydride
23	Phenanthrene-9,10-quinone	54	Indeno(1,2,3-cd)pyrene
25	x-Methyl-4-H-Cyclopenta(def)-phenanthrene-4-one	56	Benzo(ghi)perylene
26	x-Methyl-4-H-Cyclopenta(def)-phenanthrene-4-one	57	6-Nitro-Benzo(a)pyrene
28	x-Methyl-9-nitro-Anthracene	59	Coronene

Table 7.8 shows an overview of the measured exhaust gas components together with the corresponding sampling and analysis methods. The sampling was carried out from the diluted exhaust gas. The total flow method was only used for the polycyclic aromatic hydrocarbons.

Table 7.8: The sampling and analysis methods to determine some unregulated exhaust gas components

Component	Sampling Method	Analysis Method
Total Particulate Mass	Filtration	Gravimetry
Total Cyanides	Absorption	Photometry
Ammonia	Absorption	Photometry
Sulfur dioxide	Absorption	Volumetric analysis
Sulfates	Filtration	Photometry
Hydrogen sulfide	Absorption	Photometry
Total aldehydes	Absorption	Photometry (MBTH-Method)
Individual aldehydes and ketones	Absorption	HPLC (DNPH-Method)
Total phenols	Absorption	Photometry
Individual hydrocarbons	Sample bag	GC/FID
Polycyclic aromatic hydrocarbons	Filtration, Filtration with Adsorption	TLC, HPLC, GC/FID, GC/MS
Alcohols	Sample bag	GC/FID
Particle bound organic compounds	Filtration	Extraction, Thermogravimetry
Elementary composition of particulates	Filtration	Elementary analysis, Atomic absorption or x-ray fluorescence spectroscopy

DNPH - Dinitrophenylhydrazine
MBTH - 3-methyl-2-benzothiazolinonhydrazon-hydrochloride
TLC - Thin Layer Chromatography

7.4 Emission Values of Some Unregulated Exhaust Gas Components

Total Cyanides

Total cyanides which are mostly composed of hydrogen cyanide are only present in extremely low concentrations in motor vehicle exhaust gas (see Figure 7.13). In comparison to gasoline vehicles without catalytic converters, the vehicles with three-way catalytic converters or diesel engine-powered vehicles emit on an average only about 6 % (approximately 94 % decrease due to the catalytic converter concept) or 10 %. All in all, the concentrations determined are almost two orders of magnitude smaller than the isolated case determined by the EPA (a few hundred mg/mi) leading to the EPA's suspension of sale licenses.

Ammonia

The emission behavior of ammonia (Fig 7.14) differs from all other components examined here in that the highest concentrations are to be found in the exhaust gas of the gasoline-powered vehicles with three-way catalytic converters. The average amount emitted by vehicles without catalytic converters and diesel-powered vehicles is less than 3 %, as compared to the average amount emitted by vehicles equipped with three-way catalytic converters.

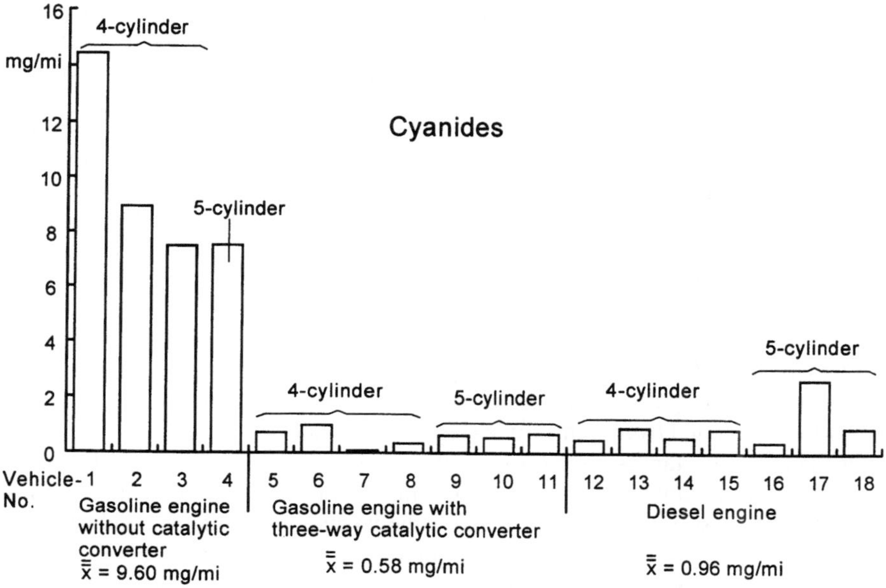

Fig. 7.13: Emission results of total cyanides, averaged over three driving cycles (EDC, FTP, SET) for different engine types, $\bar{\bar{x}}$ = overall mean

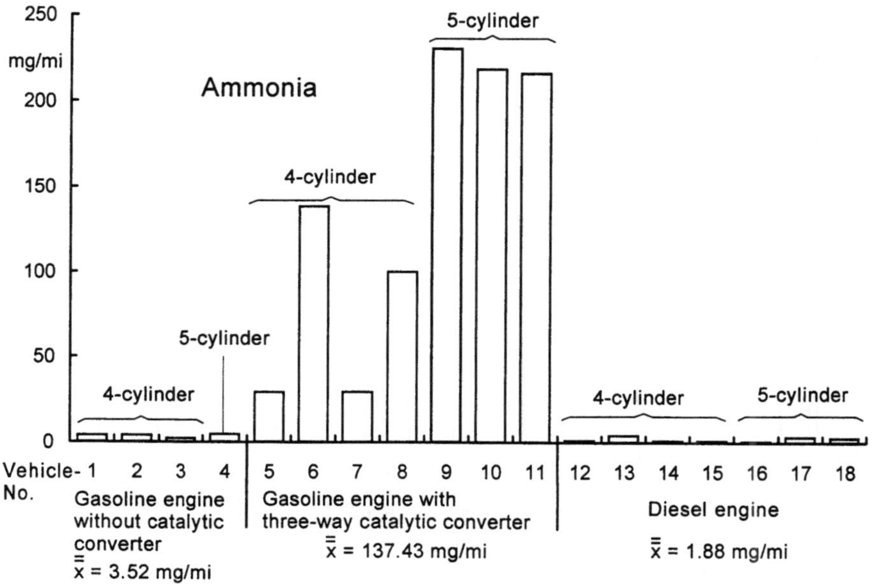

Fig. 7.14: Emission results of ammonia, averaged over three driving cycles for different engine types, $\bar{\bar{x}}$ = overall mean (see Fig 7.13)

The amount of ammonia emitted varies considerably within the catalytic converter-equipped vehicle group. However, there is a general tendency for the 5-cylinder engines (vehicles No. 9 to 11) to clearly emit more ammonia than the corresponding 4-cylinder engines (vehicles No 5 to 8). In addition to this, especially these 4-cylinder models can be divided into two distinct groups: two vehicles with relatively low emissions and two others with comparatively high emissions. Under conditions more favorable for oxidation, fewer hydrocarbons (oxidative decomposition) and less ammonia (no formation possibility) can be expected and conversely, in conditions favoring reduction, more hydrocarbons (reduced decomposition rate) and more ammonia (greater formation possibility) can be expected.

Sulfur compounds

From the group of sulfur compounds, sulfur dioxide (SO_2), sulfate deposited on the particle filter and hydrogen sulfide (H_2S) are of interest.

Of the three substances mentioned, sulfur dioxide (Figure 7.15) occurs by far in the highest concentrations. The mean emission results of the three vehicle groups (vehicles with gasoline engines, with and without catalytic converters, and vehicles with diesel engines) are approximately 13:15:100. These figures bear a direct relation to the sulfur content of the fuel used (0.02 % by weight for gasoline and 0.23 % by weight for diesel fuel). On the other hand, the catalytic converter concept has practically no influence on the sulfur dioxide content of the exhaust

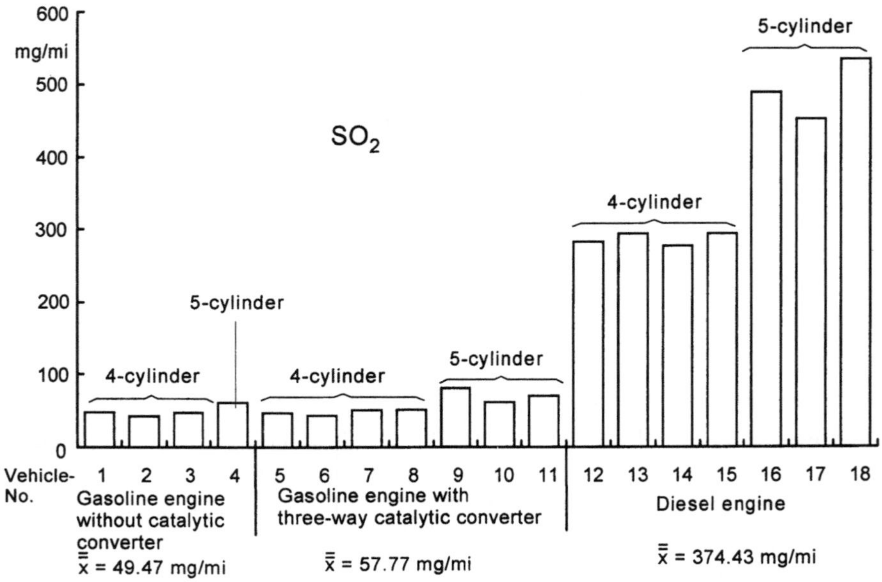

Fig. 7.15: Emission results of sulfur dioxide, averaged over three driving cycles for different engine types, $\bar{\bar{x}}$ = overall mean

gas. This is to be expected because of prevailing thermodynamic and kinetic conditions.

The sulfate emissions averaged over all measuring results came to approximately 2 % of the amount of sulfur dioxide emitted (Figure 7.16). Apart from the noticeably low emission results for vehicle No. 17, the emissions profiles of both components are remarkably similar. This analogy can be further illustrated by entering the measuring data of both components for all vehicles in pairs into a regression diagram (Figure 7.17). The regression coefficient is 0.95 for 54 value pairs.

In the case of hydrogen sulfide, the emission results under standard test conditions are in almost all cases below the detection level of 0.02 ppm and therefore not indicated here.

However, customers occasionally complain about unpleasant smells in the exhaust gases of catalytic converter-equipped vehicles, in particular of vehicles with an open-loop catalytic converter (three-way catalytic converter without λ-probe control). This indicates hydrogen sulfide emissions at least in certain operating states (after changing over from a lean fuel mixture to a rich fuel mixture). Investigations on vehicles with oxidation or three-way catalytic converters have shown that the highest emission results can mainly be observed during idling.

Figure 7.19 shows hydrogen sulfide emissions of a vehicle with a carburetor and an open loop catalytic converter (engine power 53 kW) measured in a specific

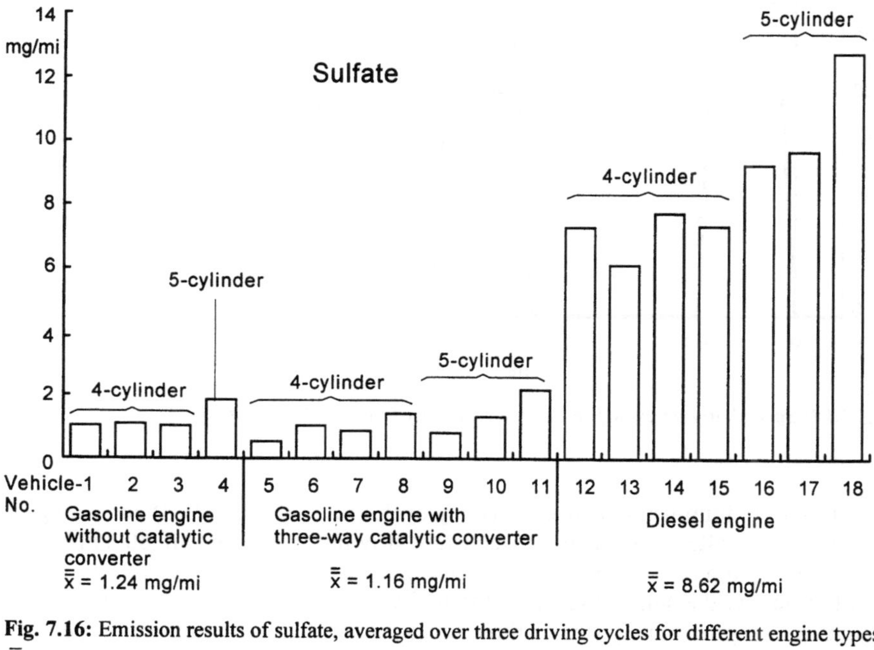

Fig. 7.16: Emission results of sulfate, averaged over three driving cycles for different engine types, $\bar{\bar{x}}$ = overall mean

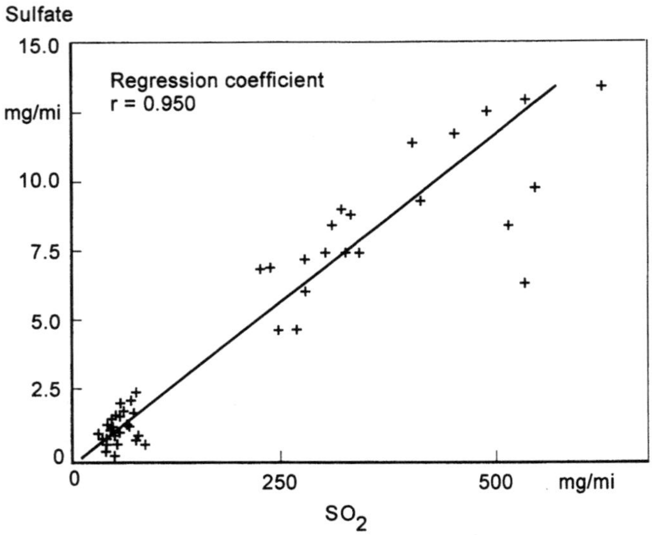

Fig. 7.17: Linear regression line analysis of sulfur dioxide and sulfate emissions for 18 test vehicles and over three driving cycles (n = 54 value pairs)

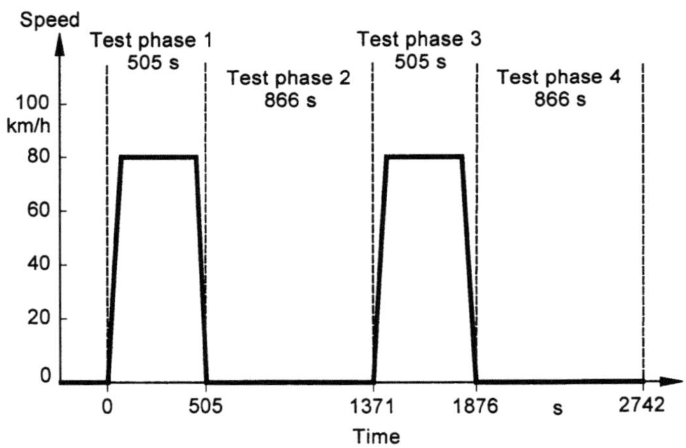

Fig. 7.18: Special driving cycle (SET) to determine hydrogen sulfide emissions

measuring program using a special driving cycle (SET) (Figure 7.18) consisting of 4 test phases (alternating between constant driving at a speed of 80 km/h and idling). Sampling took place separately during the individual test phases.

Additionally, the sulfur content of the fuel was varied in steps from 0.058 % by weight (i.e. approximately two or three times higher than for commercial fuels) to 0.034 % by weight or to approximately 0.003 % by weight.

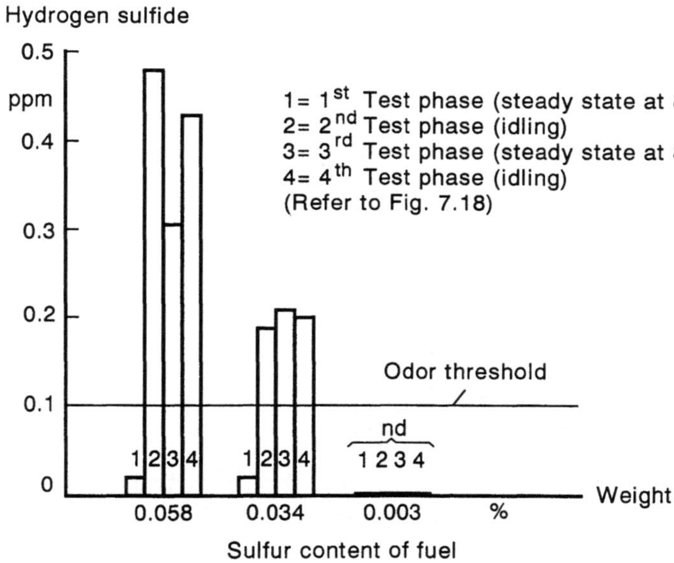

Fig. 7.19: Hydrogen sulfide emissions of a mid-sized vehicle with a carburetor and an open-loop catalytic converter system (engine output 53 kW) dependent on the sulfur content of the fuel and the test phase (special driving cycle, refer to Figure 7.18), nd = not detectable

The respective hydrogen sulfide concentrations, apart from during the first test phase, are clearly above the odor threshold of 0.1 ppm for the fuel with higher sulfur content. When the sulfur content reaches 0.003 % by weight no more hydrogen sulfide can be detected in any test phase.

7.5 Summary of the Order of Magnitude of Emission Results

The emission results (g/mi) of a representative selection of exhaust gas components are compiled in the Figures 7.20 to 7.22 for the three vehicle groups: vehicles with gasoline engines with and without catalytic converters, and vehicles with diesel engines (respective mean values of all vehicles in each group and for all three test procedures). Due to the wide range of measurement results, the respective coordinates are given on a logarithmic scale.

As can be seen in the diagrams, the emission results of the regulated and unregulated exhaust gas components of the vehicles tested cover practically 7 powers of ten. The exhaust gas components can be roughly divided into three categories with different concentration ranges.

The emission results of the catalytic converter-equipped vehicles are on an average more than one order of magnitude lower than those of vehicles without catalytic converters (refer to Figures 7.20 and 7.21). Exceptions to this rule are

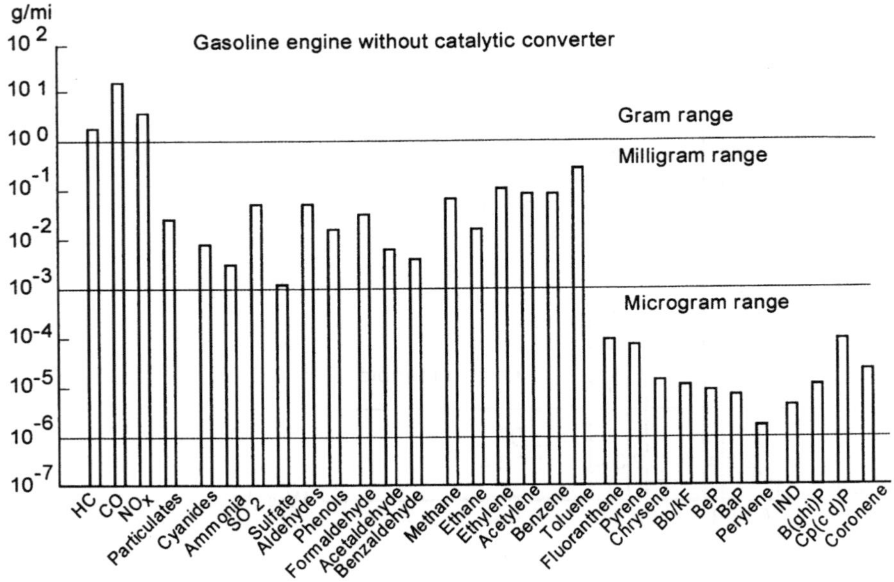

Fig. 7.20: Summary of the emission results (logarithmic scale) of gasoline-powered vehicles without catalytic converter, averaged over all the vehicles within this group and three driving cycles (EDC, FTP, SET);
(Bb/kF = total of benzo(b)- and benzo(k)fluoranthene; BeP = benzo(e)pyrene;
BaP = benzo(a) pyrene; IND = indeno(1,2,3-cd)pyrene; B(ghi)P = benzo(ghi)perylene;
Cp(cd)P = cyclopenta(cd)pyrene)

merely ammonia (the measurement results are more than one order of magnitude higher for vehicles with catalytic converters) and the sulfur compounds, sulfur dioxide and sulfate, which are present in similar concentrations in both vehicle groups.

The emission results of total hydrocarbons and carbon monoxide for the diesel engine powered vehicles are also approximately one power of ten lower than those of the gasoline-powered vehicles without catalytic converters, whereas the emissions of nitrogen oxides are only four times lower (Figure 7.22).

The inverse is true for total particle mass, sulfur dioxide and sulfate. The exhaust gas concentrations for each are approximately ten times higher. The PAH of the diesel-powered vehicles are on an average of the same concentration range as those of the gasoline-powered vehicles without catalytic converters.

Fig. 7.22: Summary of the emission results (logarithmic scale) of diesel -powered vehicles, averaged over all the vehicles within this group and three driving cycles. ▶
(See Fig. 7.20 for explanation of abbreviations; furthermore: BaA-7,12-dione = benz(a)anthracene-7,12-dione; 3-Nitro-F = 3-nitro-fluoranthene; 1-Nitro-P = 1-nitro-pyrene)

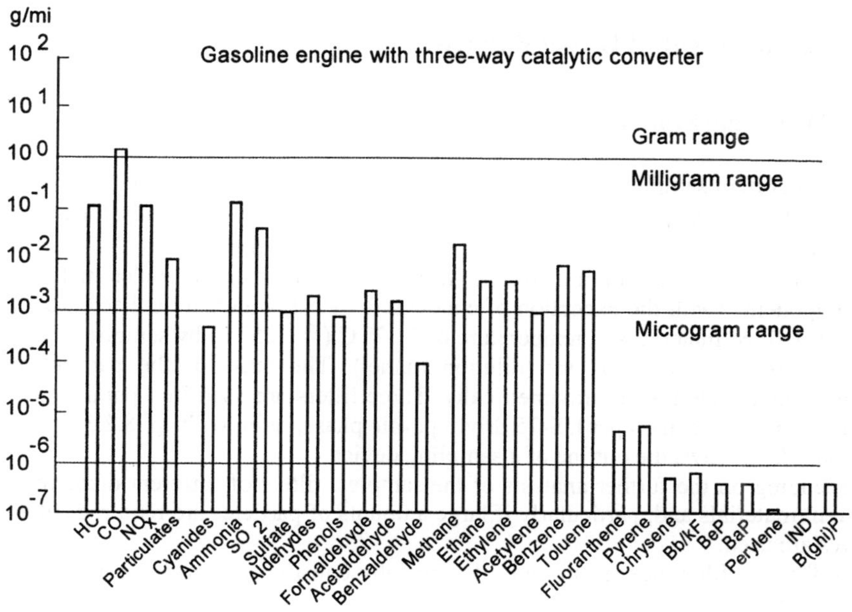

Fig. 7.21: Summary of the emission results (logarithmic scale) of gasoline-powered vehicles with three-way catalytic converter, averaged over all the vehicles within this group and three driving cycles (cyclopenta(cd)pyrene and coronene were not measured). (See Fig. 7.20 for explanation of abbreviations)

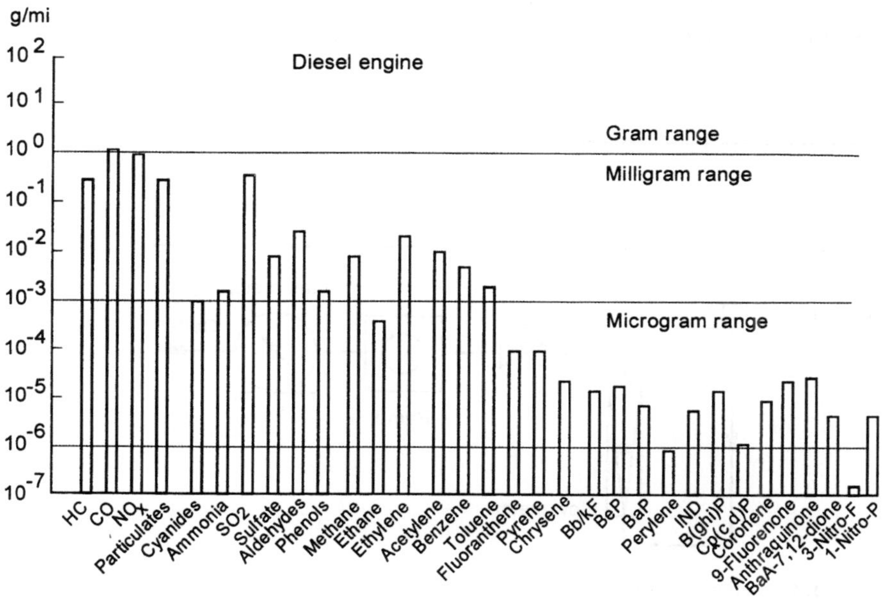

7.6 Measurement of Diesel Particles

7.6.1 Measuring Procedure

A very important sampling technique used for regulated exhaust gas components is the separation of the particulate components from the exhaust, especially for diesel engine vehicles, as they emit higher amounts of soot and sulfate.

According to the definition of the EPA, the term exhaust gas particles includes all components (with the exception of condensed water), which are collected on a filter or defined material at a temperature of 51.7° C (≙ 125° F) by sampling from the exhaust gas diluted by air (dilution tunnel). The mass of the particulate components is determined gravimetrically. Three filters are used in a parallel set-up to determine the mass per US-75 driving cycle phase, refer to Chap. 8.2.3.

Figure 7.23 shows the outline of a sampling device.

Depending on the further analysis of the samples, filters of different diameters, different materials and, particularly if large amounts of particles are required, filter sleeves are used.

The filter or filter sleeves are put into specially constructed housings, as shown in Figure 7.24.

Even measures presumed to be simple, like the gravimetrical determination of the particle mass are relatively complex, as indicated in Figure 7.25.

First of all, the filters to be laden are preconditioned for 8 to 56 hours. The temperature may vary by a maximum of ±6°C in the range of 20 to 30°C. The relative air humidity must be between 30 and 70 % with a maximum variation of ±10 %.

The weight (tare) of each preconditioned unladen filter is then determined. A scale used for weighing must have an accuracy within ±1 microgram.

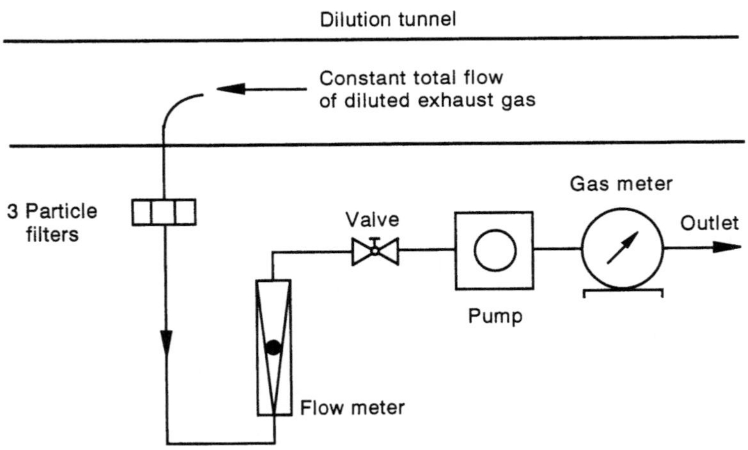

Fig. 7.23: Collection of particulate components of exhaust gas by filtration (schematic)

Fig. 7.24: Filter housing with a laden filter (below)

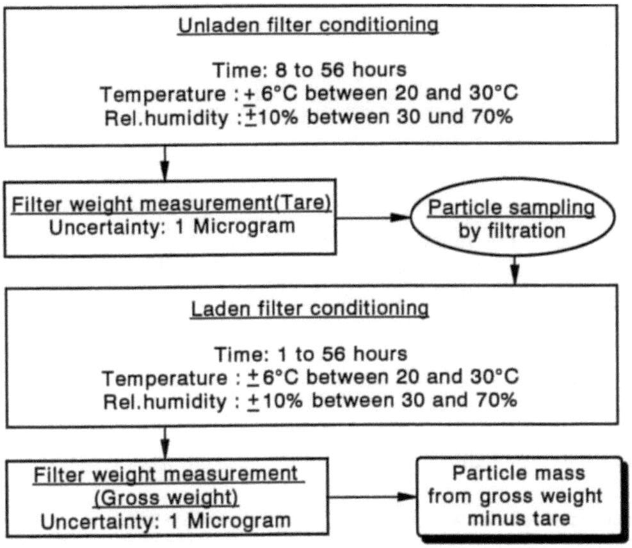

Fig. 7.25: Outline of the gravimetrical determination of the filtrated particle mass

After lading the filters during an exhaust emission test (refer to Chap. 8) they are again conditioned for 1 to 56 hours. Temperature and relative air humidity must fulfill the same requirements as for the preconditioning.

Finally, the gross weight of each laden filter is determined, again with a maximum error of ±1 microgram. Particle mass is the difference between gross weight and tare.

The filter conditioning and weighing is carried out in a special air conditioned room or in a laboratory unit in the test cell. Figure 7.26 shows a view of a laboratory unit.

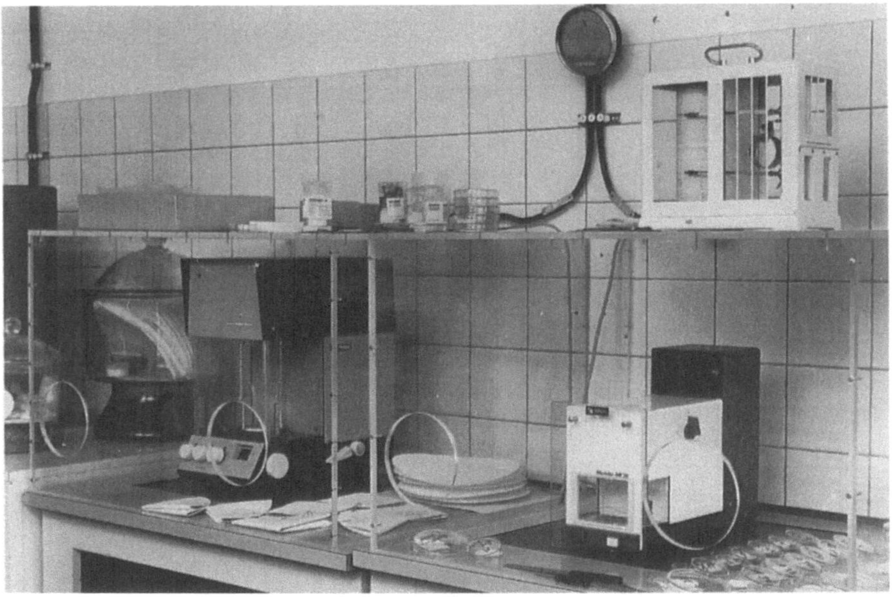

Fig. 7.26: View of a laboratory unit for conditioning and weighing of particle filters

Both the filters and the scales are housed in boxes of acrylic glass, to avoid settling of dust and disturbing air flow as much as possible. Using suitable measuring instruments the air conditioning is monitored and its data stored.

7.6.2 Measurement Results

Typical particle masses are listed in Table 7.9

Table 7.9: Typical particle masses of vehicles with different emission control systems

Engine type	Order of Magnitude of Particle Masses
Diesel engine (old)	approx. 400 mg/mi
Diesel engine (new)	approx. 80 mg/mi
Gasoline engine without catalytic converter	approx. 40 mg/mi
Gasoline engine with three-way catalytic converter	approx. 4 mg/mi

Keeping in mind the large variations of particle masses for an individual vehicle class, the figures shown in Table 7.9 indicate that the particle emissions rise each by an order of magnitude from gasoline engines with three-way catalytic converters to gasoline engines without catalytic converters to older diesel engines.

In spite of the very complex chemical composition of the particles, some important information can be gained without any additional pretreatment.

These are:

- the gravimetric determination of the total particle mass (regulated in the US and in the EU for diesel vehicles),
- the quantitative determination of carbon, hydrogen, nitrogen and oxygen contents by ultimate analysis,
- the quantitative determination of the components of heavier elements (approx. from fluorine upwards) by x-ray fluorescence analysis,
- the determination of volatile components by thermogravimetry and
- the determination of soluble components by organic extraction.

Figure 7.27 show schematically particles of diesel exhaust gas collected on a filter as an example.

Both organic as well as inorganic components settle on the filter. The soot has the greatest proportion.

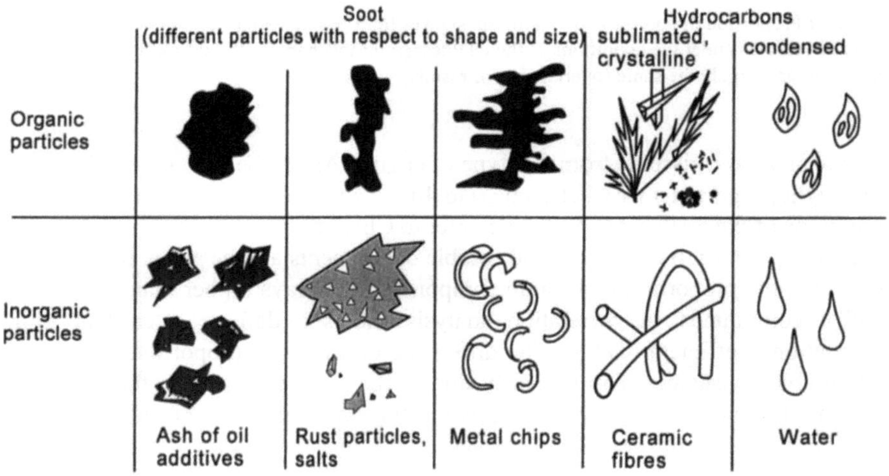

Fig. 7.27: Filtrated particles from diesel exhaust gas

7.6.3 Particle Bound Substances

In addition to determining the total particle mass, measuring the volatile and the soluble content of the particles is also important. Figure 7.28 shows the proportion of particles' content soluble in organic solvents and those volatile at a temperature of 600° C of the total mass of soot from four different diesel vehicles. Corresponding analysis results of gasoline engine vehicles are not available due to

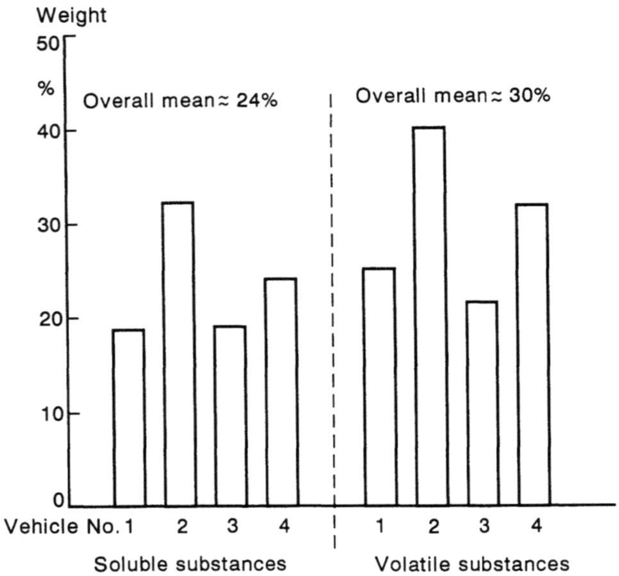

Fig. 7.28: Proportion of soluble and volatile substances (600° C) in the particles of 4 diesel-powered vehicles under the conditions of the FTP test (refer to Chap. 8). Soluble and volatile substances are largely the same (overlapping of results)

the low particle emissions from this type of engine. As can be seen in this diagram, the weight proportions vary between 20 to 40 % of the total particle mass under the conditions of the FTP test (US-75 test, refer to Chap. 8).

A great proportion of volatile and soluble components are the same. It should be noted that the proportion of volatile components is always higher than that of the soluble compounds. A correlation analysis shows a definite interrelationship (correlation coefficient = 0.97). The difference between the proportion of volatile and soluble substances may be partly due to the sulfate present in the particles. This sulfate is not soluble in organic solvents but evaporate when heated.

7.6.4 Elementary Composition of Particles

Determination of the elementary composition of particles yields important information. Elementary analysis methods were used to quantify the elements carbon, hydrogen, nitrogen, oxygen and sulfur. As a relatively large amount of particle mass is required, only measurement results for diesel-powered vehicles are given here. Iron, calcium, zinc, lead, platinum and rhodium were quantified by use of the x-ray fluorescence method.

As Figure 7.29 shows, carbon is by far the most common element in diesel particles, accounting for an average proportion of approximately 88 % by weight. The variation range of the carbon content is between approx. 82 and 94 %.

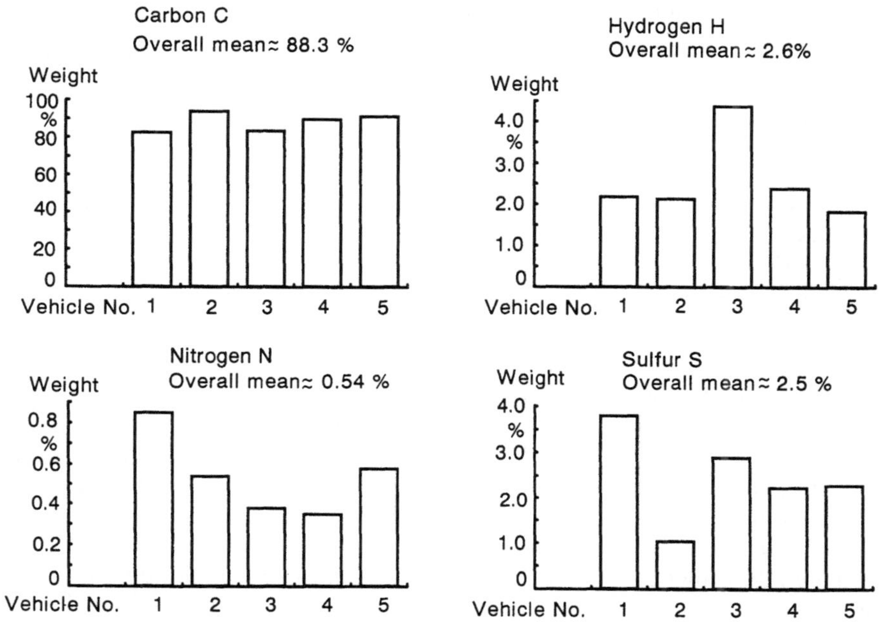

Fig. 7.29: Proportion of elements in the particles of 5 different diesel-powered vehicles

The proportions by weight of the hydrogen in the total particle mass lies between approximately 1.6 and 4.8 %, an average of about 2.6 %. A distinctly higher hydrogen content was measured on vehicle no. 3 than on the other four diesels.

The nitrogen content of the diesel particles is on an average only approximately 0.5 % by weight, the smallest proportion of all the typical organic molecule components (carbon, hydrogen, oxygen and nitrogen).

The sulfur content of the diesel particles is on an average approximately 2.5 % by weight, within the same range as the hydrogen content. However, a large difference between the two 4 cylinder test vehicles is noticeable; the particles from vehicle no. 1 contain almost 4 times as much sulfur as from vehicle no. 2. The 5 cylinder models (no. 3 to 5), on the other hand, have rather similar sulfur contents.

As the x-ray fluorescence method is more sensitive, the proportions of metals in the total particle mass could be examined for gasoline engine vehicles with and without three-way catalytic converters. Apart from lead, the measurement results lie at the detection limit; furthermore, the values vary quite considerably, and in some cases no metal could be detected at all. As the measurement results are not vehicle or test specific, Table 7.10 only shows the maximum proportions by weight found in the total particle mass.

Iron, as a corrosion product of the engine/exhaust system, is sporadically found in the particulates of all three vehicle groups. Due to the much smaller total mass of the particulates, the proportion of iron in the exhaust gas of gasoline engine vehicles is naturally higher. Calcium and zinc stem from additives in the lubricating oil and can be detected in measurements only in diesel particulates, as these

Table 7.10: Maximum metal content of particulates (nd = not detectable)

Component	Gasoline Engine without Catalytic Converter % by wt.	Gasoline Engine with Catalytic Converter % by wt.	Diesel Engine % by wt.
Iron	< 4	< 10	< 0.6
Calcium	nd	nd	< 0.3
Zinc	nd	nd	< 0.3
Lead	< 40	nd	nd
Platinum	-	nd	-
Rhodium	-	nd	-

elements are below the detection limit in the gasoline engine vehicles. As leaded fuel is only used in the ECE vehicles without catalytic converters, lead is only present in the exhaust gas of these vehicles. The abrasion products of platinum and rhodium possibly present in the particulates of the vehicles with catalytic converters could not be detected. Some other metals (e.g. aluminum, vanadium, chromium, manganese, cobalt, nickel, copper, molybdenum, cadmium and tin) lie below the detection limits in all cases.

Figure 7.30 shows again an overview of the elementary composition of diesel particulates. Carbon is the main component with about 88 % by weight, followed by oxygen, hydrogen, sulfur, the total metal content and nitrogen.

Although elementary analysis does not indicate the type of chemical bonds between the elements, rough estimations of the composition of the particulate phase of the exhaust gas are possible. A rough calculation shows that carbon cannot be present completely in a chemically bound form. If we assume a mean C:H ration of 1:1.85 for the hydrocarbons, the main group of carbon compounds in

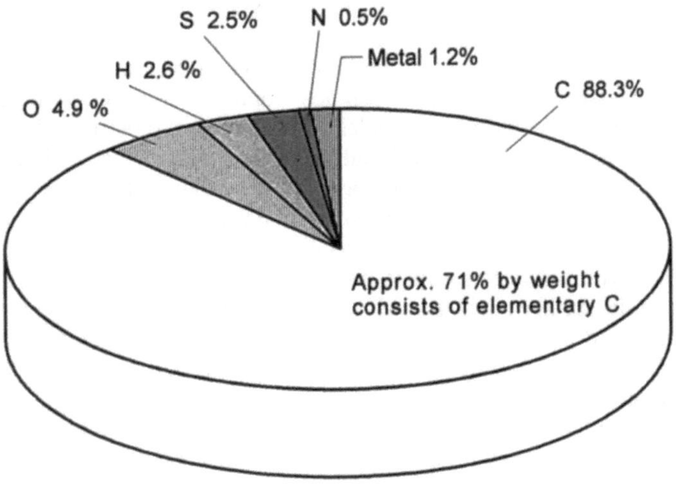

Fig. 7.30: Elementary composition of diesel particles in percent by weight

exhaust gas, one part by weight of the hydrogen is combined with 6.5 parts by weight of carbon, taking into account the atomic weights. The 2.6 % by weight of hydrogen determined in the particulates by elementary analysis can thus be combined with a maximum of about 17 % by weight of an approximate total of 88 % by weight of carbon. This means that about 71 % by weight of the total particulate mass must consist of elementary carbon in the form of soot. This value can be confirmed by experiments (thermogravimetry). Using thermogravimetric analysis, an average 70 % by weight of soot (total particulate mass minus volatile proportion) was found.

According to this estimation, approximately 20 % by weight (2.6 % by weight of hydrogen plus 16.9 % by weight of carbon) of the total particulate mass consists of hydrocarbon compounds. If we assume that almost all of the nitrogen as well as parts of the oxygen and the sulfur are bound in these substances (e.g. in the form of amino, nitro, hydroxy, carbonyl, and sulfide derivatives), the total amount of organic material will be more than 20 % by weight. The value determined experimentally for soluble substances is on an average about 24 % by weight (compare to Figure 7.28).

The remaining approximately 5 % by weight of the total particulate mass is attributed to metallic compounds and sulfates.

Figure 7.31 shows finally an overview of the *material* composition of diesel particles in percent by weight.

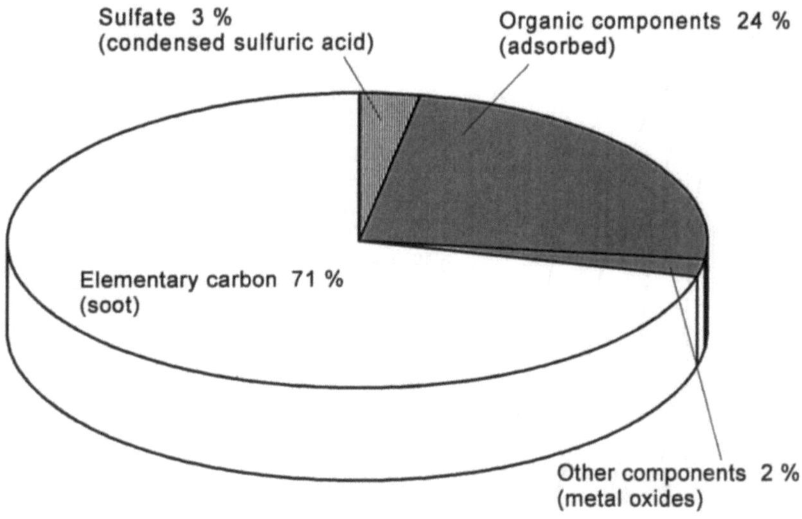

Fig. 7.31: Material composition of diesel particles in percent by weight

7.6.5 Influence of Different Parameters

The removal of particulates from the diluted exhaust gas by means of filtration is extremely sensitive to changes in the sampling parameters. For example, the particulate concentration - which is a function of the engine type and the degree of dilution - as well as the collection temperature considerably influence the amount of hydrocarbons adsorptively bound to the soot particles and the gravimetrically determined total particulate mass.

If the purpose of sampling is to measure the amount of particle bound substances (for example the polycyclic aromatic hydrocarbons), these influences must be especially taken into account. Theoretical estimates largely concur with experimental findings to show that at temperatures below 51.7° C more than 95 % of the hydrocarbons (with a molecular weight greater than ca. 200) present in the diluted diesel exhaust gas (typical particle concentration 10^{-8} g/cm^3) are adsorbed on the particulates. Figure 7.32 shows calculated adsorbed proportions of hydrocarbons as a function of temperature, with particle concentration W_p as a parameter.

At lower particulate concentrations adsorption is more and more incomplete, i.e. the hydrocarbons can no longer be accurately determined by collecting the particulates alone. Since the collection temperature cannot simply be reduced, an adsorbent must be placed behind the particulate filter in order to isolate the gaseous part of the hydrocarbons. This procedure is necessary to collect the polycyclic aromatic hydrocarbons in gasoline engine exhaust gas (particulate concentration approximately 10^{-10} g/cm^3).

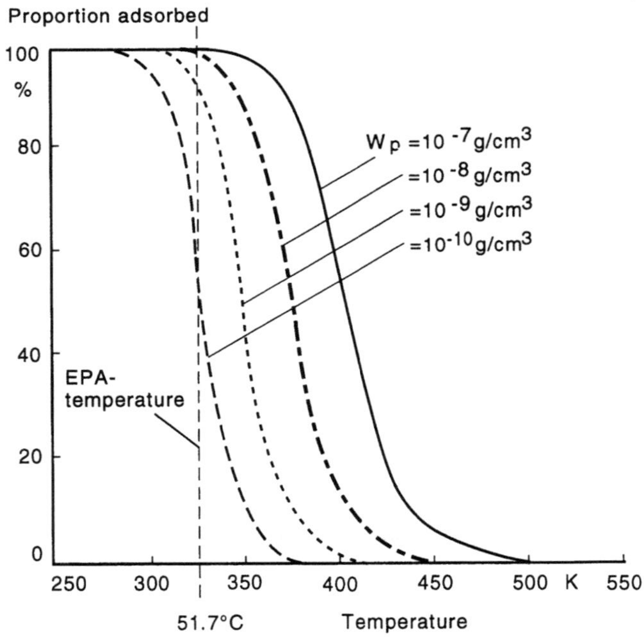

Fig. 7.32: Calculated proportion of hydrocarbons with a molecular weight greater than 200 adsorbed on diesel particulates as a function of temperature for various particulate concentrations W_p. In diluted diesel exhaust gas a value of $W_p = 10^{-8}$ g/cm^3 is typical

In undiluted exhaust gas the temperatures are higher than 500 K and therefore the hydrocarbons are still gaseous and not adsorbed to the particles. This allows the use of an oxidation catalytic converter to lower the hydrocarbon and the PAH emissions.

7.6.6 Determination of the Density Function of Aerodynamic Particle Diameters

To determine the size of particles, among others a low pressure cascade impactor is used. Its measurement principle is the size-selective deposition of particles on impact plates from an accelerated sample gas flow. This principle ist shown in Figure 7.33.

The sample from the undiluted exhaust gas flows through the cascades of the impactor. Each cascade consists of a nozzle plate, an impactor plate and a distance ring. The cascades are pressed together and sealed with temperature-resistant Teflon rings.

Each nozzle plate has several nozzles with a defined diameter, getting smaller from cascade to cascade. Corresponding to the number of nozzles for each cascade a number of aerosol streams are produced impacting upon the impact plate and deflected into the next cascade.

The impact plates are covered with aluminum foils which are conditioned as described in Figure 7.25.

Particles in the sample flow having a large enough inertia will impact upon the plate, and smaller particles will pass as an aerosol onto the next stage.

As each successive stage has higher aerosol velocities (due to nozzles with progressively smaller diameters) smaller diameter particles will be collected at

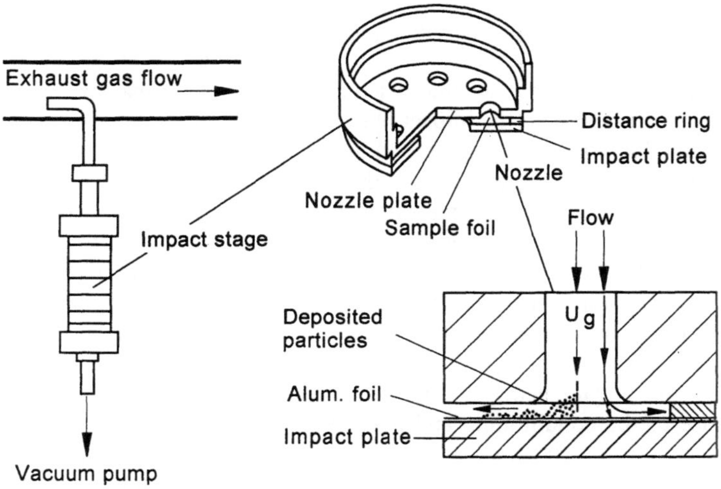

Fig. 7.33: Scheme of a Berner low pressure impactor (Source Giorgi)

each stage. For each stage, particles of a defined aerodynamic diameter range of the total aerodynamic particle diameter spectrum are deposited onto the aluminum foil covering the impact plate.

By weighing the difference of the gross/tare of the foil before and after deposition, the mass size distribution of the sampled particles are determined.
The determination of the mass of the deposited particles by weighing is comparable to the method described in Figure 7.25.

The size separation of particles depends on the inertial parameter Φ_p:

$$\Phi_p = D_p^2 \cdot \rho_p \cdot C \cdot \frac{U_g}{18} \cdot \eta_g \cdot W_g, \tag{7.1}$$

with

D_p Particle diameter
ρ_p Particle density
C Slip flow correction factor
U_g Gas velocity
η_g Gas viscosity
W_g Aerosol stream diameter

Under realistic conditions there is no distinct separation, i.e. particles with greater and smaller diameters are deposited.

In practice, the impactor manufacturer calibrates the instruments using test aerosols, thus defining the separation ranges for the aerodynamic diameters. The aerodynamic diameter is defined as the diameter of a sphere with the density of 1 g/cm^3 with the same settling speed in air at rest or in air with laminar flow as the aerosol particle to be measured.

These separation ranges defined by the instrument manufacturer are related to the standard aerosol temperatures or pressures and standard ambient conditions. At unfavorable conditions, these ranges can overlap each other and must be accepted as uncertain.

Wall losses have to be taken into account when measuring with impactors. This means particles not deposited onto the impact plates but on the other parts, e.g., on the walls. In literature loss values of approx. 3 % are given.

Additionally there is a particle bounce, i.e., the particles already deposited are again carried along by the gas flow and deposited elsewhere. In literature a coating of adhesive over the aluminum foil is recommended as a remedy for this. However, newer investigations have shown that a coating of the aluminum foil surface does not result in a difference of the deposited diesel particle mass or their size distribution, and can be dispensed with.

In Table 7.11 separation ranges of a 10 stage cascade Berner impactor are listed.

A disadvantage of an impactor is that the measurements are very time consuming because of the conditioning and placement of the foils on the plates. Furthermore, repeated measurements show great errors so that a high number of repeated measurement units (n = 25) is required for statistically more accurate results.

Table 7.11: Separation ranges of a Berner low pressure cascade impactor

Stage	D_p in µm		
1	0.015	to	0.030
2	0.030	to	0.060
3	0.060	to	0.125
4	0.125	to	0.250
5	0.250	to	0.50
6	0.50	to	1.0
7	1.0	to	2.0
8	2.0	to	4.0
9	4.0	to	8.0
10	8.0	to	16.0

Figure 2.6 in Chap. 2.1.4 shows a typical result of measurement of size distribution of particles in diesel exhaust gas, plotted on a histogram with a calculated density function and with the corresponding overall mean.

8 Vehicle Exhaust Emission Tests

8.1 Introduction

After the discussion of the

- history of exhaust emission legislation,
- exhaust emissions,
- air quality,
- effects on the environment and
- measurement procedures,

we can now turn to the exhaust emission tests.

An impression of the effort necessary is demonstrated by Figure 8.1 showing a phantom drawing of an exhaust emission test laboratory used for the mandatory exhaust emission tests following the also prescribed testing procedures described in Chap. 9.

A vehicle is placed on a chassis dynamometer behind a cooling fan. The test is performed with opened engine hood. The driver is watching a driving cycle

Fig 8.1: Phantom drawing of an exhaust emission laboratory

displayed on a screen (driver's aid) on the right hand side above the vehicle and is driving the vehicle accordingly. On the extreme right the measurement and control units are located in a separated cabin. To the left hand side at the back of the laboratory the dilution tunnel, the pump and the bags are shown. Calibration gas cylinders are placed against the back wall.

8.2 Test Techniques

8.2.1 Overview of the Test Process

The block diagram of Figure 8.2 gives a schematic overview of the important parameters and component parts of an exhaust emission and fuel consumption test.

First of all, the vehicle, the specified test fuel and the devices for simulation of driving on a road have to be mentioned:

- vehicle (tire pressure is checked),
- test fuel,
- driver,
- driving cycle,
- chassis dynamometer including fly wheel and brake.

As ambient parameters

- air humidity
- room temperature
- atmospheric pressure

have to be taken into account.

The analytical sampling, i. e., the technique of obtaining a gas probe includes the

- dilution device with dilution tunnel and CVS-unit[1] also used for measurement of the volume flow rate of the exhaust gas,
- isokinetic sampling of gas probes from the dilution tunnel,
- bags for gas probes.

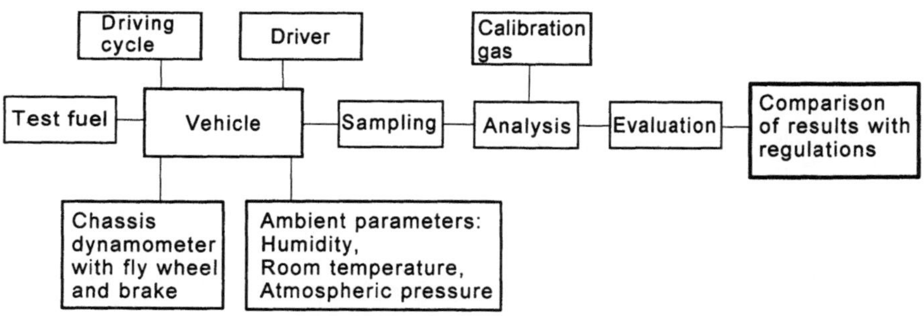

Fig 8.2: Overview of parameters and component parts, schematically

[1] CVS = constant volume sampling, see Chap. 8.2.6

The technique of analyzing exhaust emission gas components requires

- HC measurement,
- CO measurement,
- NO_x measurement,
- CO_2 measurement (necessary for indirect determination of fuel consumption and for checking),
- O_2 measurement,
- calibration of the measuring instruments using calibration gases,
- particle measurement for diesel vehicles.

The evaluation gives the corresponding measurement result of each gas component in mass per distance driven.

The comparison of the results with regulations and an interpretation complete the test. Crucial for the quality of the measurement results is the calibration of the measuring instruments using calibration gases. The fuel consumption is either gravimetrically measured (using a fuel scale) or determined by calculation based on the values of exhaust gas components.

Figure 8.3 shows a scheme of an exhaust emission test facility for gasoline and diesel engines as reguired by law.

The test can be described in detail as follows:

The vehicle is driven on to a chassis dynamometer simulating the road load on the engine, i. e., the total resistance to motion.

The driver watches a so-called driving cycle consisting of phases of acceleration, deceleration and idle and runs the vehicle according to this cycle, remaining within specified tolerances.

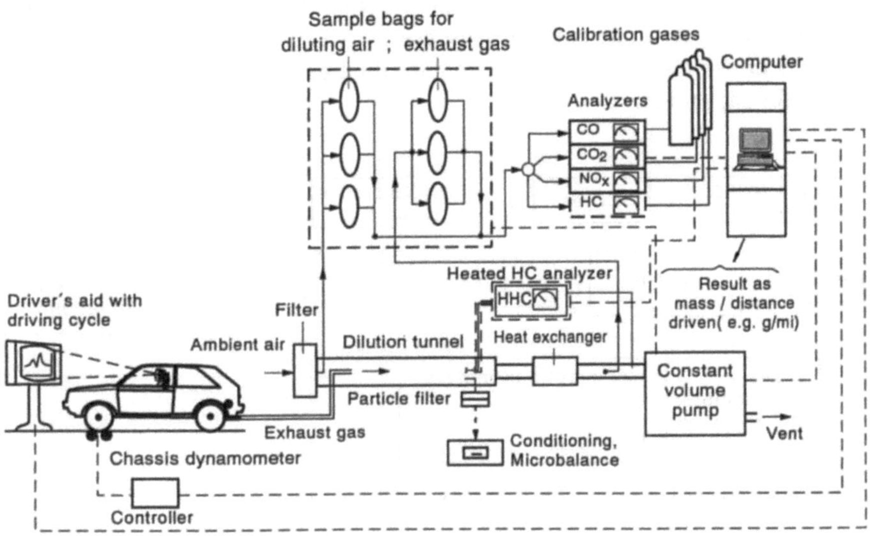

Fig 8.3: Scheme of an exhaust emission test facility, control lines are dashed

The driving cycle is meant to represent the „mean" driving behavior of many drivers running different vehicles in public traffic of different cities. Furthermore, the influence of so-called „cold starts" are to be simulated, assuming that the vehicles are started with cold engines after a longer period of non-use. The statistically estimated time for the engine to warm up to the operational temperature has to be taken into account.

The escaping exhaust gas is diluted by air. At known volumetric delivery of the mixture of air plus exhaust gas (dilution rate ≈ 1 : 8) the total volume flow rate is thus determined. Part of the exhaust gas is used to fill three bags, one after each of the three sections of the US-75 driving cycle. According to the European driving cycle, the exhaust gas is filled into one bag. Furthermore, part of the air used for diluting the exhaust gas is also filled into three further bags (background).

The concentrations of the corresponding components are measured with the analyzers described in Chap. 6. The concentration values of the components of the air probe (background) are subtracted from the corresponding concentration values of the exhaust gas components. Multiplication of the resulting concentration differences with the flow rates integrated over the same driven distance establishes the respective masses per distance driven, e. g. g/km or g/mi.

For diesel vehicles and HC-measurement, the diluted exhaust gas probe is lead though a heated line to a heated FID. The particles are measured according to a procedure described in Chap. 7.6.

The total US-test is called Federal Test Procedure (FTP) or US-75 test. The test valid for Europe is called here the European test.

Figure 8.4 schematically shows an additional so-called modal bench unit enabling time resolved measurements of concentrations of the regulated exhaust

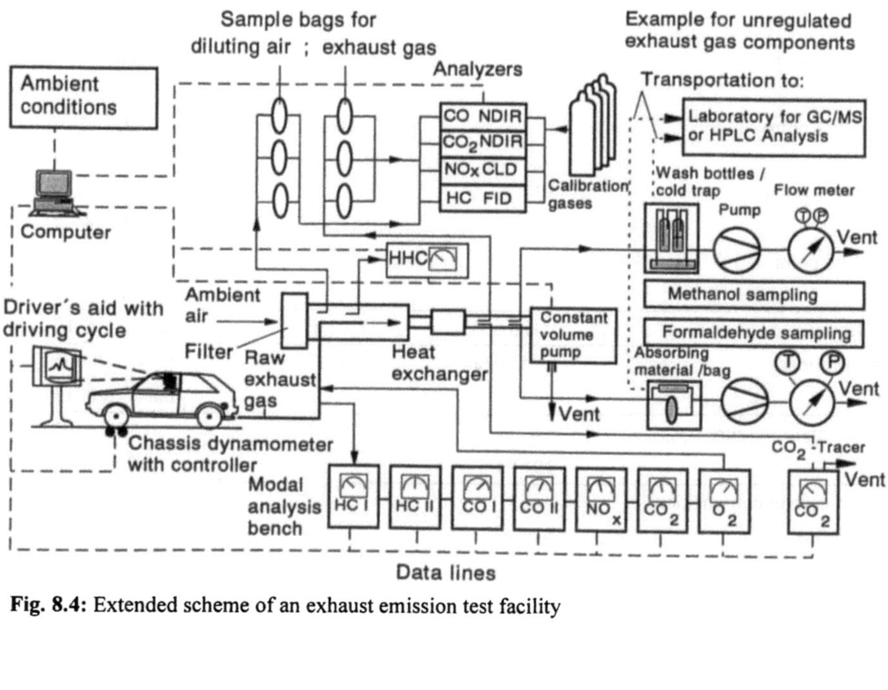

Fig. 8.4: Extended scheme of an exhaust emission test facility

gas components by sampling from the raw exhaust gas (undiluted exhaust gas). The volume flow rate is here determined by a so-called tracer procedure comparing the CO_2-concentrations of the raw exhaust gas with those of the diluted exhaust gas. For CO two analyzers, CO I and CO II, are necessary for different measuring ranges. Although such a time resolved measurement of the mass emissions of regulated exhaust gas components is not regulated, it is important for optimizing reduction of exhaust emissions. Furthermore, in Figure 8.4 the measurement of unregulated exhaust gas components is indicated, see Chap. 7.

8.2.2 Specified Test Fuels

Specified test fuels must have characteristic parameters as shown in Table 8.1 as an example for the US-75 test.

Table 8.1: Specified characteristic parameters of Otto-cycle test fuel (unleaded) for the US-75 test (1990)

Item	Unit	ASTM test method No.	Value
Octane, research		D2699	min. 93
Sensitivity			min. 7.5
Lead (organic)	g / US gal	D3237	max. 0.050
	g / l		max. 0.013
Distillation range:			
Initial boiling point	°F		75 - 95
	°C		23.9 - 35
10 % point	°F		120 - 135
	°C		48.9 - 57.2
50 % point	°F	D86	200 - 230
	°C		93.3 - 110
90 % point	°F		300 - 325
	°C		148.9 - 162.8
End point	°F		max. 415
	°C		max. 212.8
Sulfur	weight-%	D1266	max. 0.10
Phosphorus	g / US gal	D3231	max. 0.005
	g / l		max. 0.0013
Reid vapor pressure	psi	D3231	8.7 - 9.2
	10^3 Pa		60.0 - 63.4
Hydrocarbon composition:			
Olefins	Vol.-%		max. 10
Aromatics	Vol.-%	D1319	max. 35
Saturates			Remainder

8.2.3 Driving Cycle

8.2.3.1 Remarks

Exhaust emission tests are supposed to furnish quantitative data concerning the exhaust emissions and fuel economy to be expected from a vehicle in operation on the road without the necessity of having to perform measurements during an actual drive on the road. For this reason, all exhaust emission test procedures are based on the principle of simulation, i.e., an effort to imitate operating conditions on the road on a dynamometer. This is based on the assumption that the emissions or the fuel consumption will be the same on a dynamometer and on the road, provided that the sequence and mean magnitude of forces and speeds acting on the vehicle are the same as well.

Under these conditions, exhaust emission and fuel economy tests necessitate the use of driving cycles whose speeds and acceleration phases come as close as possible to duplicating the mean of actual driving conditions on the road. Driving cycles are therefore curves of speed via time with specified tolerances.

There are two distinct categories depending on their development history:

a) driving cycles derived from records made of actual drives on the road, such as those of the US-75 test (Figure 8.5)
b) cycles which have been synthetically constructed from a number of constant acceleration and constant speed phases also based on records of actual driving on the road. Most of the driving cycles considered here belong to the latter category, e.g., the European driving cycle with the high speed phase (Figure 8.6).

The US driving cycle consists of three phases, as shown in Figure 8.5.

To determine the exhaust emissions according to the US-75 test or to the European test, the vehicles are tested in a „cold start" after conditioning them in a room at a temperature of approximately 20 °C for 12 h (for the US test) and 6 h

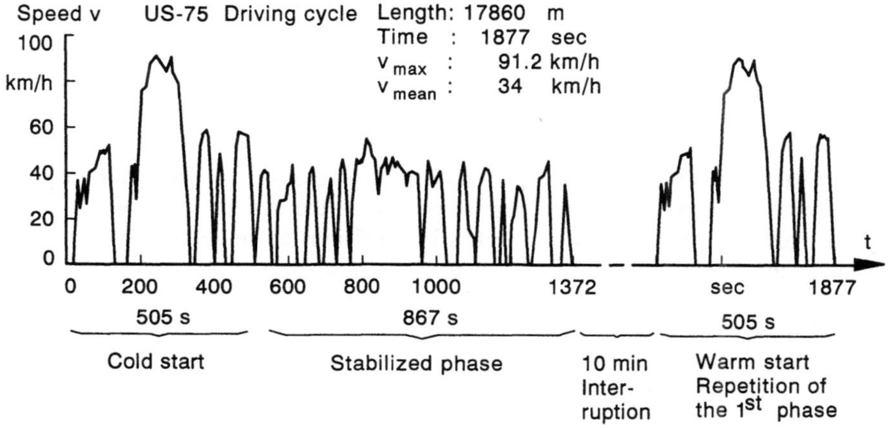

Fig 8.5: Driving cycle of the US-75 test; speed via time t

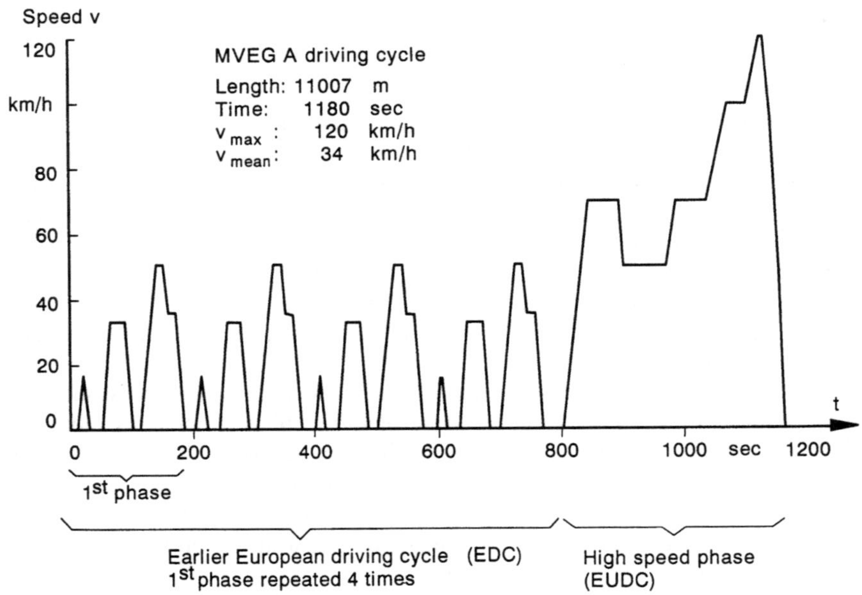

Fig 8.6: New European driving cycle (MVEG: Motor Vehicle Emission Group)

(for the European test). To determine the fuel consumption, no conditioning is required and the vehicles are started with a hot engine (warm start).

The new European driving cycle valid for the European Union (EU) from 1993 on includes as a new phase, a high speed phase to enable simulation of drives on extra-urban roads and highways. This high speed phase has been added to the former European driving cycle which consists of 4 identical phases.

For the US-75 test, unlike the new European driving cycle, a separate driving cycle is regulated for the high speed phases, the so called Highway Driving Cycle (HDC). Its curve is shown in Figure 8.7. Again, a warm start is used. In Figure 8.7 details of the test procedure are listed.

8.2.3.2 Development of Driving Cycles

Let us first discuss the development of a driving cycle based on a selection of parts of records of actual drives. The first step towards the development of a driving cycle is always the definition of the traffic conditions to be simulated. Thus, for instance, the earlierUS-72 cycle (first and second phases of the curve in Figure 8.5) is specifically designed to duplicate the traffic conditions prevailing during the morning rush hour in Los Angeles.

The US-75 driving cycle differs from the US-72 by an additional "hot" phase which is identical to the phase of the first 505s. The hot engine is restarted after a stop of 10 minutes and the exhaust emissions are measured during these further 505s (see Figure 8.5) so as to include the warm start behavior.

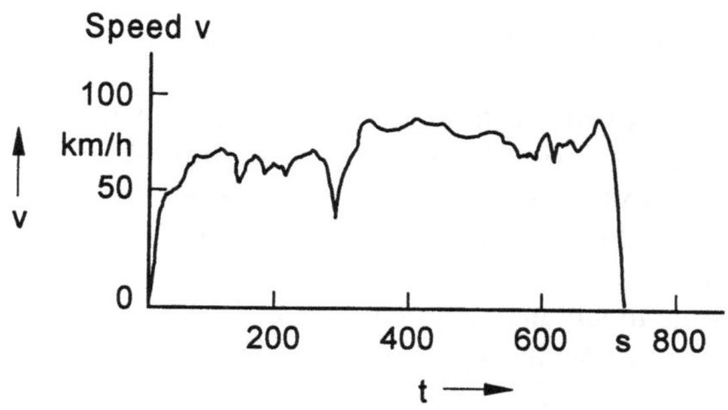

Highway driving cycle (HDC)

Length: 16.5 km (10.22 mi)
Time: 765 s
v_{max} : 96 km/h
v_{mean} : 77 km/h

Conditions:

1. First cycle for engine warm up
2. Second cycle for measurement
3. Continuous sampling into 1 bag
4. Based on an actual stretch of US highway

Fig. 8.7: US Highway driving cycle (HDC) (high speed driving cycle)

The selection of the traffic conditions to be simulated or the corresponding roads has a decisive influence on the exhaust emissions measured in an exhaust emission test. What conditions are actually to be simulated should depend on the impact on air quality caused by exhaust emissions on certain types of road as well as the number of persons affected by it. Thus, for instance, city centers as well as intra-urban streets might be selected for a driving cycle. Once a route has been selected, driving tests are run on this route in order to get recordings of driving speeds as a function of time. These test drives are made with vehicles of various classes as well as by drivers of various driving behavior.

From the multitude of driving cycles thus recorded, a single curve is selected which is as representative as possible. This is done by means of visual assessment of speed curves as well as by using a number of quantitative criteria summarily characteristic of the driving cycle in question. Among the criteria most frequently used are average speed and the proportions of time spent idling, accelerating, decelerating and at a constant speed. The curve thus selected is then modified as required, for instance to accommodate certain rates of acceleration and deceleration which represent the performance limits of the dynamometer used. On the whole, the US-75 driving cycle was arrived at in this fashion.

As an alternative, it is possible to draw up a "synthetic" polygonal driving cycle. This is done by subdividing the driving curves recorded into a number of driving modes according to one of several different procedures, and establishing their frequency and in some cases, even their typical sequence. Some of these individual driving modes are then selected according to their frequency and combined into a representative driving cycle. Although much computing is required for this procedure, it is mainly independent of the length of the test track, and the resultant driving cycle can be varied within wide limits, yet simplification and schematization should not be taken too far. The European driving cycle was developed along these lines.

Moreover, the two methods - selecting and constructing - may be combined.

Constructing driving cycles for the categories described from records of values measured during drives in traffic allows theoretically infinite possibilities of selections and thus of the patterns of driving cycles; consider, for instance, the sequence of selected driving cycle phases.

Furthermore, it is nearly impossible to simulate extreme traffic conditions, such as traffic jams, by means of only one driving cycle because only statistically determined mean values of driving behavior, data from drivers, traffic flow, cities, roads and road inclinations can be used.

8.2.3.3 Assessment and Comparison Criteria

As a mere representation of speed as a function of time is insufficient to assess and compare driving cycles, other criteria are required for this purpose. A number of assessment criteria are used which are listed in the following Table 8.2. For some criteria, lower limits have to be defined according to the accuracy of the data measured.

Table 8.2: Assessment criteria for comparing driving cycles

No.	Criterion	Abbre-viation	Unit
1	Average speed of the entire driving cycle	$\overline{v}_1$	km/h
2	Average speed during actual driving (idle time excluded)	$\overline{v}_2$	km/h
3	Average acceleration of all acceleration phases ($\dot{v} > 0{,}1$ m/s^2)	$\overline{\dot{v}}_+$	m/s^2
4	Average deceleration of all deceleration phases ($\dot{v} < -0{,}1$ m/s^2)	$\overline{\dot{v}}_-$	m/s^2
5	Mean duration of a driving period (from start to standstill)	τ	s
6	Average number of acceleration-deceleration changes (and vice versa) within one driving period	M	
7	Proportion of idle time ($v \leq 3$ km/h, $\lvert\dot{v}\rvert \leq 0{,}1$ m/s^2)	S	(%)
8	Proportion of acceleration time ($\dot{v} > 0{,}1$ m/s^2)	B	(%)
9	Time proportion at constant speed ($\lvert\dot{v}\rvert \leq 0{,}1$ m/s^2)	K	(%)
10	Proportion of deceleration time ($\dot{v} < -0{,}1$ m/s^2)	V	(%)

The text in this table includes the corresponding definitions. The most important criterion here is the average speed $\overline{v}_2$ with the idle time excluded.

Applying the assessment criteria to the known driving cycles developed for city traffic and shown in Figure 8.10 , one arrives at the figures shown in Table 8.3.

From Table 8.3 it is apparent that different driving cycles are prescribed in different countries or areas. Worldwide the US-75 driving cycle is predominant.

Comparing the European driving cycle without the high speed phase (EDC) with the Japanese 10-Mode driving cycle or with the other driving cycles, some criteria

Table 8.3: Assessment criteria figures of some well known driving cycles, refer to Fig. 8.10

	Driving cycle						
Criterion	EDC	Former Californian	Former US-72	US-75	Japan 10-Mode	Japan 11-Mode	Mean
$\bar{v}_1$	18.8	35.6	31.5	34.1	17.7	28.9	**27.8**
$\bar{v}_2$	26.4	41.7	38.1	41.2	23.7	38.4	**34.9**
$\bar{\dot{v}}_+$	0.64	0.64	0.58	0.59	0.63	0.52	**0.60**
$\bar{\dot{v}}_-$	0.71	0.65	0.69	0.70	0.62	0.58	**0.66**
τ	46.3	117	63.0	67.4	50.3	95.0	**73.2**
M	1.0	3.0	3.9	4.2	2.0	5.0	**3.2**
S	28.7	14.6	17.3	17.4	25.4	24.8	**21.4**
B	21.5	32.8	34.0	33.7	25.9	33.3	**30.2**
K	30.3	21.2	20.0	20.5	22.2	11.9	**21.0**
V	19.5	31.4	28.6	28.5	26.4	30.1	**27.4**

- such as average speed, length of the driving cycle, reversal of load or proportion of idle time - show distinctly different values.

For the first rough assessment of a driving cycle it is best to use its average speed $\bar{v}_1$. The driving cycles of the former Californian test as well as those of the US-72 and US-75 tests are „fast" (higher $\bar{v}_1$), whereas those of the European and the Japanese 10-Mode tests are decidedly „slow". Their lower average speed is not merely due to generally slower driving, which of course depresses the mean speed $\bar{v}_2$ of a driving phase, but also to prolonged idle periods S. The European cycle is the one in which both idle and constant speed periods, S and K, are by far the most frequent.

Suitable graphical illustrations to demonstrate the differences between the driving cycles are (relative) frequency distributions of the speed and accelerations. Figure 8.8 shows as an example the frequency distributions of speed and acceleration of driving cycles in a threedimensional graph. The relative frequency is given by closed area, i.e., by areas given in percentage. These areas define the percentage of time where acceleration or speed do not exceed certain values which are given here in rough steps of 0-50 %, 50-75 %, and 75-90 %.

Graphs like this one illustrate the differences between cycles derived from actual test runs, such as the US-72 and US-75 cycles, and constructed driving cycles, such as the European, the Californian, the Japan 10-Mode and Japan 11-Mode. Whereas the structures of the frequency distributions derived from the US-72 and US-75 cycles are wide rounded areas, those of the „synthetic" cycles consist of lines and dots which have acquired a certain spread only because a smoothing program was used in drawing up the graph.

Furthermore, it is noticeable that the accelerations and decelerations occurring in the US-72 and US-75 driving cycles at speeds up to 30 km/h are higher than those of the other driving cycles. The differences of the *mean* acceleration and deceleration are, however, small.

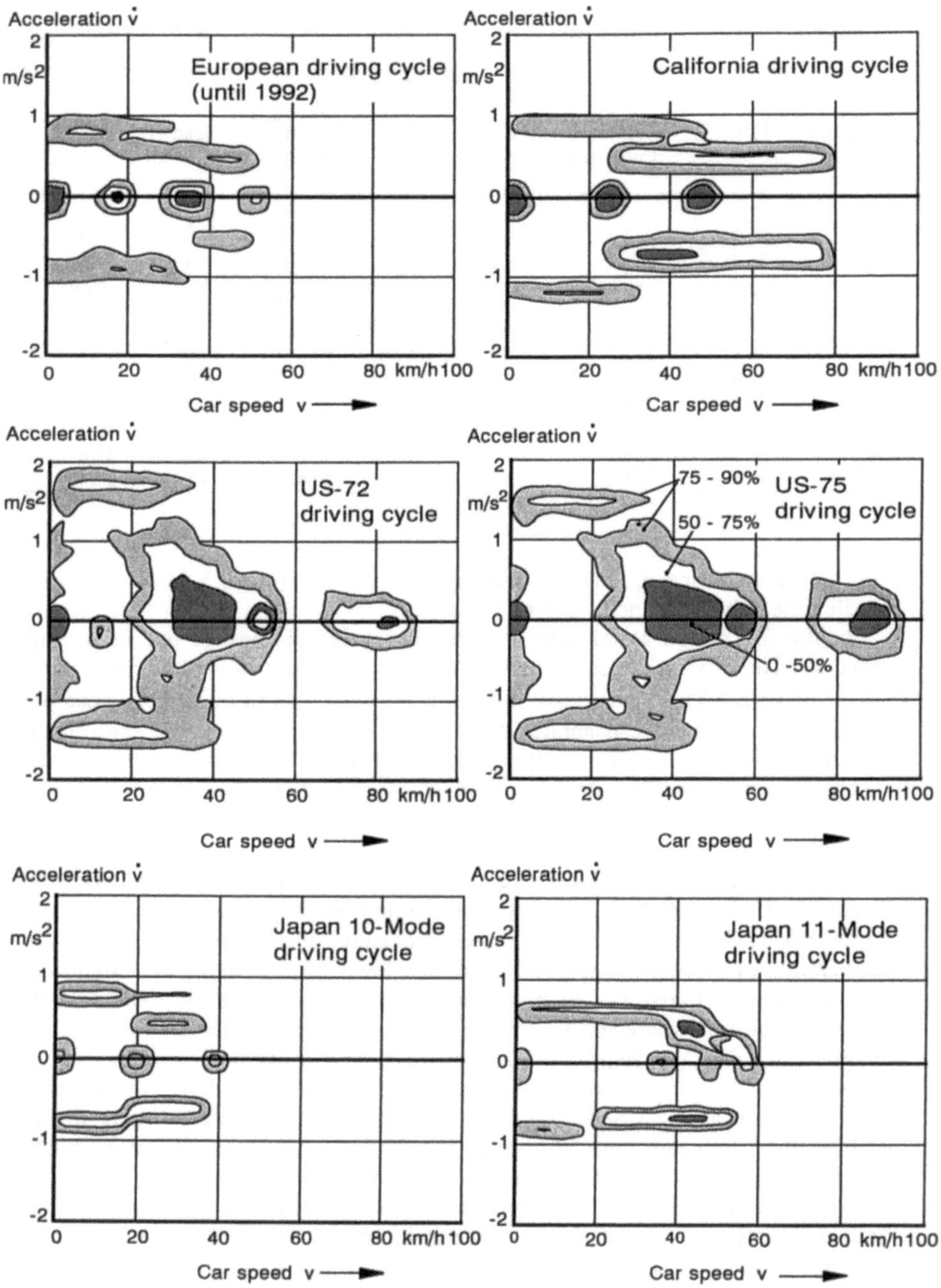

Fig. 8.8: Relative frequency distributions of speed and acceleration of different driving cycles in steps of
0 - 50 %, 50 - 75 % and 75 - 90 %

8.2.3.4 Results of Test Runs

The tests evaluated here were run by several European automobile manufacturers in a number of European cities as a part of a traffic noise study. For this purpose, it was necessary to analyze typical urban traffic. The test routes were selected by the manufacturers without prior consultation, a fact which made basic differences in driving behavior probable.

To compare the results from intra-urban routes to the regulated driving cycles, the criteria listed in Table 8.2 are used. Figure 8.9 shows these criteria for test routes and driving cycles.

If we look first of all at the average speeds inclusive of idle periods ($\overline{v}_1$, Figure 8.9 top) we find that the driving cycles used in the US-72 and in the Japanese 11-Mode tests correspond most closely to the overall mean value MV derived from the test runs. On the other hand, the driving cycles of the European and the Japanese 10-Mode tests, which are far below this mean, agree rather well with the city center test route 3. Differences occurring in the average speeds $\overline{v}_1$ of individual test routes and driving cycles are about equally due to variations in the average speeds not inclusive of idle periods $\overline{v}_2$ and to the variations in the relative proportions of idle time S.

What is to be said about $\overline{v}_1$ corresponds nearly exactly to what can be said about $\overline{v}_2$.

If we then proceed to consider the relative proportion of time spent at idle S, both the European and the Japanese 10-Mode driving cycles deviate most markedly as far as the mean time at idle is concerned.

The duration of acceleration and deceleration is more or less the same on all test routes. The same applies to all mandatory driving cycles, the sole exceptions being the US-72 and US-75 driving cycles, where a slightly longer time is spent accelerating than decelerating.

On the other hand, the constant speed periods of the mandatory driving cycles are markedly longer than those of the test routes, with the sole exception of the Japan 11-Mode driving cycle. Compared to this, the proportion of the European driving cycle which is spent at constant speed is extremely high.

As far as the average duration of a driving segment (τ) is concerned, the US-75 driving cycle corresponds quite well with the mean MV drawn from the various test routes. Both the European and the Japan 10-Mode tests are decidedly below the mean MV but still higher than the average duration of a driving phase on the city center test route 3.

Concerning M (number of acceleration-deceleration changes), driving cycles and test routes averages do not correspond well. Not a single driving cycle does so much as approximate the mean MV derived from the test runs; the Japan 11-Mode driving cycle comes closest, whereas the European cycles is farthest removed. As M expresses irregularity of speed, we may say that all driving cycles are smoother than is warranted by actual driving on the road. Of all criteria hitherto described, M was considered to be the least significant.

The mean accelerations of the driving cycles $\overline{\dot{v}}_+$ are slightly lower than those drawn from the test runs.

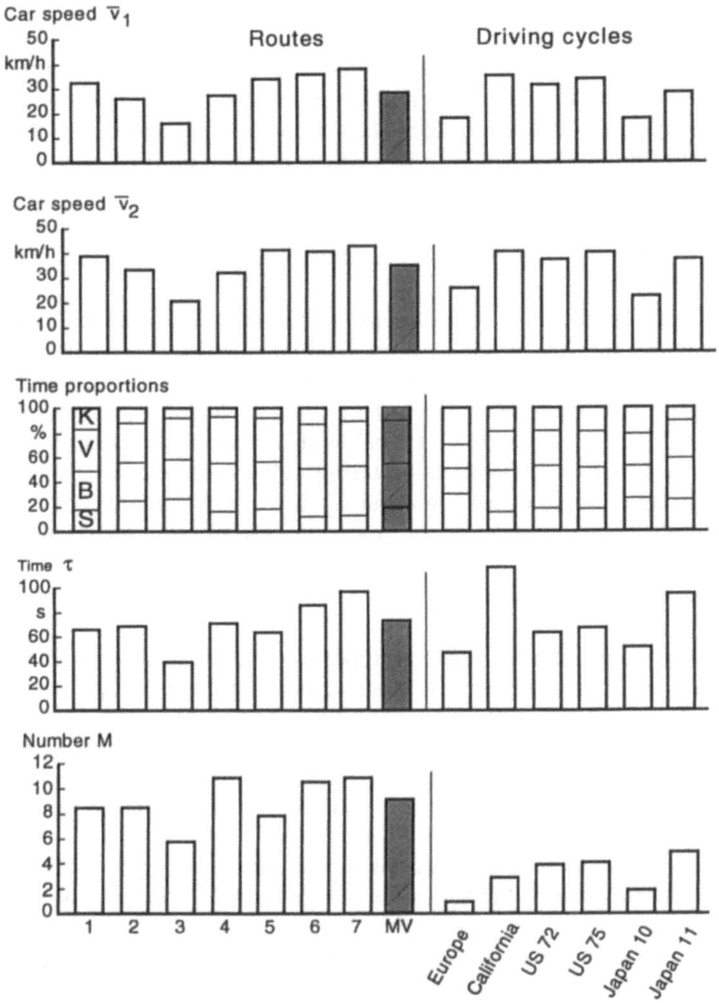

Fig. 8.9: Assessment of test routes 1 through 7 and of the mandatory driving cycles according to criteria of Table 8.2. MV are the corrected overall mean values of the respective criteria for the routes

Summing up, we may say that as far as nearly all the criteria under consideration are concerned, the US-72 and the Japan 11-Mode driving cycles approximate closest the mean values MV drawn from the test runs. In contrast, the European driving cycles deviate widely from the mean values MV of nearly all significant criteria. Its closest resemblance is route 3, which represents dense traffic in an old city center, and this is what it was supposed to simulate originally. Yet this set of traffic conditions is becoming increasingly insignificant as streets in city centers are turned to pedestrian zones or the traffic flow is improved.

As a general statement, it can be concluded that the average traffic prevailing in big European cities is reflected quite well by the US-72 driving cycle. The US-75 is a good approximation whereas the former European driving cycle (without the high speed phase) is inadequate for this purpose, as it reflects former traffic conditions in the inner parts of European cities.

Since the Japan 11-Mode driving cycle was developed to reflect average traffic conditions prevailing in big Japanese cities, the US driving cycles were developed for conditions in big US cities and since both reflect the situation in big European cities as well, we may conclude that the US driving cycles may be used worldwide for all big cities, whereas the former European driving cycle is adequate for the inner cities.

Different driving cycles for big cities are not necessary from the scientific point of view. They were regulated for political reasons alone and cause unnecessary expense for the automobile manufacturer operating worldwide.

8.2.3.5 Proposal for a New Worldwide Standard Driving Cycle

There are many proposals to standardize the driving cycles for worldwide use. Figure 8.10 shows a new driving cycle developed by VW for worldwide use (bottom curve) compared to the known driving cycles.

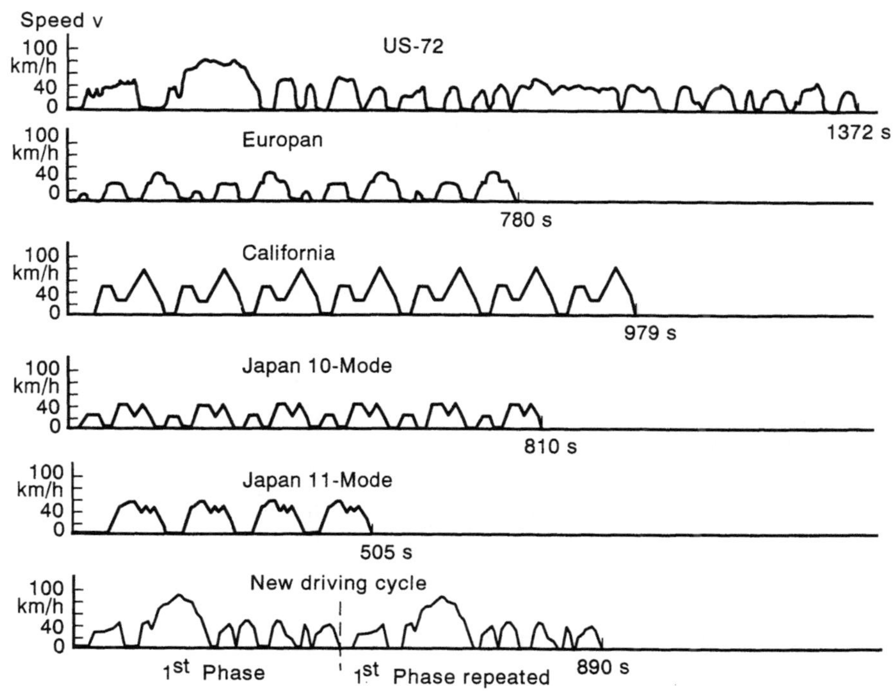

Fig 8.10: Newly proposed driving cycle (bottom) compared to known driving cycles

This new driving cycle was developed on the basis of the US-72 cycle and consists of two identical sections. Its development procedure is described in literature.

Unfortunately, for political reasons it is nearly impossible to introduce such a proposal worldwide. Discussions on this proposal in the UN did not lead to a consensus of opinion among the representatives of all countries.

8.2.4 Driver

The second factor discussed here influencing the exhaust emission test results is the driver, i.e., the human being as a part of the closed loop circuit between driving cycle and vehicle. This closed loop is shown schematically in Figure 8.11.

The human driver is a significant factor in emission tests. During the dynamometer run the driver observes the running display of a driver's aid (computer printout of the nominal (mandatory) driving cycle) and compares it with the actual driving cycle that he produces. Close tolerances have to be observed, otherwise the test is terminated. The vehicle should be operated with the least amount of gas pedal movement possible.

The tolerances for speed and time are geometric means. For the former European driving cycle ±1 km/h and ±0.5 s and for the US test, ±2 mph (3.2 km) and ±1 s are allowed.

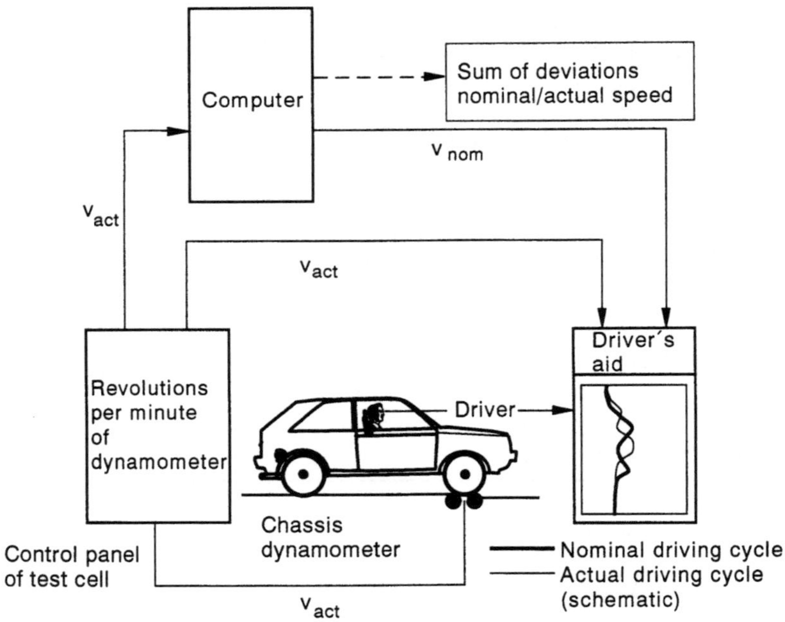

Fig. 8.11: Schematic of a closed loop circuit: driving cycle-driver-vehicle

Mechanized Driver (Automatic Driver)
Efforts are being made to introduce an automatic driver. Individual vehicles with automatic transmission have been used with automatic drivers earlier for round robin tests (refer to Chap. 9). This is also true of vehicles which are tested for durability and driven on a dynamometer according to other specifications than the driving cycles.

For mandatory exhaust emission tests an automatic driver must function properly even for manual transmission. The most important requirement is that the installation and removal of the automatic driver should take only a few minutes at the most.

Figure 8.12 shows an automatic driver developed together with Volkswagen installed in a vehicle. The system is mechanically attached to the seat.

It has three operating elements for the clutch, brake and gas pedals. In addition, there is an adjustable arm which can move on all three axes, enabling the shifting of a manual transmission. Each of the six axes of this mechanical system are driven by a direct current servo motor with an integrated tachometer and a coupled angle coder. To transfer the rotation of the motors into the linear movement required, suitable step-down gearing is carried out via toothed-belt disks using threaded spindles.

To adapt the automatic driver to the space requirements there are several simple individually adjustable mechanisms. The system can be vertically adjusted and shifted laterally, i.e., parallel to the transverse axis. The clutch pedal, gas pedal and z-axis can be shifted vertically and horizontally, thus enabling adjustment to the geometry of the vehicle's pedals.

To recognize the reference point for the basic adjustment, each axis is equipped with a magnetic proximity switch. In addition, the clutch and gas pedals and the

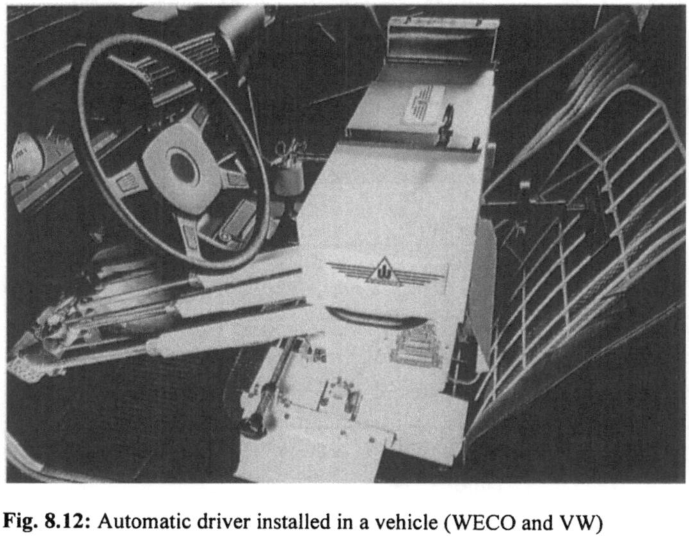

Fig. 8.12: Automatic driver installed in a vehicle (WECO and VW)

z-axis are also equipped with a proximity switch which acts only when a defined elastic force is reached. For the brake pedal a force sensor placed at the „foot" provides information on the contact point and the maximum pedal position. The reference values are stored in a computer.

For checking the operation and for necessary „learn" phase, a hand driven control unit is available. The electronic unit and the electronic control technique and software will not be discussed here.

Test Results

To be able to pass judgment on the automatic driver with regard to its usefulness for exhaust emission testing and for measuring fuel consumption, a number of tests were performed according to prescribed procedures (European, US-75) with different types of vehicles (compact to mid size) and various engines (carburetor, fuel injection, turbo charged, diesel and turbo diesel). Installation and removal of the automatic driver was performed repeatedly without any problems in approximately 7 minutes.

Figure 8.13 gives an impression of the actual speed deviations v_{act} from the set values v_{nom}.

In the lower part of this figure the speed versus time for a part of the US-75 driving cycle is shown. The upper part of the figure shows the regulated tolerance ranges and the corresponding recorded errors of an automatic driver and a human

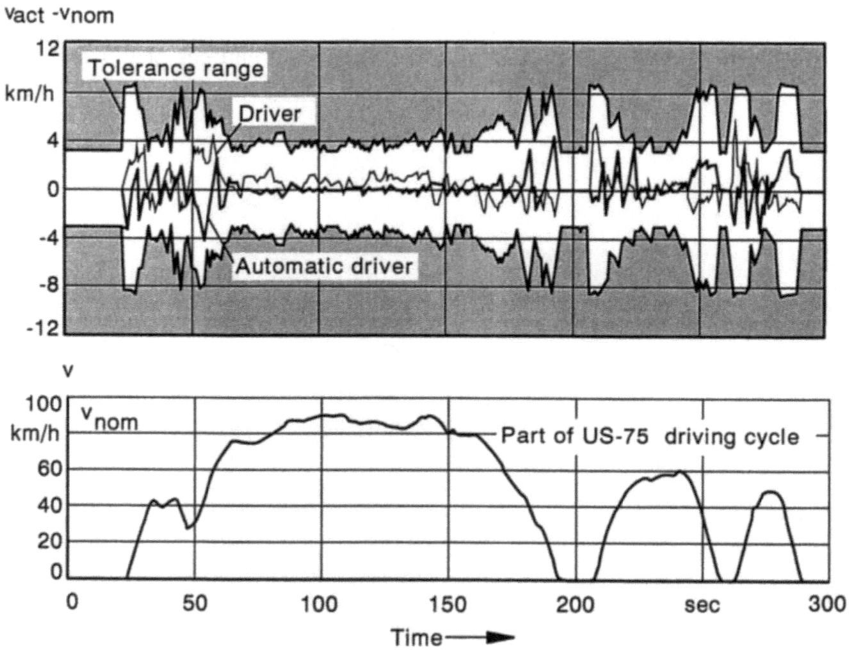

Fig. 8.13: Speed deviations from the set value for an automatic driver and a human driver

driver. For both the automatic driver and the human driver the actual speed stays within the tolerance range. For the areas of constant speed the deviation is smaller for an automatic driver than for the human driver. Only in areas of large speed changes are the deviations between set and actual values smaller for a human driver.

Figures 8.14 to 8.17 show as examples results of exhaust emission tests and fuel consumption measurements performed with an automatic driver in comparison to the results of several human drivers. Number of repeated measurements n = 10.

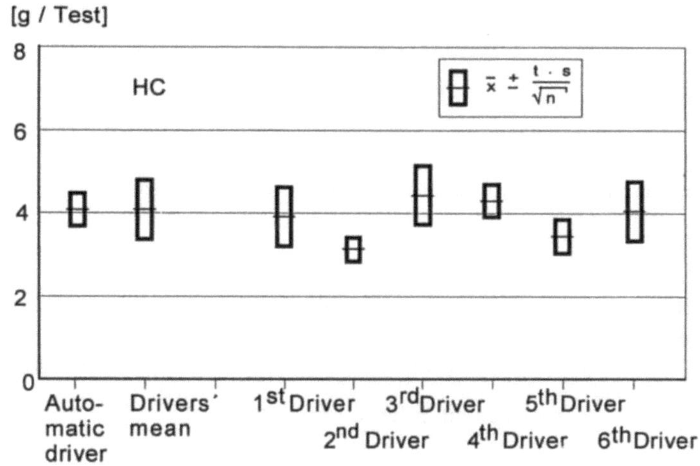

Fig. 8.14: Comparison of automatic driver to human drivers for HC (European test)

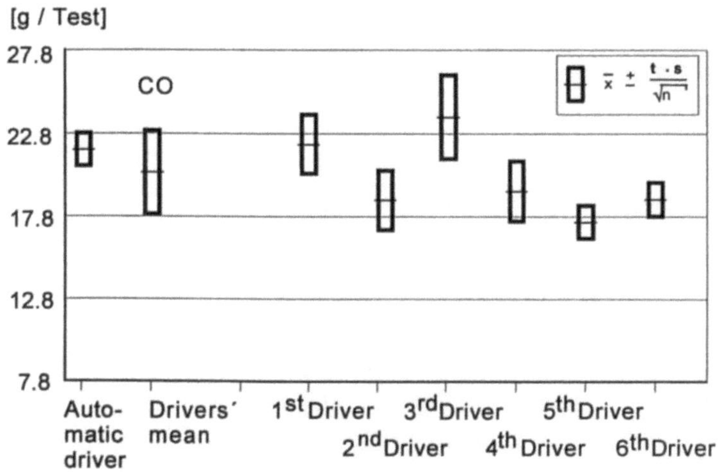

Fig. 8.15: Comparison of automatic driver to human drivers for CO (European test)

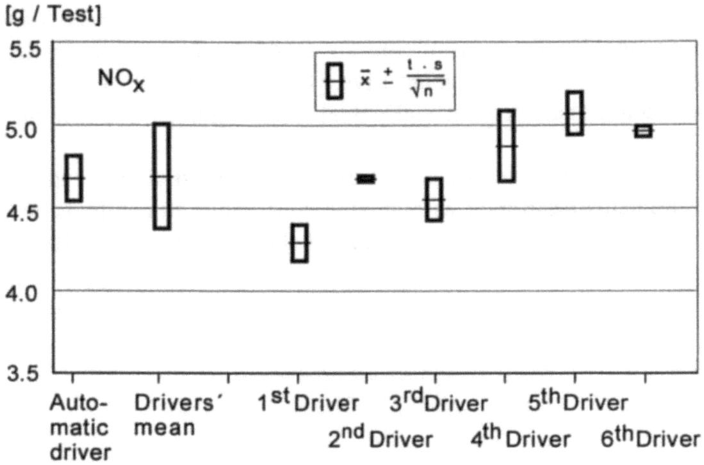

Fig. 8.16: Comparison of automatic driver to human drivers for NO_X (European test)

The results of the human drivers vary considerably depending on the condition of the drivers. The mean value $\bar{x}$ of the results of several drivers corresponds sufficiently to the mean values of the results of an automatic driver when taking the confidence ranges $\pm \frac{ts}{\sqrt{n}}$ into account, refer to Chap. 9.

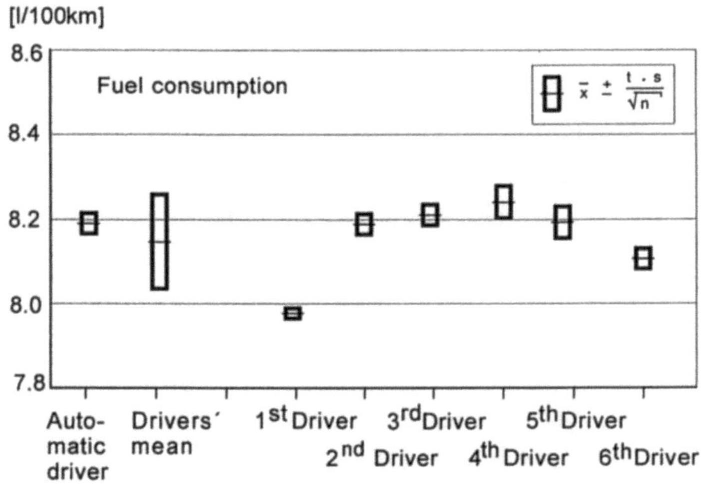

Fig. 8.17: Comparison of automatic driver to human drivers for fuel consumption (European test)

8.2.5 Chassis Dynamometer

8.2.5.1 Remarks

An important factor of the test is the chassis dynamometer with a suitable brake, e.g. a water brake, an eddy current brake or a direct current brake. To produce driving conditions similar to those on the road on a chassis dynamometer, the inertial weights (weight of the vehicle and weight of the vehicle's rotating parts) are simulated by coupled gyrating masses (fly wheels) and the total resistance to motion is simulated by a defined braking of the dynamometer with a suitable brake. Figure 8.18 shows the principal configuration of such a dynamometer.

This figure shows two rolls on the left with the live axle with two drive wheels above them. On the right the fly wheels and the brake are outlined.

The established chassis dynamometers are equipped with digital control circuits, a controlled DC electrical device as a brake, adjustable mechanical fly wheels, and electronic simulations of small gyrations between the fly wheels.

In the following, the equations for the torques are derived as

$$M(t)_{Road} \quad \text{and} \quad M(t)_{Dyno}$$

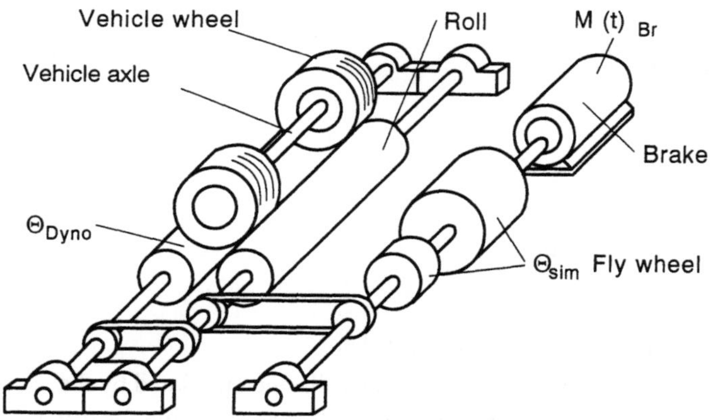

Fig. 8.18: Principal configuration of a chassis dynamometer. M = torque, Θ = inertial moment, m = mass

8.2.5.2 Resistive Forces on Vehicles on the Chassis Dynamometer

Driving a vehicle on a dynamometer results in a total resisting force of (refer to Figure 8.18):

$$F(t)_{Dyno} = \frac{M(t)_{Dyno}}{r} = \frac{M(t)_{Br} + \Theta \dfrac{d\omega(t)}{dt}}{r} \tag{8.1}$$

with

$F(t)_{Dyno}$ tractive force on the drive wheels of the vehicle on the dynamometer,

$M(t)_{Dyno}$ corresponding torque on the drive wheels,

$M(t)_{Br}$ resistive moment of the dynamometer brake,

Θ total moment of inertia of the chassis dynamometer,

$\frac{d\omega(t)}{dt}$ angular acceleration,

r dynamic rolling radius of the vehicle drive wheels.

The quantity r is practically constant within the speed range of the driving cycles, refer to Chap. 8.2.3. This was proven with an error of less than 1 % by results of measurements of r up to a speed of 100 km/h. Therefore, instead of the force $F(t)_{Dyno}$ the torque $M(t)_{Dyno}$ alone needs to be considered. Thus (8.1) becomes simplified to:

$$M(t)_{Dyno} = M(t)_{Br} + \Theta \frac{d\omega(t)}{dt} \tag{8.2}$$

or, because of

$$\Theta \frac{d\,\omega(t)}{dt} = \left(\Theta_{Dyno} + k\,\Theta_{sim}\right) \frac{d\,\omega(t)}{dt} \tag{8.3}$$

with

Θ_{Dyno} inertial moment of the dynamometer rolls,

Θ_{sim} additional inertial moment acting via the rolls and simulated by the flywheels or simulated electronically (which is not possible for water brakes)

k factor to take into account a possible transmission between the roll axle and the fly wheel axle (in most cases $k = 1$)

and for $k = 1$ to:

$$M(t)_{Dyno} = M(t)_{Br} + \left(\Theta_{Dyno} + \Theta_{sim}\right) \frac{d\,\omega(t)}{d\,t} = M(t)_{Br} + \frac{\Theta_{Dyno} + \Theta_{sim}}{r_{Dyno}} \frac{\mathrm{dv}}{\mathrm{dt}}\,, \tag{8.4}$$

with

r_{Dyno}	Radius of the rolls
v	tangential speed of wheels in direction of F_R and in the contact point (area) of the wheels on the rolls

8.2.5.3 Resistive Forces for Vehicles driven on the Road

The resistive moment when driving the vehicle on the road is (with r being practically constant again) (refer to Figure 8.19):

$$F(t)_{Road}\, r = M(t)_{Road} = r\,[R_{Ro}(t) + R_G(t) + R_A(t) + R_I(t)] \tag{8.5}$$

with

$F(t)_{Road}$	tractive force on the drive wheels on the road
$M(t)_{Road}$	corresponding torque on the drive wheels on the road
R_{Ro}	rolling resistance road/wheels,
R_G	gradient resistance,
R_A	aerodynamic drag,
R_I	inertial force of rotating masses of the vehicle parts (axles, wheels, crankshaft, etc.),

or

$$M(t)_{Road} = r\left[F_{Ro}\, mg + mg\sin\alpha + \frac{\rho}{2} c_w A \mathrm{v}^2 + \lambda m^* \frac{d\mathrm{v}(t)}{dt}\right] \tag{8.6}$$

with

F_{Ro}	coefficient of rolling resistance, which is approx. constant in the speed range of driving cycles but is in general: $F_{Ro} = F_{Ro}(\mathrm{v})$,
m	vehicle mass,
g	gravitational acceleration,

Fig. 8.19: Total resistive force F_{Road} acting on a driven vehicle and air drag R_A

α gradient angle of the road,
ρ standardized air density,
c_W drag coefficient,
A cross sectional area of vehicle,
v tangential wheel speed in direction of F_R in the contact area of the wheels on the road,
λ vehicle parameter to take into account the rotating masses in the vehicle (e.g. axles, crankshaft),
m^* effective vehicle mass.

8.2.5.4 Procedure for Calibrating the Test Stand

Decisive for the simulation of driving a vehicle on a road by using a dynamometer is that the dynamometer can be adjusted so that the resistive forces on the vehicle driven on a road or a dynamometer are equal. This means that for each instant the difference between the torques dyno - road is equal to zero, hence:

$$\Delta M(t) = M(t)_{Road} - M(t)_{Dyno} \overset{!}{=} 0 \tag{8.7}$$

In this case the operation of a vehicle on the road is simulated totally, i.e., in complete accordance with the laws of physics.

A presupposition is however that the dynamometer has a suitable construction. Suitable is a dynamometer which has flywheels and a brake, both of which can be precisely adjusted.

For the presupposition that driving on a road can be replaced by driving on a dynamometer, namely $\Delta M(t) = 0$, we derive from the equations for the moments on the dynamometer (8.4) and on the road (8.6):

$$\begin{aligned}\Delta M(t) = M(t)_{Br} + \frac{\Theta_{Dyno} + \Theta_{sim}}{r_{Dyno}} \frac{dv}{dt} \\ - r\left[F_{Ro}\, m g + m g \sin\alpha + \frac{\rho}{2} c_w A v^2 + \lambda m^* \frac{dv(t)}{dt}\right] = 0\end{aligned} \tag{8.8}$$

To perform realistic exhaust emission tests and fuel consumption measurements on test stands with a chassis dynamometer, i.e., to achieve a simulation of the resisting forces on the road as accurately as possible, these resistive forces have to be measured first of all. Then the test stand has to be adjusted to them. To do this, the use of a suitable torque measurement system is best. Then the driving torque is measured in a first step as a function of a step-wise constant speed (speed time constant procedure). In a second step the driving torque is measured as an integral over the torque versus time during a defined driving cycle (speed time dependent procedure). This separation into two steps is done for practical reasons.

Speed Time Constant Procedure

The speed time constant procedure, the first step, gives the characteristic torque function, torque as a function of the step-wise constant speed. In these steps, the torque is constant and

$$\frac{dv}{dt}=0, \qquad \frac{d\omega}{dt}=0, \qquad v = \text{const}, \quad \omega = \text{const}. \tag{8.9}$$

Hence, for both moments in (8.2) and (8.6) the terms with differentiation come to zero. Furthermore, the vehicle is driven on a road without a gradient ($\alpha = 0$), i.e. $R_G = 0$.

Thus we get for constant v and ω from the equations (8.5) or (8.4) for M_{Road}, or M_{Dyno}:

$$M_{Road} = r\,(R_{Ro} + R_A) \qquad \text{and} \qquad M_{Dyno} = M_{Br} \tag{8.10}$$

and from the required equality of the moments:

$$M_{Br} = r\,(R_{Ro} + R_A) = r\,(a_0 + a_2\,v^2) \tag{8.11}$$

with $a_0 = R_{Ro}$ $\qquad a_2 = R_A$.

M_{Road} is measured on the two wheels of the drive axle and on a non-gradient road ($R_G = 0$) at constant speeds between 20 km/h and 100 km/h in steps of $\Delta v = 10$ km/h. The torques (M_{20}, M_{30}, ... M_{100}) corresponding to these constant speed levels are obtained. The parameter r is also taken as a constant in this speed range, otherwise, at higher speeds, $r = r(v)$.

From these measured torque values the torque characteristic function is calculated corresponding to (8.11) using the least square method with

$$a_0 = R_{Ro} = F_{Ro}\,m\,g \qquad \text{and} \qquad a_2 = R_A = \frac{\rho}{2} c_w A \tag{8.12}$$

Using the same test vehicle the curve thus determined (partial road load curve) is then reproduced on the dynamometer so that for $v = 20$ km/h to $v = 100$ km/h, $M_{Dyno} = M_{Road}$, as illustrated in Figure 8.20. To this end, the brake load on the dynamometer (brake characteristic in the case of a DC electric device) is changed by changing M_{Br} until it matches the partial road load curve.

The geometrical dimensions of the test stand, such as the roll diameter, etc., are automatically taken into account when using this calibration procedure based on the torque measurement. Thus these dimensions no longer have any influence.

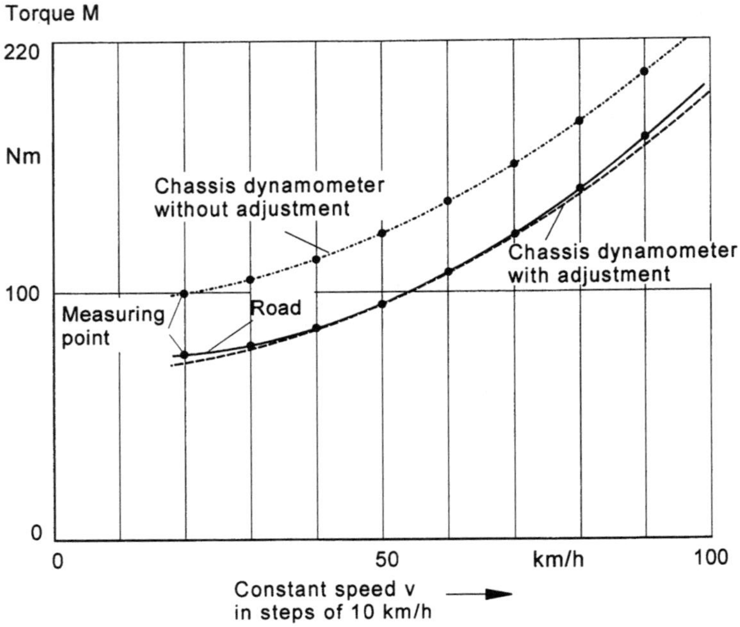

Fig. 8.20: Adjustment of the brake characteristic to the partial load curve. Measurement is performed in steps of 10 km/h, steady-state driving

Speed Time Dependent Procedure

In the time-dependent procedure the acceleration and deceleration effects are included by driving according to a part of a driving cycle, i.e.

$$\frac{d\,\mathrm{v}(t)}{dt} \neq 0 \quad \text{and} \quad \frac{d\,\omega(t)}{dt} \neq 0. \tag{8.13}$$

The torque difference $\Delta M(t)$ between the torque when driving on a non-gradient road and the torque when driving on a dynamometer must again be equal to zero. From (8.1) to (8.8) follows:

$$\begin{aligned} \Delta M(t) &= M(t)_{Road} - M(t)_{Dyno} \\ &= \underbrace{\left[r(R_{Ro}+R_A) - M(t)_{Br}\right]}_{\substack{=0 \\ \text{after adjustment according to} \\ \text{speed time constant procedure}}} + \left[r\lambda m^* \frac{d\,\mathrm{v}}{dt} - \frac{(\Theta_{Dyno}+\Theta_{sim})}{r_{Dyno}}\frac{d\,\mathrm{v}}{dt}\right] \end{aligned} \tag{8.14}$$

The term $r\,(R_{Ro}+R_A)$ is equal to M_{Br}, because the time constant adjustment of the test stand is carried out in the first step.

Then one obtains since the sign of ΔM and $\frac{dv}{dt}$ change with acceleration or deceleration in the same direction:

$$|\Delta M(t)| = \left[r\,\lambda\, m^* - \frac{(\Theta_{Dyno} + \Theta_{sim})}{r_{Dyno}} \right] \cdot \left| \frac{d\,\mathrm{v}(t)}{d\,t} \right| \quad . \tag{8.15}$$

From the requirements for adjusting the test stand characteristics to the road characteristics

$$|\Delta M| \rightarrow 0 \quad \text{or} \qquad \left| M(t)_{Dyno} \right| \overset{!}{\approx} \left| M(t)_{Road} \right| \tag{8.16}$$

this condition follows:

$$\frac{\Theta_{Dyno} + \Theta_{sim}}{r_{Dyno}} \overset{!}{=} r\,\lambda\, m^* \quad . \tag{8.17}$$

The dynamic adjustment takes place by selection of a suitable fly wheel simulation (inertial moment Θ_{sim}). If this adjustment is to be carried out experimentally, it is not recommended to match every point of the time function $M(t)_{Dyno}$ to $M(t)_{Road}$. It is better to use the mean torque as follows, calculated over a sufficient time period, for example over any given part of a driving cycle:

$$\overline{M} = \frac{1}{t_2 - t_1} \int_{t_1}^{t_2} M(t)\; dt \quad . \tag{8.18}$$

$\overline{M}$ is the magnitude of the calibration for the dynamic procedure with a positive sign for acceleration and a negative sign for deceleration.

To determine or evaluate the adjustments, (8.15) is integrated and multiplied by $\frac{1}{t_2 - t_1}$. The following is thus obtained for the mean deviation between the torques $\overline{M}_{Road}$ and $\overline{M}_{Dyno}$

$$|\Delta \overline{M}| = \left[r\,\lambda\, m^* - \frac{(\Theta_{Dyno} + \Theta_{sim})}{r_{Dyno}} \right] \cdot \underbrace{\left| \frac{1}{t_2 - t_1} \int_{t_1}^{t_2} \frac{d\,\mathrm{v}(t)}{d\,t}\, dt \right|} \tag{8.19}$$

or

$$|\Delta \overline{M}| = \left[r\,\lambda\, m^* - \frac{\Theta_{Dyno} + \Theta_{sim}}{r_{Dyno}} \right] \cdot |\,\overline{\mathrm{v}}\,| \quad . \tag{8.20}$$

The conditions for adjustment are, as in (8.16)

$$\left|\Delta \overline{M}\right| \rightarrow 0 \quad \text{or} \quad \overline{M}(t)_{Dyno} \overset{!}{=} \overline{M}(t)_{Road} \;, \tag{8.21}$$

which means that again

$$r\,\lambda\,m^{*} = \frac{\Theta_{Dyno} + \Theta_{sim}}{r_{Dyno}} \;, \tag{8.22}$$

refer to (8.17).

The time-dependent adjustments occur via the selection of the suitable fly wheel simulation Θ_{sim}.

For the adjustment of the test stand in relation to the road load, in the second step it is sufficient to determine the mean torque $\left|\overline{M}(t)_{Dyno}\right|$ and $\left|\overline{M}(t)_{Road}\right|$ and match them together through measuring until their difference is zero.

The driving cycle part and the length of the time interval (t_1, t_2) can be selected for any $\overline{M}$. To get a clear measuring effect, it is recommended to select a driving cycle with enough high acceleration, like the US driving cycle or parts of it.

Torque Measurement Unit

The determination of the torque on the road or the dynamometer is performed by means of a suitable measurement unit. Figure 8.21 shows schematically the important elements of such a system.

The computer to collect and process the data, together with the torque and wheel-speed sensors and the fuel consumption instrument, are packaged into a unit for use in the vehicle operated on the dynamometer and on the road.

The torque sensor (Figure 8.22 shows an exploded view) consists mainly of two disks which are welded at suitable points.

The vehicle wheel is on the left. The torque sensor is mounted on the brake disk by an adapter. The anterior torque sensor disk is furnished with spokes equipped with strain gauges.

The measuring effect is produced by the bending of the spokes caused by the moments acting on the wheels while the vehicle is in motion on a road or on a dynamometer. Adapters allow the adaptation of the torque sensor disk to different types of vehicles. Calibration of the torque sensors is performed in a calibration laboratory by using a force measurement machine equipped with a lever.

Figure 8.23 shows the view of a mounted torque sensor and Figure 8.24 shows a view of the measurement instrumentation mounted into a vehicle with a driver driving on a road in a test area. The driver looks at part of a driving cycle changing on a display and drives accordingly or drives at constant speeds v. $\overline{M}$ and $M_{v_1}...M_{v_2}$ are measured and stored.

Figure 8.25 shows as an example of a measurement result the torque function $M(t)$ and the resulting positive and negative mean torque $\overline{M}(t)$ calculated by integration when driving according to part of the former European driving cycle.

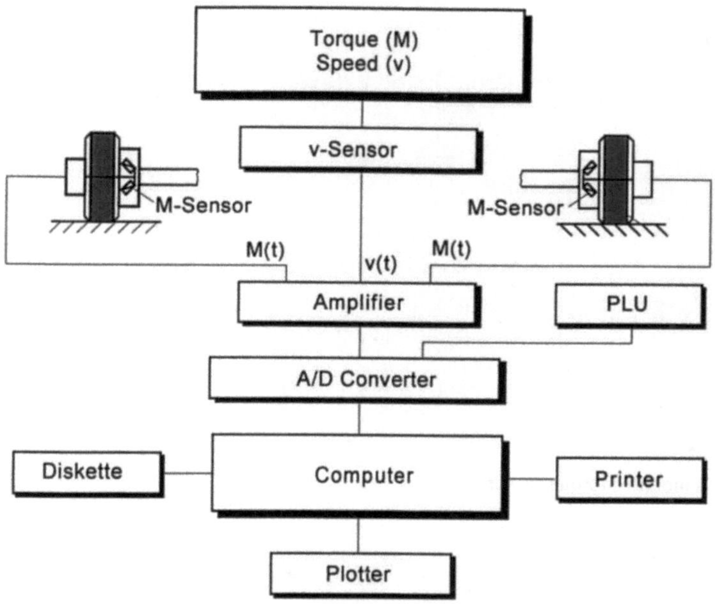

Fig. 8.21: Scheme of a packaged torque measurement unit (suited for on-board measurement) (VW), PLU = fuel consumption instrument of Pierburg

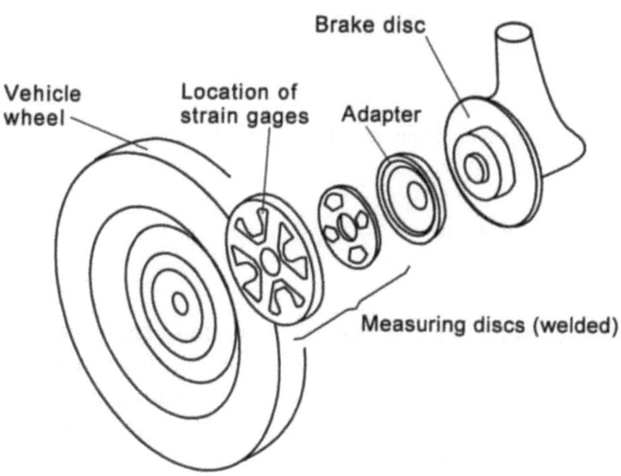

Fig. 8.22: Torque sensor with strain gauges, exploded view

Fig. 8.23: View of a mounted torque sensor

Fig 8.24: View of a torque measurement instrumentation mounted into a vehicle

8.2.6 Exhaust Emission Sampling System

8.2.6.1 Remarks

According to regulations the exhaust gas probes are sampled, analyzed and evaluated while driving the vehicle on a chassis dynamometer following the European or US driving cycle. For the US driving cycle, the sampling, analyzing and evaluation are performed separately for the three phases. Thus, at the end of the test, integral measured values are available.

The sampling of the exhaust gas emitted by a vehicle is performed by using a CVS unit. CVS stands for Constant Volume Sampling, i.e. sampling with the total volume staying constant. ($V_{exh.gas} + V_{air} = V_{total}$ = constant.)

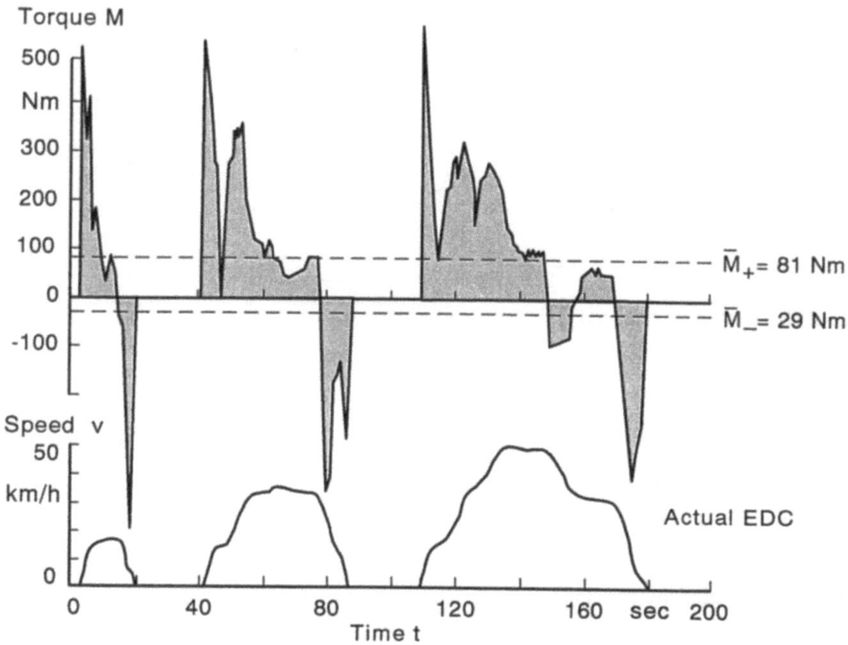

Fig. 8.25: Mean torque $\overline{M}$ for a part of the former European driving cycle

Figure 8.26 shows again the principle of the procedure prescribed by the US Environmental Protection Agency (EPA), refer to Figure 8.3.

The total volume of the exhaust gas produced during the test is pumped through a dilution tunnel using a positive displacement pump (PDP) in case of the CVS procedure. The pump has a pumping capacity distinctly higher than the maximum exhaust gas volumetric flow rate of the vehicle engine at full load.

The volume difference between the exhaust gas volumetric flow rate and the pumping capacity is compensated for by sucking filtered ambient air into the dilution tunnel mixing it with the exhaust gas. The exhaust gas is thus diluted and mixed in continuously changing ratios (e.g., more in idle than in full load of the vehicle engine).

During the total sampling time the revolutions of the pump are added up, the result giving - after multiplication with the constant of the unit and a correction factor -the total volume of exhaust gas and air per test or test phase.

Isokinetic sampling into bags is done by continuously drawing off a partial stream out of the dilution tunnel after proper mixing. (The homogeneity of the exhaust gas - air mixture is a requirement for proper sampling into the bags.) The bags are emptied through exhaust gas analyzers to measure the concentrations of the respective components, refer to Chap. 6. This value of the total volume together with the concentration allows the calculation of the mass of an exhaust emission component and of the fuel consumption (refer to Chap. 8.2.9). The air volume has to be measured as a function of time (Volume flow rate, dV/dt) simultaneously with the respective concentrations.

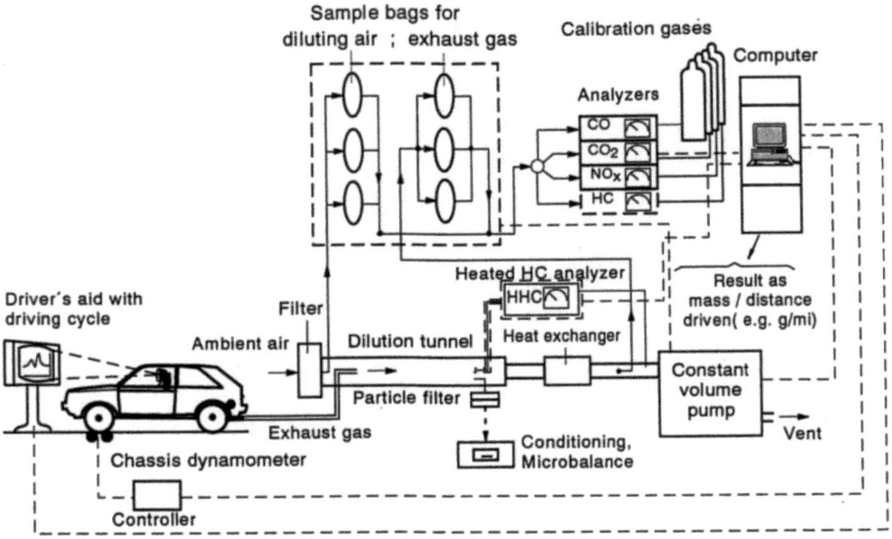

Fig. 8.26: CVS procedure (principle)

Before the test starts the pump is calibrated using an independent procedure, thus determining the pumped volume per revolution, the unit constant and the correction factor.

Remarks

Instead of the forced discharging pump which is a measuring instrument in itself, a set up can be used with a venturi tube (CFV = Critical Flow Venturi) for measuring the volumetric flow rate. For pumping, for instance, blowers, ventilators or jet pumps can then be used. A venturi tube allows the determination of the volumetric flow rate by measuring the pressure difference across a contraction as shown in Figure 8.27.

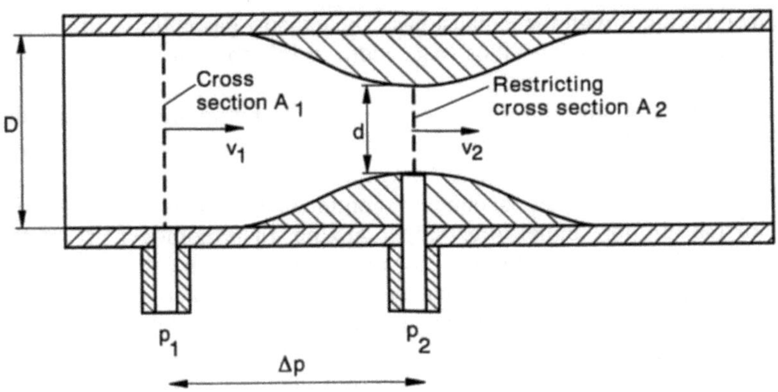

Fig. 8.27: Volumetric flow rate measurement according to differential pressure flow metering

The following equations are valid for the venturi tube:

$$\frac{dV}{dt} = \alpha \varepsilon A_2 \sqrt{\frac{2\Delta p}{\rho}} \qquad (8.23)$$

with

$\frac{dV}{dt}$	volumetric flow rate,
A_2	restricting cross section,
α	flow coefficient,
ε	coefficient of expansion,
Δp	differential pressure ($= p_2 - p_1$),
ρ	gas density.

Δp is measured, all other quantities are known.

The following properties of the CVS procedure should be mentioned:

1. Immediately after the end of the test, the results can be calculated. The determination of the volume takes place simultaneously with the test run.
2. By dilution, condensation and the chemical reactions within the exhaust gas probe are strongly reduced.
3. Considerable space is required for the sampling system, but technically it is movable to other locations if need be.
4. The circuit of the analyzer unit can be relatively simple. Problems of dead volume time and resolutions are less severe as compared to continuous measurements.

The disadvantages of the CVS procedure are:

1. The distribution of the exhaust emissions according to the operating conditions of the engine (time resolution) is only possible with additional instrumentation. This distinction is not allowed for certification measurements, but is useful for engine development.
2. Further investigations, for example of different hydrocarbons or so called unregulated components are more difficult because of the lowered concentrations caused by diluting the exhaust gas.

Besides the sample outlets shown in Fig. 8.26, more sample outlets for specific hydrocarbons and particles are located at the dilution tunnel, refer to Chap. 7.

8.2.6.2 CVS Calibration

For calibration of the constant volume pump the procedure prescribed in the US Federal Register, uses a Laminar Flow Element (LFE) as a measuring gauge. A similar procedure is prescribed in Europe and described in the instruction 70/220/EEC.

Figure 8.28 shows the set up required. The differential pressure measured with the LFE is a measure of the volumetric flow rate in case of laminar flow, which is produced with a honeycomb construction with approximately 500 channels per square centimeter (not shown here).

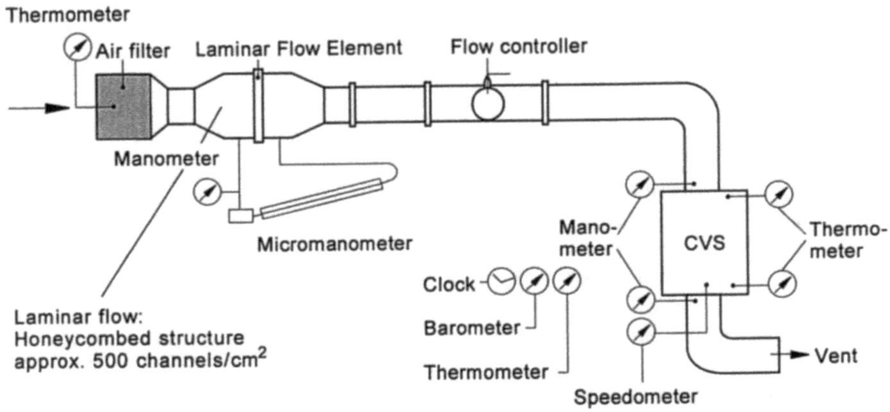

Fig. 8.28: Scheme of the CVS calibration set up. (Laminar Flow Element: honeycomb structure with approx. 500 channels/cm^2)

The basic equation is:

$$\frac{dV}{dt} = A \cdot \Delta p_{LFE} \cdot K_p \cdot K_T \tag{8.24}$$

with

$\frac{dV}{dt}$	volumetric flow rate,	
A	unit constant,	
Δp_{LFE}	differential pressure at the laminar flow element,	
K_p	pressure correction	with regard to standardized conditions of the LFE.
K_T	temperature correction	

Besides the gauge and a flow controller between LFE and pump, all the measuring instruments listed in Table 8.4 are required, with measuring ranges and tolerances prescribed.

Table 8.4: Instruments for CVS calibration

Instrument	Measuring range		Tolerance
Micromanometer	2.5	10^3 Pa	± 1 Pa
Manometer	0...2.5	10^3 Pa	± 25 Pa
Manometer	0...15	10^3 Pa	± 25 Pa
Manometer	95...110	10^3 Pa	± 30 Pa
Thermometer	0...50	°C	± 0.05 K
Thermometer	0...150	°C	± 0.20 K
Stop watch	0...10,000	s	± 0.25 s
Event counter	0...10,000	min^{-1}	± 0.01 min^{-1}

The calibration procedure is based on the determination of the absolute values of the parameters of the pump and the flow controller. Prerequisites for accurate calibration are the following:

1. The pressure gauges used for pressure correction should be mounted on the pump in the center directly at the inlet and outlet and not in the tubes ahead of or behind the pump. Only in this way the actual chamber pressure can be determined
2. During calibration the temperature must stay stable because the Laminar Flow Element gives changing values in case of temperature fluctuations. Stepwise changes of ±1 K are acceptable if they take place over several minutes.
3. All connections between pump and Laminar Flow Element must be absolutely gasproof.

With the set up shown in Figure 8.28 the following data is determined for all available pump speeds and at least six settings of the flow controller for each speed (steps of approx. 10^3 Pa at the pump inlet):

1. Ambient air temperature at the LFE inlet,
2. Pressure ahead of the LFE,
3. Differential pressure across the LFE,
4. Temperature at the pump inlet or ahead of the venturi nozzle,
5. Temperature at the pump outlet (PDP only),
6. Pressure at the pump inlet (or ahead of the venturi nozzle),
7. Pressure at the pump outlet (or behind the venturi nozzle),
8. Sum of all revolutions of the pump (PDP only),
9. Time per measuring period,
10. Ambient air pressure,
11. Ambient air temperature.

Each measuring period requires at least 120 s and for stabilization between the measuring periods a period of 180 s is required. Before starting the calibration the total system should be warmed up by running for 20 minutes.

These calibrations should be performed before using the unit for the first time and then in a sequence of 50, 100, 200, 400, etc. operational hours.

One possibility for absolute calibration is the use of a calibrated rotating lobe gas meter (measuring range of up to 10 m^3/min) which can be placed ahead as well as behind the blower of a compact CVS unit thus enabling the elimination of influences of leaks. As with the LFE, the parameters mentioned above under 4. to 11. as also additional temperature and pressure at the gas meter have to be determined. For accurate measurements a sufficiently long and undisturbed length of advancing flow is required.

8.2.6.3 Checking the Overall System

The US regulations and also the European directives require interim checks of the constants of the sampling system. This check is to be done by introducing known quantities of propane (C_3H_8) or carbon monoxide (CO) into the CVS system.

The CVS system is operated in the same way as for exhaust emission testing. The values measured must, within given tolerances, be in accordance with predefined values. In case of larger deviations, the sources of the error (e.g. leakage, faulty measurement of pressure, temperature or speed) should be discovered and eliminated. If necessary, the constants should be determined again by complete calibration.

The metering of the pure gases can be done for propane using a gravimetric method or for CO by a volumetric input via a gas meter, refer to Chap. 8.2.8.

Another procedure to check the CVS system in stationary operation is the use of calibrated nozzles. Either propane or CO is introduced at a critical velocity which corresponds to the ratio of the pressures ahead of and behind the nozzle where the volumetric gas flow rate through the orifice reaches a maximum. The volume of the pumped gas is, in this case, a function of the pressure ahead of the nozzle. The back pressure plays no role here. The basic equation is:

$$\frac{dV}{dt} = \frac{A + B + p}{\sqrt{T}} \tag{8.25}$$

with

$\frac{dV}{dt}$	volumetric flow rate ,
A, B	calibration constants,
p	inlet pressure,
T	gas temperature

and the limiting conditions for the pressure ratio: $\frac{p_{behind}}{p_{ahead}} \approx 0.5$.

The working pressure of approx. $6 \cdot 10^5$ Pa must be measured very precisely, requiring sensitive instruments ($\pm 10^3$ Pa $\hat{=}$ ca. $\pm$ 9 % volume error ΔV). The accuracy of the gas temperature measurement also influences the result considerably ($\pm$ 0.5 K $\hat{=}$ ca. $\pm$ 9 % ΔV).

Sufficiently accurate measurements are only possible with such a unit by exercising utmost care during operation and proper maintenance.

Because of the toxicity of CO and propane used at high concentrations (refer to Chap. 5) care schould be taken to prevent health hazards.

8.2.7 Quantitative Analysis of the CVS Gas Samples

The following diagram (Figure 8.29) shows an example of a circuit and pneumatic network of the analyzing system when sampling according to the CVS procedure. The pneumatic network is very complicated.

To determine HC, CO_2, CO and NO_x, a number of analyzers are required which are defined with regard to the measuring principles and to their interconnection.

Analyzers based on other measuring principles may be used with the prior permission of the environmental agency as long as they give results comparable to the defined analyzers.

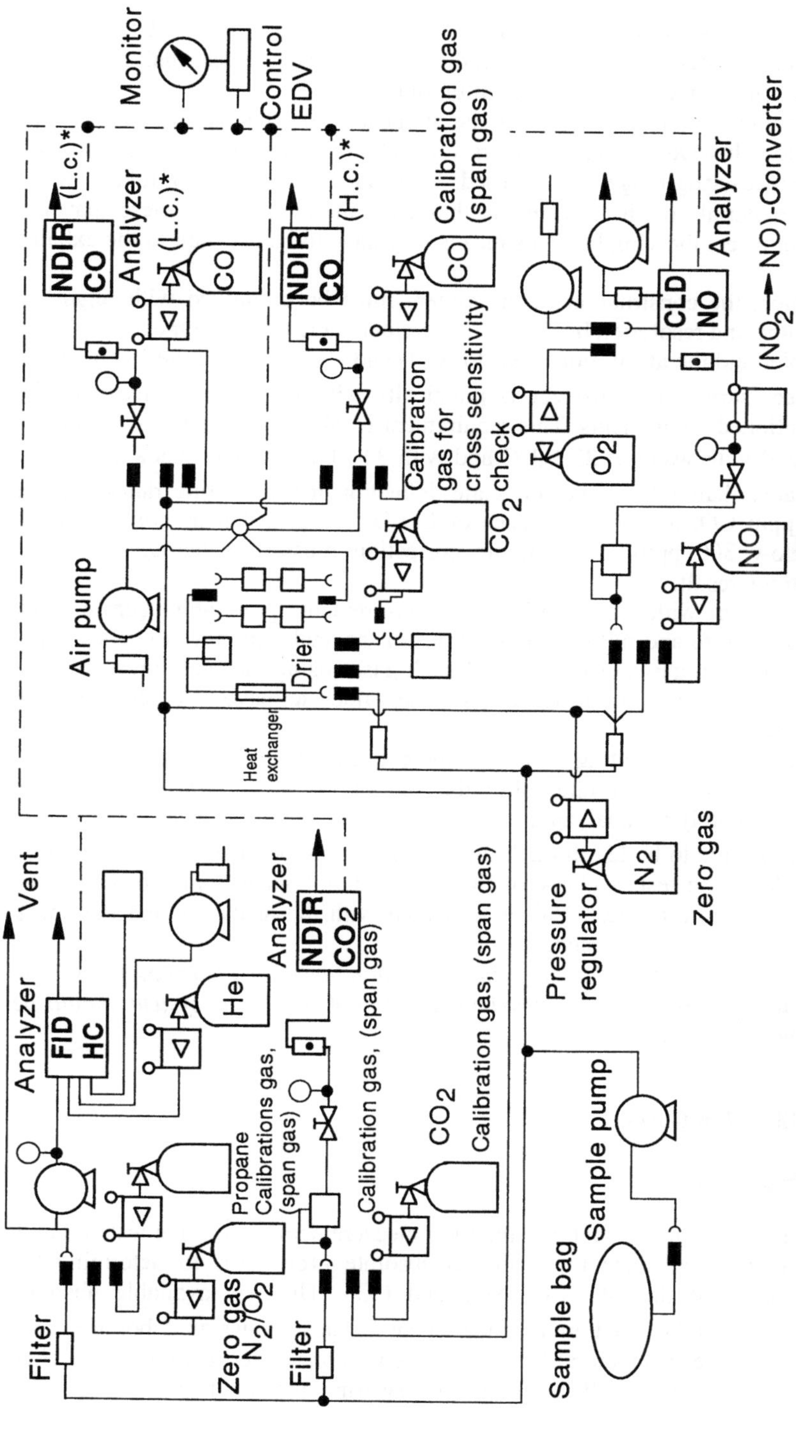

* L.c. = Low concentration
H.c. = High concentration

Fig. 8.29: CVS analyzing system with calibration gases

The gas samples are pumped out of the bags (1., 2. and 3. for the US test) into the pneumatic system and led to the individual analyzers. Each analyzer has its own subsystem for measurement and calibration.

The diagram is not binding; it is supposed to characterize a possible configuration. Instead of two analyzers for CO, for instance, one analyzer with adjustable measuring ranges can be used. With regard to the pump, tubes, valves and other materials of the system, the basic requirement is that these may not influence or affect the samples. This means they have to be resistant to the exhaust gas.

In practice, the construction of the system must be optimized with regard to response time and resolution by timing flow velocity and dead volume.

The cooler and the absorption tubes for CO_2 and H_2O ahead of the CO analyzer are not necessary if the analyzer is practically without cross-sensitivity to the above-mentioned components. This requirement is taken to be met if a calibration gas saturated with water at 23 ± 3°C and with 3 % CO_2 content has a measuring signal smaller than 1 % of the full scale deflection for measuring ranges higher than 300 ppm CO, or if the influence of CO is less than 3 ppm, for measuring ranges below 300 ppm. The volume correction must then be ignored when calculating the result.

In connection with each series of measurement - preferably before each measurement - zero and full scale deflections of the analyzers have to be set using suitable calibration gases or mixtures of such gases using the same pressures and flow rates which are used during the test. The values set have to be documented, refer to Chap. 8.2.8.

The samples of the diluting air and of the exhaust gas air mixture both have to be analyzed within 20 minutes after the end of the test. If the analyzers are connected separately the analyzing should begin with NO_x.

The samples are to be led through the analyzers until a constant value has been achieved. After the test, a check of the zero and full scale deflections must be performed as before the test. Values measured with the analyzers and the check values must also be documented.

A problem that arises is that the system is contaminated by the exhaust gas if the system is used continuously. Approximately every 4 weeks the system should be cleaned and calibrated.

8.2.8 Calibration Gases

8.2.8.1 Principles

For exhaust emission and fuel consumption measurements in routine tests relative procedures are used since there are no absolute procedures to determine the emitted masses of the HC, CO, NO_x and CO_2. The few available absolute procedures cannot be used in practice in an exhaust emission laboratory. An exception is the determination of the particle emissions (refer to Chap. 7.6). They are determined gravimetrically with certain measuring uncertainties.

The basis of the usual procedures to determine concentrations within an exhaust emission analysis is a comparison of the exhaust gas samples with mixtures of gases, the compositions of which are known. These gas mixtures are used to measure the calibration curves, i.e., to determine the measuring signals which are assigned to certain gas concentrations. Furthermore, they are used for routine tests of linearity and the setting of zero and full scale deflections of the analyzers during operation and to cancel or to consider drifts.

According to the agreement VDI-Guideline no. 3490 these gas mixtures are classified "Prüfgase" (calibration gases). In German there is no differentiation between calibration gases and span gases, although the accuracy and quality levels for these gases are defined. Span gases are gases with lower accuracy used in exhaust emission laboratories acc. to US-regulations.

A calibration gas consists of a basic gas containing admixtures of one or more other gases. The basic gas can be a pure gas or a gas mixture and it is usually the main part of the calibration gas in proportion to the quantity. In general, it is not to affect the analyzer. In certain cases (e.g., to analyze HC with a FID) it may be necessary to admixture other components along with the component to be measured in order to produce conditions which are similar to the exhaust emission sample or to compensate for cross-sensitivities, refer to Chap. 6. The admixture to the basic gas is the gas or vapor content which must be precisely known and measured. This admixture is the basis for calibration and checking the measuring instruments.

The concentration values of the calibration gas mixtures used for exhaust emission testing are given as a ratio of the volume of the admixture compared to the volume of the basic gas in Vol.- % or ppm.

8.2.8.2 Legislation

The exhaust gas regulations of different countries permit deviations from the set values of the admixtures concentrations of ±1 % for the calibration gases used to measure the calibration curve, or ±2 % for full scale deflection calibration gases. Furthermore, US regulations set the limit for maximum impurities for zero scale deflection calibration gases at:

1	ppm HC (C1)
1	ppm CO
0.1	ppm NO_x
400	ppm CO_2

Both requirements cannot be adequately fulfilled for span gases with the available analytical instruments. The analytical accuracy required of the measuring instruments is not sufficient. When considering the high total uncertainties of the exhaust emission tests (refer to Chap. 9) it becomes clear that it is economically and technically pointless to demand high accuracy for calibration gases as one of the parameters of the exhaust emission testing as the effort involved ist disproportionately high. However, systematic deviations of the calibration gases directly influence the final result of a test.

To check the linearity of analyzers in the US regulation (FTP), for each measuring range calibration gas mixtures with concentrations of 15, 30, 45, 60, 75 and 90 % of the full scale deflection are required.

The span gases should have concentrations of approximately 80 % of the full scale deflection. These requirements are similar for European and Japanese tests.

The following overview (Table 8.5) gives an impression of the resulting demand for calibration gases for a European automotive company, operating worldwide.

Table 8.5: Number of necessary calibration gas mixtures

Propane in air	Hexane in N_2	CO in N_2		CO_2 in N_2	NO in N_2
from 8 ppm to 18.000 ppm in 18 steps	from 50 ppm to 1000 ppm in 8 steps	from 7.5 ppm to 2500 ppm in 22 steps	from 0.3 % to 10.0 % in 11 steps	from 0.08 % to 16.0 % in 18 steps	from 5 ppm to 9.000 ppm in 15 steps

More than 100 different calibration gas mixtures are used as span gases for this example, with 5000 gas cylinders per year filled according to procedure and transported to the different laboratories.

8.2.8.3 Calibration Gases in Cylinders

The calibration of the exhaust emission analyzers must be done under the same conditions as for the measurement of the exhaust gas samples. This results in the considerable demand for calibration gases mentioned above. Due to of the high quantities of calibration gases used during a test and because of the number of tests per day, this demand can only be met today by using gas mixtures in cylinders. Depending on the size of the cylinder, the cylinder pressure and the type of mixture, between 0.5 and 10 m^3 of calibration gas can be provided per cylinder.

In principle, the gas mixtures in cylinders can be produced gravimetrically but the effort required is considerable. Even to approximately achieve the required accuracy, a scale with very high precision is necessary. This procedure is impractical for producing large quantities of span gases and full scale deflection gases, because the gravimetric method requires too much time. A practical way to continuously produce higher quantities of calibration gases is to dose roughly via partial pressure, following it up with an analysis of the homogenized mixture. At ambient pressures, liquid substances (hexane, water) are dosed with injection syringes, vaporised and then handled as gases. For the vapors however, care should be taken that certain maximum filling pressures should not be exceeded because there is a danger of condensation and segregation at higher cylinder pressures.

For permanent gases, the mixture of the calibration gases is carried out from gas cylinders via controlled valves, where the partial pressure of each mixture component is defined (partial pressure procedure). After the mixing, the composition of the gas is roughly known. The exact determination of this mixture achieved by using specially calibrated and maintained gas analyzers which are similar to the exhaust emission analyzers but are only used for the analysis of calibration gases.

From the point of view of the measuring technique it is very unusual to use the same kind of instruments for both measurement and calibration.

The total sequence of the production of the span gases is outlined in a sequence chart in Figure 8.30, and is as follows:

1. Preparation of the gas cylinders: Appointment with the TÜV, checking for rest pressure, purification, addition of Teflon balls.
2. Introduction of calibration gas: Filling with the gas component according to the partial pressure procedure.
3. Filling: Introduction of the back gas from liquid nitrogen via a liquid gas pump.
4. Homogenization: Thorough mixing by rotating the cylinders; the Teflon balls in the cylinders intermix the gases.
5. Analysis: Analysis of the gas mixture with analyzers previously calibrated with very precise gases.
6. Labeling and Release: The cylinders are labeled with bar codes containing the dates, concentration, etc.
7. Delivery.

The calibration of the calibration-analyzers finally is carried out with calibration gases of a higher accuracy class as shown schematically in Figure 8.31.

The calibration gas analyzer is checked using a gas standard M before and during each analysis of a calibration gas „X“ produced with the partial pressure procedure. This gas standard M has been verified by comparison with gases of higher accuracy.

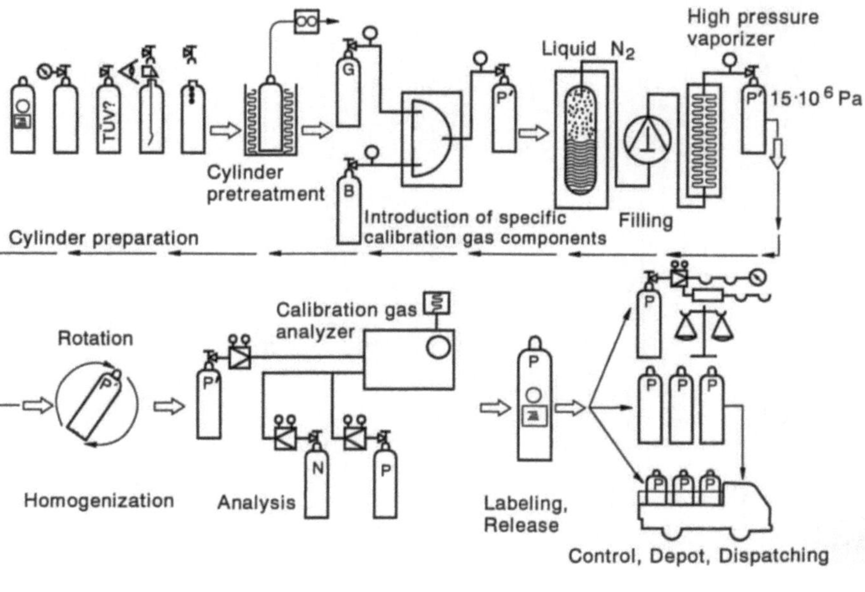

Fig 8.30: Sequence chart for calibration gas production

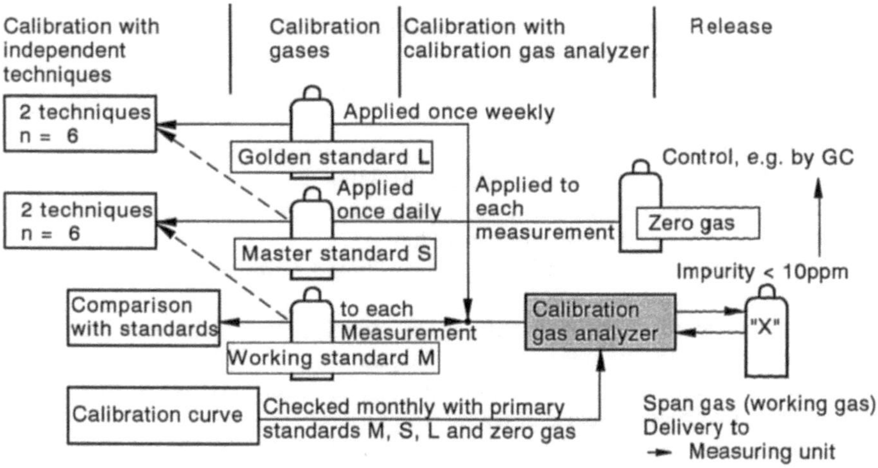

Fig 8.31: Schematic of calibration gas analysis

Furthermore, there is a daily calibration control with the gas standard S, the concentration of which is measured earlier using an independent technique. The calibration analyzer is checked weekly using the gas standard L which has been verified earlier using at least two independent technique for analysis.

Fig. 8.32: View of the measuring instruments to check span gases

Finally, once a month the calibration curve of the calibration analyzers is checked with S or L standards.

Figure 8.32 shows a view of the measuring instruments for checking span gases.

The following Table 8.6 shows a list of the principles of independent procedures to ensure the higher accurate calibrations standards (L, S, and M standards). With the exception of one example, further details will not be described here.

Table 8.6: Independent procedures for calibration gas assurance

Admixture	Independent Procedures	Routine Instruments (see Chap. 6)
CO	- dynamic, volumetric (mixing pump, nozzle) - gravimetric I (Voland scale) - gravimetric II (Combustion $\rightarrow CO_2$) - chem. analysis (oxid. J_2O_5)	NDIR
CO_2	- dynamic, volumetric - gravimetric I - gravimetric II (adsorption)	NDIR
NO	- dynamic, volumetric - gravimetric I - chem. analysis	CLD
Hexane	- dynamic, volumetric - gravimetric I	NDIR
Propane	- gravimetric II - gravimetric III	FID

One of the most important and most reliable procedures today for controlling calibration gases is the gravimetric filling of gas mixtures which is described more in detail here as an example of the independent procedures.

The standardized calibration gases used by the US National Bureau of Standards (NBS) for controlling „Standard Reference Materials" and approved by the EPA as a basis for calibration gases are, for example, produced by using a precision scale.

In Germany the Federal Agency for Material Testing (Bundesanstalt für Materialforschung und -prüfung - BAM) is the agency responsible for control of calibration gases.

When using a precision scale, steel cylinders with a volume of 50 liters and a mass of up to 100 kg can be weighed with an error of ± 20 to 30 mg. Thus the relative error is $\frac{\Delta m}{m} = 2 \cdot 10^{-7} \ldots 3 \cdot 10^{-7}$ with a repeatability of $\frac{\Delta m}{m} = 3 \cdot 10^{-8}$. In this way fillings of gas mixtures down to approximately 10 ppm are possible in one step.

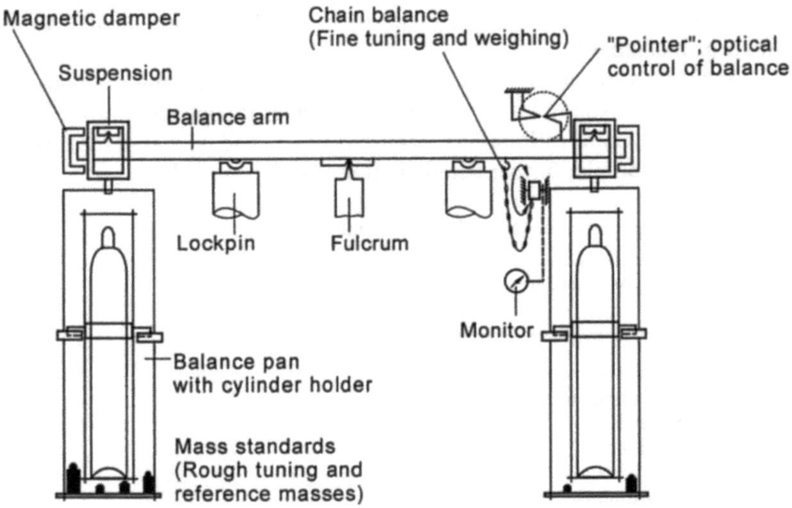

Fig. 8.33: Principle of the calibration gas cylinder scale

The operating principle of such a scale, which must be installed in a thermostatic control room, is shown in Figure 8.33.

To determine the precise composition of the gas mixtures (standards), the gas is weighed after each admixture during the filling. From the weights the composition can be determined according to amount.

The adjustment of the scale was carried out together with the Federal Institute for Metrology, Braunschweig-Berlin (Physikalisch-Technische Bundesanstalt - PTB). The first comparison tests showed errors of -20 to +30 mg for 100 kg. This corresponds to an accuracy of $-2 \cdot 10^{-7}$ to $+3 \cdot 10^{-7}$. For repeated weighing the standard deviations were 3 to 31 mg for 100 kg ($3 \cdot 10^{-8}$ to $1 \cdot 10^{-7}$) depending on the filling procedure. The masses of the admixtures vary from 1.2 g (100 ppm-gases) to 1.2 kg (10 %-gases). The error is then approximately 2 to 3 % (100 ppm) or 1.2 %. For the 100 ppm range, this error is still higher than the regulated 1 to 2 %. Gas chromatic controls of these gravimetrically produced mixtures resulted in deviations of approximately -3 % for CO (900 ppm - gas) and CO_2 (4 %-gas) compared to the earlier laboratory standards. One needs to take into account that measurements with a gas chromatograph are less accurate than the measurements with the scale.

Figure 8.34 shows a view of such a scale

Advanced technologies and new materials as well as special handling are the necessary requirements to achieve this precision. Operating the scale requires special care and a lot of training.

8.2.9 Final: Calculation of the Measurement Results

The last step in the process of the individual exhaust and fuel consumption testing procedures is the calculation of the measurement results.

Fig. 8.34: View of a calibration gas cylinder scale (VOLAND)

The mass emissions of the exhaust gas components and the fuel consumption will be calculated by the following equations for the measurement values for concentration, volume, temperature, pressure, humidity as well as the relevant constants for each driving cycle.

8.2.9.1 US-75 Test

The end results of the US-75 test are composed of three phases (acc. to the three driving cycle phases) which are evaluated differently:

$$m_i = 0.43 \cdot \frac{m_{i\,CT} + m_{i\,S}}{s_{i\,CT} + s_{i\,S}} + 0.57 \cdot \frac{m_{i\,HT} + m_{i\,S}}{s_{i\,HT} + s_{i\,S}} \tag{8.26}$$

with

m_i	mass of the component i	in	g / mi,
$m_{i\,CT}$	emitted mass of i in cold start phase (1st phase)	in	g,
$m_{i\,HT}$	emitted mass of i in warm start phase (2nd phase)	in	g,
$m_{i\,S}$	emitted mass of i in stabilized phase (3rd phase)	in	g,
$s_{i\,CT}$	length of driving cycle, cold start phase	in	mi,
$s_{i\,HT}$	length of driving cycle, warm start phase	in	mi,
$s_{i\,S}$	length of driving cycle, stabilized phase	in	mi.

The masses emitted in the individual phases are calculated for a CVS with PDP according to the following equation:

$$m_{ij} = c_{ijk} \cdot \rho_i \cdot \frac{V_o \cdot n_j \cdot p_{Pj} \cdot 2.893}{T_{Pj}} \tag{8.27}$$

The following equation is applicable for a CVS unit with venturi nozzles (CFV):

$$m_{ij} = c_{ijk} \cdot \rho_i \cdot V_{jk} \tag{8.28}$$

with

i	exhaust gas component (i = HC, NO_X, or CO)		
j	driving cycle phase (US-75) (j = *CT*, *S* or *HT*)		
m_{ij}	mass of the component i in phase j	in	g,
c_{ijk}	concentration of the component i in the phase j after correction (k) acc. to (8.29) ... (8.33)	in	10^{-2} %,
	or	in	10^{-6} ppm,
ρ_i	density of the component i	in	g/m^3,
V_{jk}	total volume (exhaust-air mixture) in phase j, corrected under normal conditions (k)	in	m^3
V_o	volumetric flow rate of the probe sampling unit	in	m^3 per revolution
n_j	total blower revolutions in phase j,		
p_{Pj}	absolute pressure ahead of the pump in phase j	in	10^3 Pa,
T_{Pj}	temperature ahead of the pump in phase j	in	K,

If the CO measuring unit shows cross sensitivity to CO_2 or H_2O, correction calculations need to be carried out when using reagents or cooling to remove CO_2 and water vapor, refer to Chap. 8.2.7. A correction of the influence of humidity is necessary for NO_X. Additional corrections are also required when CO, HC or NO_X is present in the diluted air probe.

The correction of the concentration value c_{ijk} is carried out according to the following equations:

– for HC:

$$c_{HC\,jk} = c_{HC\,jm} - c_{HC\,jd}(1 - 1/DF_j), \tag{8.29}$$

– for NO_X :

$$c_{NOx\,jk} = K_{NOxj}\,[c_{NOx\,jm} - c_{NOx\,jd}(1 - 1/DF_j)], \tag{8.30}$$

– for CO (as long as H_2O and CO_2 are excluded from the gas flow by reagents or cooling):

$$c_{COjk} = c^*_{COjk} - (1 - K_{Hj}) \cdot c_{COjd} \cdot (1 - 1/DF_j)$$
$$c^*_{COjk} = c_{COjm} - (K_{2j} + K_{Hj}) \cdot c_{COjm} \tag{8.31}$$

with

c_{ijm} measured concentration value (m) for component i in phase j in 10^{-2} %

c_{ijd} measured value of the component i in phase j in the diluted air (d), or in 10^{-6} ppm

DF_j dilution factor for phase j :

$$DF_j = \frac{13.4}{c_{CO_2 j} + (c_{HC\,jm} + c^*_{CO\,jk}) \cdot 10^{-4}}, \tag{8.32}$$

K_{NOxj} humidity correction factor for NO_x in phase j :

$$K_{NO_x j} = \frac{1}{1 - 0.0329(H - 10.71)}$$
$$H = \frac{6.211 \cdot R_{aj} \cdot p_s}{p_a - p_s \cdot R_{aj} \cdot 10^{-4}} \tag{8.33}$$

H absolute humidity in g water per kg of dry air,

R_{aj} relative humidity in phase j in %,

p_s saturated vapor pressure at the ambient temperature in 10^3 Pa,

p_{aj} ambient air pressure in phase j in 10^3 Pa,

K_{2j} volume correction after exclusion of CO_2 in phase j

$$K_{2j} = 1.925 \cdot c_{CO_2 j} \cdot 10^{-2},$$

$c_{CO_2 j}$ concentration of CO_2 in phase j in %,

K_{Hj} volume correction after exclusion of water vapor in phase j

$$K_{Hj} = 3.23 \cdot R_{aj} \cdot 10^{-4}.$$

8.2.9.2 Calculation of the Fuel Consumption

From the carbon balance (considering the fact that all emissions containing carbon must derive from the fuel), the fuel amount used can be calculated from the mass emissions of HC, CO and CO_2 emitted during the test drive. In the USA, the following formulas are given for the calculation of fuel economy according to 40 CFR 600.113-88:

- for vehicles with gasoline engines:

$$FE = \frac{5174 \cdot 10^4 \cdot CWF \cdot SG}{\left[(CWF \cdot m_{HC}) + (0.429 \cdot m_{CO}) + (0.273 \cdot m_{CO_2})\right] \cdot \left[(0.6 \cdot SG \cdot NHV) + 5471\right]} \tag{8.34}$$

- and for vehicles with diesel engines:

$$FE = \frac{2778}{(0.866 \cdot m_{HC}) + (0.429 \cdot m_{CO}) + (0.273 \cdot m_{CO_2})} \tag{8.35}$$

with

FE	fuel economy value	in	mi/gal,
CWF	carbon proportion (Carbon Weight Fraction) according to ASTM D 3343,		
NHV	net heating value according to ASTM D 3338,	in	Btu/lb
SG	fuel density (specific gravity) according to ASTM D 1298.	in	g/ml

Here m_{HC}, m_{CO} and m_{CO_2} are inserted in g/mi.

The following calculation formulas are given for Europe in the EC regulation 93/116/EEC:

- for vehicles with gasoline engines:

$$FC = \frac{0.1154}{\rho_{TF}} \left[(0.866\ m_{HC}) + (0.429\ m_{CO}) + (0.273\ m_{CO_2})\right], \tag{8.36}$$

- for vehicles with diesel engines:

$$FC = \frac{0.1155}{\rho_{TF}} \left[(0.866\ m_{HC}) + (0.429\ m_{CO}) + (0.273\ m_{CO_2})\right], \tag{8.37}$$

where the following applies

FC	fuel consumption	in	l/100 km,
m_{HC}	hydrocarbon emissions measured	in	g/km,
m_{CO}	carbon monoxide emissions measured	in	g/km,
m_{CO_2}	carbon dioxide emissions measured	in	g/km,
ρ_{TF}	density of the test fuel	in	kg/l .

The factor before the brackets contains the carbon content *and* the driving cycle length per test. When changing these parameters , the corresponding value must be corrected. In the European test according to the old European driving cycle (length: 4.052 km), when applying the CVS procedure and using the currently prescribed test fuel with a density of $\rho_{TF} = 0.755$ kg/l, the following equation results for the calculation of the driving cycle length fuel consumption:

$$FC = \frac{0.866 \cdot m_{HC} + 0.429 \cdot m_{CO} + 0.273 \cdot m_{CO_2}}{26.51} . \tag{8.38}$$

As the proportion of m_{CO_2} in the exhaust is larger than 90 %, the CO_2 predominates (more than 70 %) when investigating the fuel consumption. The error in this calculation is mainly caused by the error in the determination of CO_2.

8.2.10 Measuring the Fuel Evaporation Emission in the SHED Test

Within the framework of a complete US emission test (FTP), the measurements for hydrocarbon emissions are also carried out in the SHED test (SHED - Sealed Housing for Evaporative Determinations). Figure 8.35 shows an open SHED chamber with a vehicle. In the closed chamber, the fuel is heated in the vehicle's tank. The hydrocarbon concentrations in the ambient air of the chamber are measured with a FID.

Fig. 8.35: View of an open SHED chamber

An overview of the fuel evaporation emission test (SHED test) valid until 1995 is shown in sequence in Figure 8.36.

A new fuel evaporation emission test is phased in from 1995 on. The corresponding overview of this test is shown in sequence in Fig. 8.37.

The fuel vapor losses measured in the SHED diurnal breathing and hot soak tests are calculated by the following formula:

$$m_{HC} = k \cdot V_n \cdot 10^{-4} \cdot \left(\frac{c_{HCf} \cdot p_{af}}{T_f} - \frac{c_{HCi} \cdot p_{ai}}{T_i} \right) \tag{8.39}$$

with

m_{HC} mass of hydrocarbon vapor in g,

k = 1.2 (12 + H/C)

V_n	net volume of the SHED (for the test vehicle, the lump sum of 1.42 m^3 is subtracted from the gross volume; unless the exact vehicle volume has been determined beforehand and approved by the authorities)	in	m^3,
c_{HC}	HC-concentration	in ppm C_1	at each beginning(i)
p_a	atmospheric pressure	in 10^3 Pa	and end (f) of the
T	SHED room temperature	in K	measuring phase,
H/C = 2.33	hydrogen to carbon ratio for the diurnal breathing (DB) test,		
H/C = 2.20	hydrogen to carbon ratio for the hot soak (HS) test.		

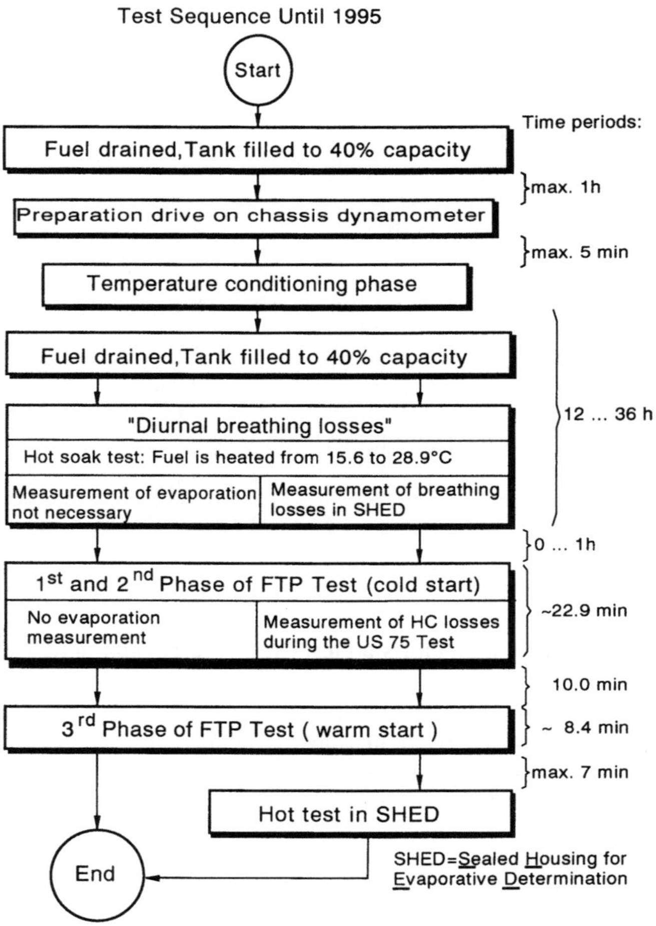

Fig. 8.36: Sequence plan for the measurement of fuel evaporation emissions via the HC-concentration in the ambient air in the SHED chamber with an intermediate, complete US-75 exhaust emission test

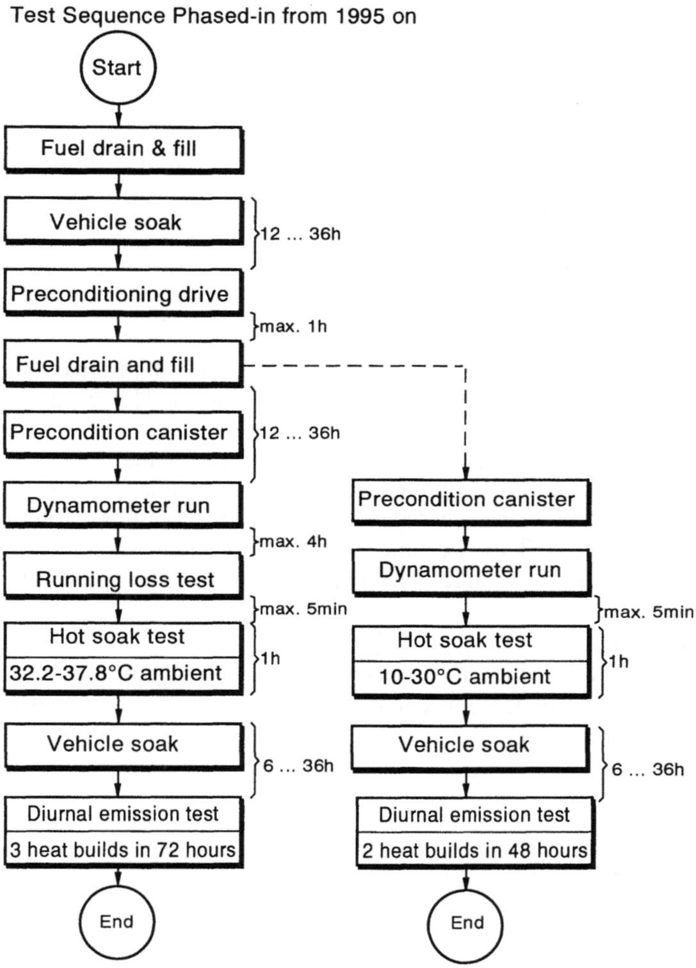

Fig. 8.37: Sequence plan for the new fuel evaporation emission test, phased in from 1995 on

The total evaporation emission mass m_{total} shall be computed by the following formula:

$$m_{total} = m_{DB} + m_{HS} \,. \tag{8.40}$$

8.3 A New Proposed Exhaust Emission Test System

Future, more stringent regulations (refer to Chap. 2) require from the point of view of measuring techniques a totally new measuring and testing system, because of

the low concentrations to be measured and other disadvantages of the current test system. The disadvantages of the current test system are as follows:
- complex and expensive sampling (CVS),
- the use of the difference of the concentrations of diluted exhaust samples and diluting air sample,
- the great number of different instruments based on different measuring principles for analysis,
- expensive calibration procedures,
- high expense of maintenance,
- separate measuring benches, one for integral measurements, the other for time resolved measurements

Desirable is a proposed system as shown schematically in Figure 8.38 inserted into the scheme of the current system (refer to Figure 8.3) and shaded. This proposal will be described below.

After sampling and conditioning, including the measurement of the flow rate, the raw exhaust gas is analyzed with regard to the regulated and some unregulated components with an instrument which is based on just one physical principle. Both the integral measurement as well as the time resolved measurements of many components are simultaneously possible. The requirements for this measuring instrument are the following:
- great dynamic measuring range from ppm to Vol.-%,
- real time measurement of the concentrations with a time resolution of 1 second,
- simultaneous measurement of the concentrations of 25 to 30 chosen exhaust emission components,

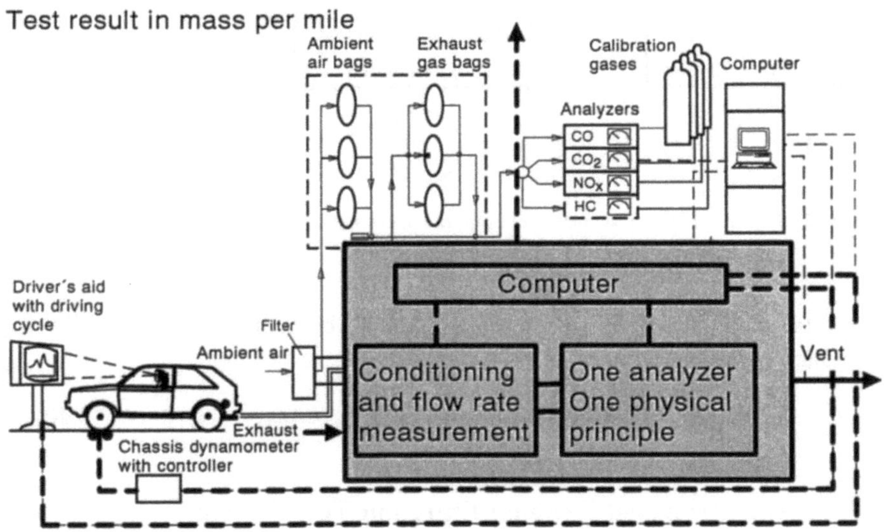

Fig. 8.38: Scheme of a new proposed test system (shaded) inserted into the scheme of the current system (refer to Fig. 8.3)

- data output not longer than 10 minutes after the end of the test,
- calibration intervals > 1 month
- availability > 95 %.

The FTIR described in Chap. 6.1.7 is such an instrument. A corresponding test system has also been developed, as shown schematically in Figure 8.39.

The measurement is performed using undiluted exhaust gas (raw gas).

To measure the flow rate of raw undiluted exhaust gas a vortex sensor is used. The principle of this sensor is based on vortices which are produced in the flow of the exhaust gas by a disturbing body and which pass through a focused ultrasonic beam perpendicular to the exhaust gas flow. The time for passing through the beam is measured to determine the flow rate of the exhaust gas. A part of the undiluted exhaust gas is led through a sample conditioning unit and then through the measuring cell of the FTIR measuring instrument and finally back into the undiluted exhaust gas flow. The temperature of the exhaust gas sample to be analyzed in the measuring cell of the FTIR is kept constant at 185°C ± 5 K. To measure the particles the undiluted exhaust gas is then led into a small dilution tunnel. A sample of the diluted exhaust gas is pumped through particle filters to determine the particle mass (as in the current procedure). The small dilution tunnel can be additionally used to control the flow rate. To measure the concentration of oxygen, an additional conventional measuring instrument which is not shown here is required (refer to Chap. 6.4).

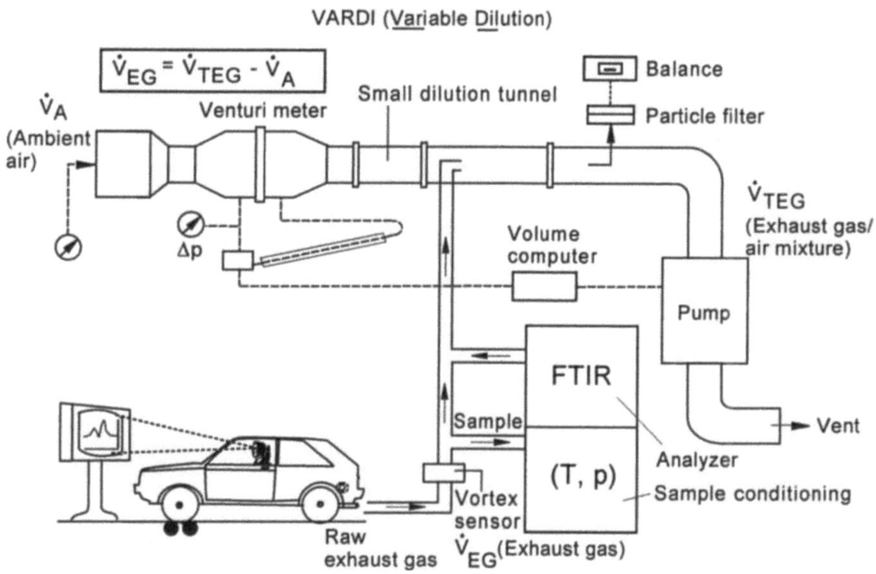

Fig. 8.39: Scheme of a future exhaust emission test system; $\dot{V}_{EG}$ = Flow rate of undiluted exhaust gas, $\dot{V}_{TEG}$ = flow rate of total exhaust gas, $\dot{V}_A$ = flow rate of diluting air

The advantages of this new test system are the following:

- The sensitivity limits for the exhaust gas components are higher by a factor of 8 to 10 because there is no dilution. In this way the low ULEV values, refer to Chap. 2.3, can be determined more easily.
- It is not necessary to take the difference of concentrations of diluted exhaust sample-air sample (this is important for low concentrations → ULEV).
- The adsorption of organic components with high boiling points (e.g. PAH) to soot is prevented or lowered. This means that these components are gaseous.
- The condensation of water is prevented. Thus polar hydrocarbons such as aldehydes can also be measured.
- With diesel exhaust gas, the high temperature of 185°C prevents the formation of sulfur compounds.

For the introduction of the new system it is important that the measurement results are identical to those measured with the current systems. This has already been proven by measurements on many vehicles.

Many specialists, consider to be this system the system of the future. Together with the new proposed driving cycle (Chap. 8.2.3.5) this would facilitate testing and would lower the expense considerably.

9 Vehicle Exhaust Emission Testing Procedures - Overview and Criticism

9.1 Regulated Procedures

Fig 9.1 provides an overview of the mandatory tests to ensure that the automotive manufacturer and the vehicle owner comply with the standards for limited exhaust emission components.

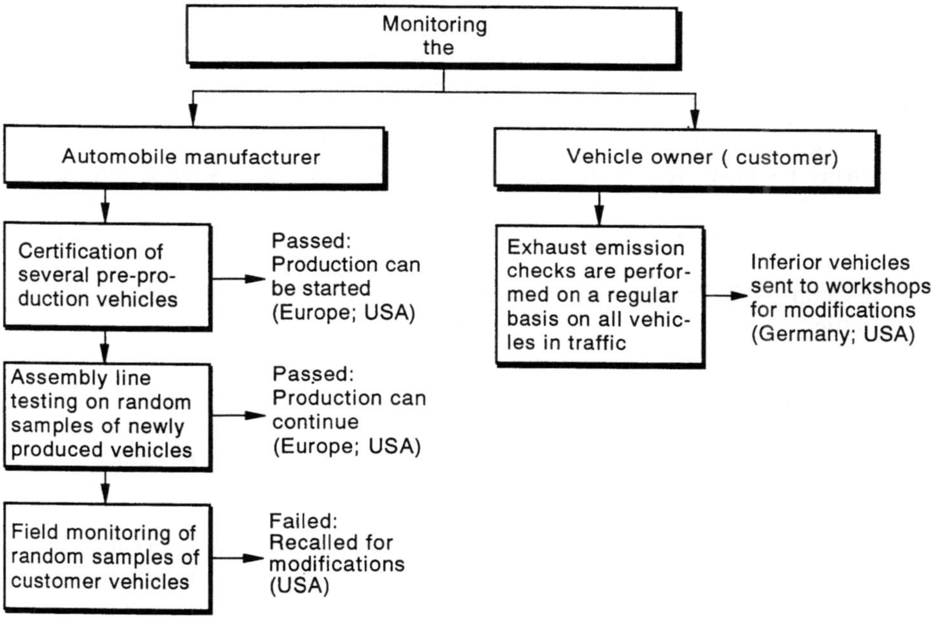

Fig 9.1: Monitoring the automotive manufacturer and the vehicle owner

9.1.1 Monitoring the Automobile Manufacturer

The certification of new vehicles is prescribed as the first check applying to the automotive manufacturer. The results of this test determine whether or not a newly developed vehicle or a vehicle containing significant changes may be produced.
New vehicles are tested as they leave the assembly line by taking random samples and measuring the exhaust emission components. The results of these measure-

ments determine whether production can continue. Random sampling methods are different in Europe to those in the USA, particularly in California. In Europe, assembly line testing is rarely performed, if at all.

In the US a random sample of used vehicles is tested in so called field tests. These tests ascertain if the exhaust emission values still meet the standards after 50,000 or 100,000 miles respectively. The results of these tests determine whether the vehicles can be certified to have met standards or if they need to be recalled for modifications. Recalls are issued by the manufacturer and modifications are paid for by the manufacturer.

With these three checks, the automotive manufacturer's compliance with regulations is relatively closely monitored.

9.1.2 Monitoring the Automobile Owner

The situation is quite different for monitoring the automobile owner as it is often more difficult for technical and political reasons. In principle, the vehicle owner is supposed to see to maintenance. Otherwise, the efforts of the automobile manufacturer would be wasted and the exhaust emission regulations would be meaningless. The sum of all exhaust emissions from vehicles on the road would be higher than intended by the legislator. This problem was discussed in the US and in Germany for many years before legislation was passed. Details are described in Chap. 9.9.

A regular inspection of vehicles on the road is required in nearly all the states in the US.

In Germany the exhaust emission check, called AU, is regulated for all vehicles in traffic.

9.1.3 Lack of Standardized Procedures

In different countries, all three checks for monitoring the automobile manufacturer are based on one test procedure. This procedure, however, has significant differences in Japan, Europe and the USA so that a comparison of exhaust emission standards or of measured values is not easily possible, refer to Chap. 8.

This lack of standardized test procedures poses a big problem for the automotive manufacturers operating worldwide because the vehicles need to be developed differently for different markets. From the viewpoint of preventive environmental measures, such differences are pointless.

9.2 Statistical Errors

Inherent in all the tests is the problem that basically each exhaust emission value measured contains an unavoidable random error (uncertainty). The total random error is made up of the statistical variation of the vehicle's or engine's exhaust emissions and the measuring uncertainty.

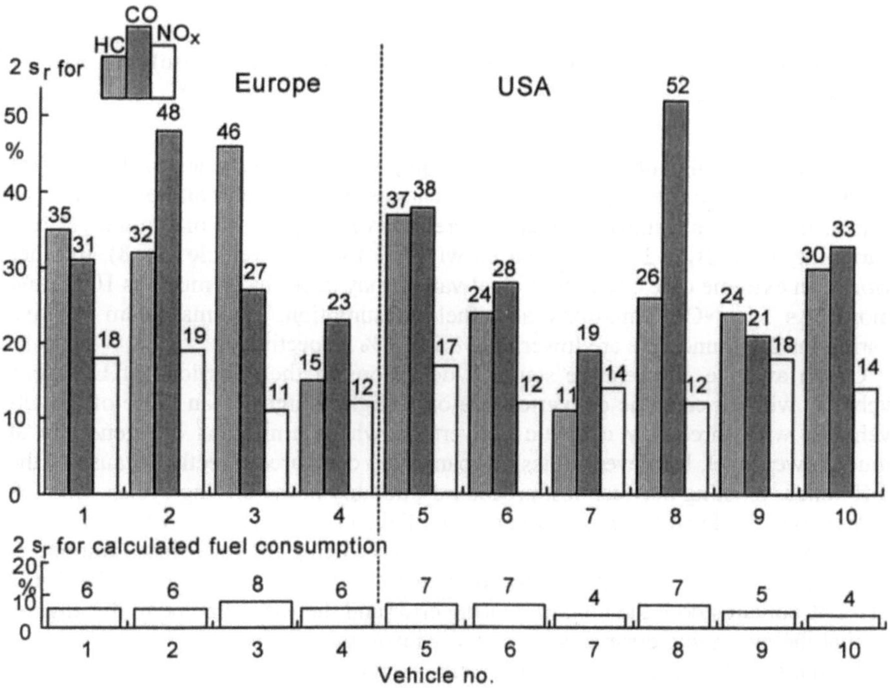

Fig. 9.2: Relative variations ($2s_r$, P = 95%) for HC, CO and NO_x emission data and the fuel consumption of 50 repeated former European and US-75 cold start tests on European and US vehicles. Vehicles 1-4 without catalytic converter, vehicles 5-10 with three-way catalytic converter

Figure 9.2 shows a view of relative variations in percent from 50 repeated measurements on the same vehicle for ten vehicles produced in Europe and USA. The European vehicles were measured according to the former European test and the US vehicles according to the US-75 test, both in cold start, refer to Chap. 8.

Figures above the bars each indicate the relative variability ($2s_r$):

$$2s_r = 2\frac{s}{\bar{x}} \cdot 100 \text{ in } \%, \tag{9.1}$$

s_r = Relative standard deviation
s = Standard deviation
$\bar{x}$ = Mean

} Probability $P = 95$ %.

Variability is less extensive than usual because all vehicles were serviced with extreme care, the chassis dynamometers and analyzers were calibrated carefully before the tests began, and all the tests were run by the same driver and on the same chassis dynamometer using the same analyzers. In addition, mean variations of the environmental parameters were small:

Atmospheric pressure	± 1.2 %	
Relative humidity	± 41 %	($2\,s_r$, $P = 95$ % in each case).
Air temperature	± 0.7 %	

Despite the careful calibration of the test equipment, the constancy of the ambient conditions and the use of one driver, one chassis dynamometer and one analyzing bench, the relative standard deviations are still very high. The maximum relative variability of $\pm\, 2s_r = \pm\, 52$ % occurred with CO for a US vehicle (no. 8). In other words, in extreme cases, two individual values may differ by as much as 100% and more. As for NO_X emissions and fuel consumption, the maximum relative variations $2s_r$ found here are lower: 19 % and 8 % respectively.

On an average, the relative standard deviations of the emissions of European vehicles without catalytic converters are only slightly higher than those of the US vehicles with three-way catalytic converters, whose emissions are generally at much lower level. However, emissions cannot be compared directly because of the differences existing between test procedures, notably in the driving cycles. One US vehicle (no. 7) showed the lowest variation of all, 11 % for HC and 19 % for CO.

When considering all variations of the US vehicles for HC, these amount to 37 % for vehicle no. 5 and 11 % for vehicle no. 7. In this specific case, the figure of 11 % contains both the total measuring error and the variation due to the engine, so that the measuring error alone is actually lower than 11 %.

Assuming that the measuring errors pertaining to the two vehicles are comparable, it is possible to estimate the HC measurement error due to the engine of vehicle no. 5 alone, refer to (9.2). This is physically reasonable because both were tested by the same driver on the same test facility. According to the Gaussian law of error propagation, and on the simplifying assumption that the total variation of vehicle no. 7 virtually contains only measurement errors, neglecting in this case errors due to the engine, the following estimation of the upper and lower limit of the error due to the engine is obtained:

$$\pm 37\,\% \geq \pm 2s_{r,Eng.} \geq \pm\sqrt{(2s_{r\,total})^2 - (2s_{r\,Meas.})^2} \approx \pm\sqrt{37^2 - 11^2} \approx \pm 35\,\% \tag{9.2}$$

The error due to the engine is between 35 % and 37 % and determines the total relative standard deviation $2s_r$.

As is well known, the results of repeated measurements can be plotted as histograms to which density functions can be fitted. Figure 9.3 shows as an example histograms and fitted density functions for the results of 50 repeated HC and CO measurements on one and the same US vehicle with a three-way catalytic converter from a European manufacturer.

Three density functions were investigated for their ability to reflect the empirical histograms of exhaust emission values:

- normal density function,
- log-normal density function,
- Weibull density function.

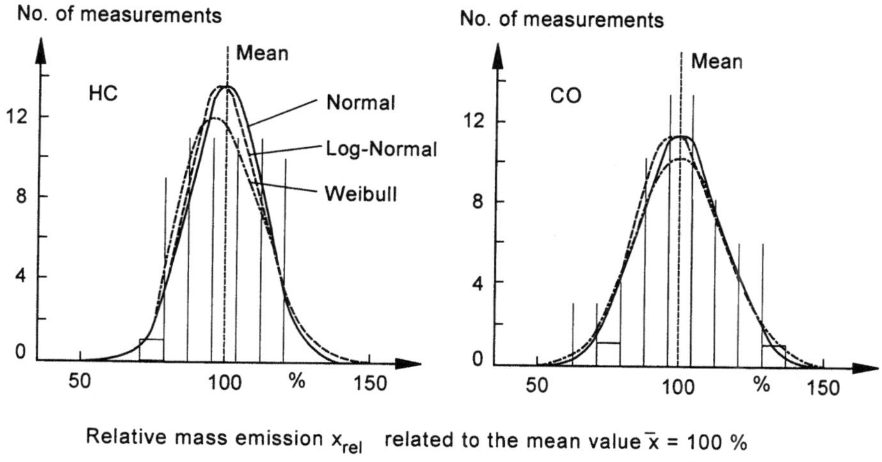

Fig. 9.3: Histograms and fitted density functions for HC and CO (50 repeated measurements on vehicle no. 6)

Table 9.1 shows the formulas (9.3) to (9.11) and the most important parameters of these three density functions. Figures 9.4 to 9.6 show their curves.

Unlike the normal density function, which is defined by mean and standard deviation, both the Weibull and the log-normal density functions are defined by three parameters.

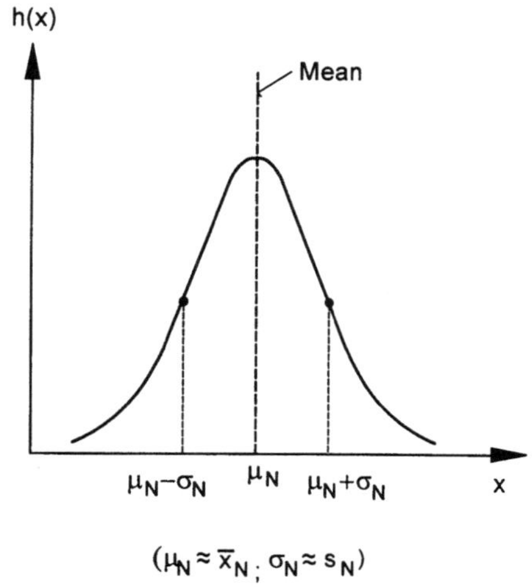

Fig. 9.4: Normal density function h(x)

Table 9.1: Density functions

	Normal	Log-Normal	Weibull
Plot	Fig. 9.4	Fig. 9.5	Fig. 9.6
Density function	$h(x)=\frac{1}{\sqrt{2\pi}\cdot\sigma}\cdot e^{-\frac{1}{2}\left(\frac{x-\mu}{\sigma}\right)^2}$ (9.3)	$h(x)=\frac{1}{\sqrt{2\pi}\,\sigma_L}\cdot\frac{1}{x-\gamma}\cdot e^{-\frac{1}{2}\left(\frac{\ln(x-\gamma_L)-\mu_L}{\sigma_L}\right)^2}$ (9.6)	$h(x)=\frac{\beta}{\vartheta}\cdot\left(\frac{x-\gamma_W}{\vartheta}\right)^{\beta-1}\cdot e^{-\left(\frac{x-\gamma_W}{\vartheta}\right)^{\beta}}$ (9.9)
Mean	$\mu\approx\bar{x}=\frac{1}{n}\sum x_i$ (9.4)	$\mu_L=\gamma_L+e^{\mu_L+\frac{\sigma_L^2}{2}}$ Maximum at: $e^{\mu_L-\sigma_L^2}$ Median value at: e^{μ_L} (9.7)	$\mu_W=\gamma_W+\Gamma\left(\frac{1}{\beta}+1\right)\cdot\vartheta$ (9.10)
Variance	$\sigma^2\approx s^2=\frac{\sum(x_i-\bar{x})^2}{n-1}$ (9.5)	$\sigma_L^2=(e^{\sigma_L^2}-1)\cdot e^{(2\mu_L+\sigma_L^2)}$ (9.8)	$\sigma_W^2=\vartheta^2\cdot\left\{\Gamma\left(\frac{2}{\beta}+1\right)-\Gamma^2\left(\frac{1}{\beta}+1\right)\right\}$ (9.11)
Parameter	μ - Mean $\approx\bar{x}$ σ - Standard deviation $\approx s$	γ_L - Minimum response parameter μ_L - Mean of log-values ln $(x_i - x_o)$ σ_L - Standard deviation of log-values ln $(x_i - x_o)$	γ_W - Minimum response parameter β - Shape factor ϑ - Scale factor Γ - Gamma function μ_W - Mean σ_W - Standard deviation

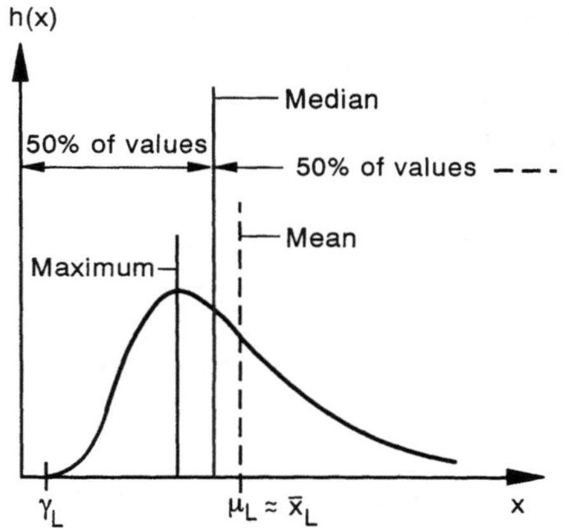

Fig. 9.5: Log-Normal density function. Here the median value (50 % of the values lie to the left and to the right of the median value) differs from the mean

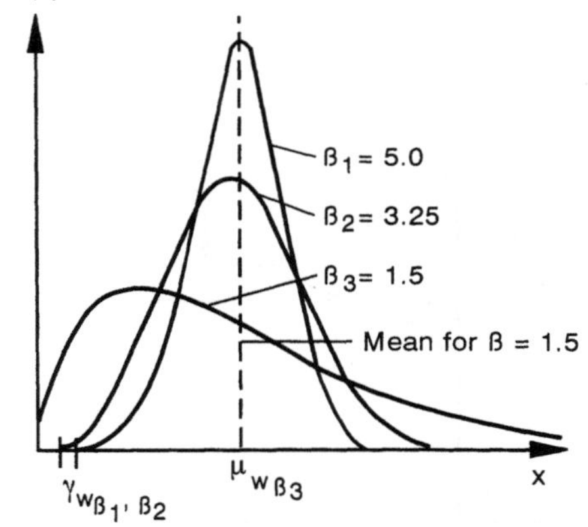

Fig 9.6: Weibull density function h(x) with different parameters β. This is the most variable of the three density functions

Applicability to measuring results

Testing the density function of a series of measurements can be done if the array of values is a random one and if the series itself is homogeneous, so that there are no discontinuities, clusters or trends to be observed. Only in this case is it permissible to assume that the sample given by the series of measurements constitutes a true reflection of a parent population. But even if the homogeneity of the series is ensured by the selection of the vehicles involved, there may be outliers in the exhaust emission values. These outliers must be disregarded in fitting tests. As they are not identifiable by any universally applicable criterion, any value which might be an outlier must be checked individually.

The given series of 50 repeated test runs cleared of any outliers on each of the 10 vehicles were checked by means of computations and graphs to establish which of the three density functions considered here would fit the empirical values best. The procedures used for this test are shown in Table 9.2.

Table 9.2: Computational and graphic methods for determining a suitable density function

Density function	Computational method	Graphical method
Normal	χ^2 - test Test of the 3rd and 4th Moment	Plot of normal probability graph
Log-Normal	χ^2 - test Test of the 3rd and 4th Moment	Plot of logarithmic probability graph
Weibull	χ^2 - test	Plot of Weibull probability graph

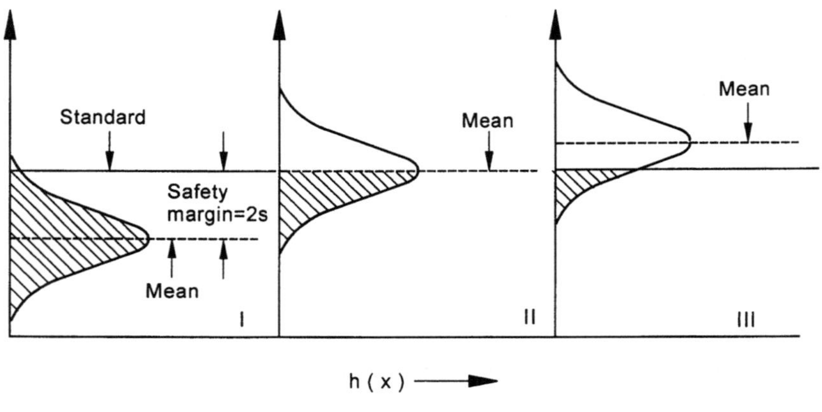

Fig 9.7: Three different positions of the density functions of mass emissions readings relative to the standard

Three cases have to be distinguished between:

I: There is a sufficient safety margin of 2*s* between mean and standard, and the probability of readings being below the standard is P = 95 %.

II: Mean and standard coincide, so the probability being either above or below the standard is uniformly P = 50 %.

III: The mean exceeds the standard, yet there is sufficient probability that individual readings will still be below the standard.

Generally, the investigations showed that

> *data from repeated measurements on one and the same vehicle are sufficiently approximated by normal density functions.*

From the large standard deviation, i.e., the large variations of test data, it is statistically obvious that

> *single measurements are pointless for exhaust investigations. Only the mean data from several measurements is a usable value to judge exhaust mass emissions. The mean value approximately corresponds to the actual emission value if the number of repeated measurements n is large enough.*

Since the regulated certification procedures (refer to Chap. 9.6) in Europe or in the US only allow a limited number of repeated measurements, there is only a limited probability of a vehicle passing the certification test even if its actual emissions are lower or equal to the standards. On the other hand, there is also the risk of a vehicle passing the certification test whose true emissions exceed the standards. Figure 9.7 illustrates this point (systematic deviation is still disregarded - refer to Chap. 9.4) using three different cases.

9.3 Operational Characteristics

This problem that proper vehicles may be rejected and inferior vehicles may be accepted makes it understandable that manufacturers as well as government authorities are very interested in knowing the probability of acceptance or rejection of a vehicle. The situation is further complicated by the fact that the HC, CO, and NO_X emission standards have to be met simultaneously so that the total probability of meeting all three standards decides the question. Curves representing probability as a function of the mean related to the standard are called *operational characteristics* and they are well suited for judging test procedures.

Figure 9.8 shows examples of such operational characteristics computed for one test ($n = 1$) with the relative standard deviation $\sigma_r (\approx s_r)$ known from former repeated measurements m, as parameter. The abscissa show the relative mean values $\mu_r \approx \bar{x}_r$ related to the standard (which is set equal to 100 %) of the parent population of measured values.

Values μ_r relativ to the standard give the probability to pass the test. In the case of μ_r = standard = 100 % the probability of passing or failing is equal and equal to 50 % for all σ_r (see middle part of Figure 9.7).

The risk is reversed.

With the increasing σ_r, the probability of passing decreases or of failing increases, respectively.

Only in case of the theoretically ideal step curve of a certification does the probability of passing amount to 100 % as long as μ_r is smaller than or equal to the

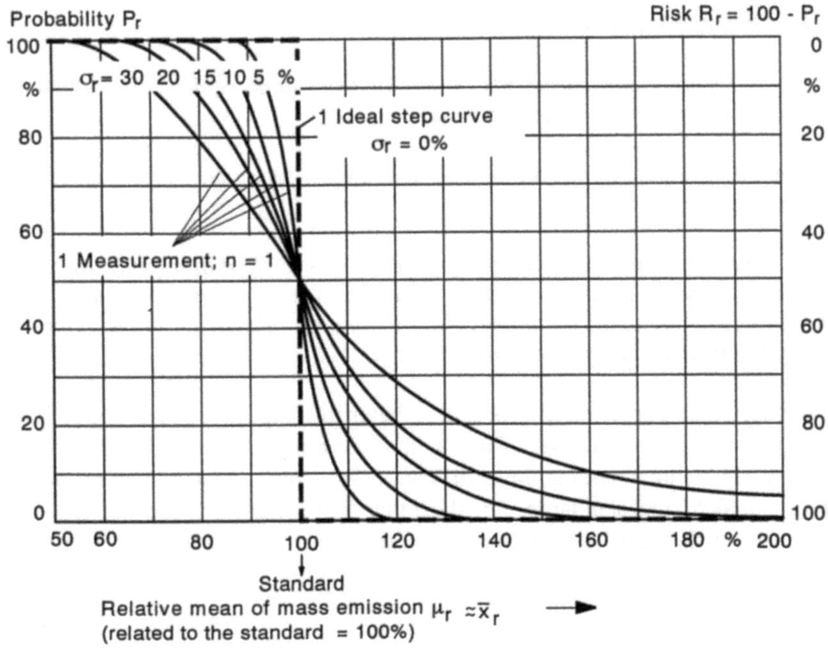

Fig. 9.8: Operational characteristics for $n = 1$, and $m \Rightarrow \infty$ (step curve), and for different σ_r as parameter, known from former $m \geq 25$ repeated tests

standard and come down to zero as soon as the μ_r increases beyond the standard, the risk then being 100 %.

All certification tests are based on the assumption that a manufacturer who presents a vehicle for certification already knows its emissions of the exhaust gas components to be tested. Consequently a number of tests (m) is performed by the manufacturer before the certification testing; another predetermined number of tests (n) is run by the authorities for certification. Mathematically, a computation of the probability of passing an exhaust emission test on the basis of one single measurement ($n = 1$; measured value $= x$) is equivalent to forming an integral over the density function $h(x)$ between $-\infty$ and the standard x_G, thus:

$$P\,(x \leq x_G) = \int_{-\infty}^{x_G} h(x)\, dx \qquad \text{with} \qquad 0 \leq P(x) \leq 1 \tag{9.12}$$

As there are only upward limits to mass emissions, we are here only interested in the probability $P(x \leq x_G)$ for the measured value x remaining below the standard x_G.

Normally, it is assumed that the data obtained from repeated tests approximately follow a normal distribution, and this assumption has been confirmed experimentally. This being so, $h(x)$ is completely described by the true mean value μ and the standard deviation σ of the parent population, which means that for the computation we need to know both the true mean value μ and the standard deviation σ.

Under practical conditions, these two quantities are approximated by the mean $\bar{x}$ and the standard deviation s which are found, for instance, by the manufacturers performing a number (m) of tests before applying for certification.

For the means of n tests, the probability of passing the certification is governed by the same formula, with x being replaced by $\bar{x}$ and s by $\frac{s}{\sqrt{n}}$.

For other criteria than the ones considered here - one actual test ($n = 1$) and the mean $\bar{x}$ drawn from n values - the probability of passing the certification test is computed from the probability ratings of all criteria combined. Thus, for instance, the probability of passing the US test, which involves a simple re-test ($n = 2$) (see Figure 9.9), is equal to the sum of the probability of passing the first test (P_1) and the probability of passing the re-test (P_2) multiplied by the risk of the first test ($1 - P_1$), so that

$$P_{US} = P_1 + P_2 \cdot (1 - P_1) . \tag{9.13}$$

In this case, $P_1 = P_2$, i.e.

$$P_{US} = 2\,P_1 - P_1^2 \tag{9.14}$$

$$P_{US} = 1 - (1 - P_1)^2 . \tag{9.15}$$

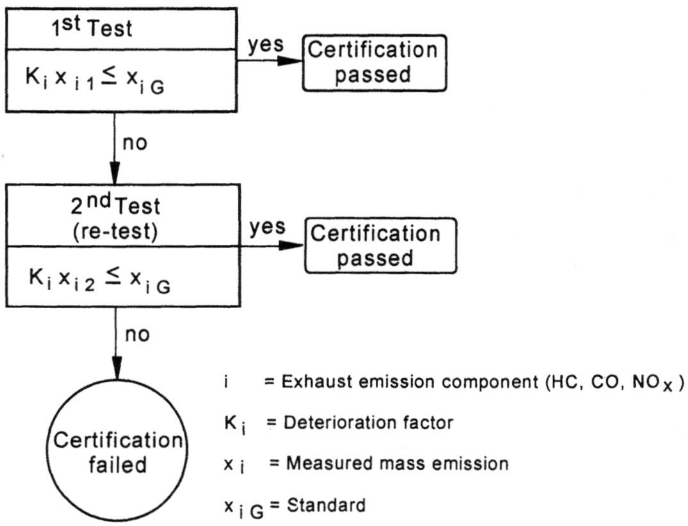

Fig. 9.9: Flow chart of the US certification test procedure. One re-test is permitted. If it is carried out, only $x_{i,2}$ is valid

This line of reasoning may be applied analogously also to other composite criteria, such as those for the European test.

On the other hand, it is possible to compute for a given probability P(x) the maximum emissions x_{MAN} or the maximum mean emissions $\bar{x}_{MAN}$ which automobile manufacturers have to meet before entering a vehicle for certification testing. (Systematic errors are not taken into account.)

Monte Carlo method

These equations, however, cannot always be solved explicitly. For such cases and whenever other types of density functions occur, it is simpler to apply the Monte Carlo method. This method has already been described in Volume A: Acoustics [Akustik], Chap. 9.2.3. The computation is based on a given density function $h(x)$, for instance a normal density function with true mean value μ and standard deviation σ, or in other words, a normalized probability function $P(x)$: A computer is used to generate a number of random numbers z_i which are uniformly distributed between 0 and 1. These numbers z_i are used as P values and then the x arguments are determined (Figure 9.10), which in turn, are distributed randomly according to the given density function $h(x)$.

This procedure can also be used in connection with other density functions, because the values P_i are by definition always uniformly distributed between 0 and 1.

The individual figures making up the sample generated in this fashion (no less than 1,000 individual figures in our case) are regarded as data obtained from repeated tests, and are distributed around μ according to the given density function. This fictional data is then computer-processed according to legal or theoretical test criteria. For this purpose, the test criteria are made into flow chart diagrams (see

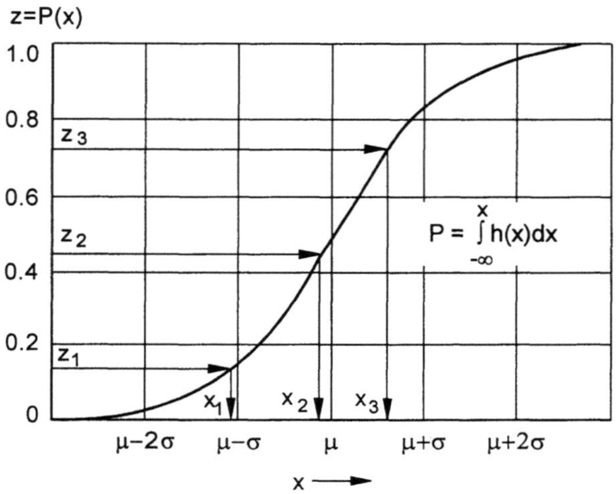

Fig. 9.10: Selecting individual figures (x_i) from a parent population defined by a cumulative distribution, using uniformly distributed random numbers (z_i) with $0 \leq z_i \leq 1$

e.g. Figure 9.9, which is a flow chart of the US certification test) and appropriately designed computer programs are generated. Each individual test mentioned in the official test criteria, for instance each of the two tests which make up the US procedure, must be run on an independent sample. Let us use the particularly simple US test procedure to explain the process of computation. Assuming that there is a total of 2,000 figures, 1,000 of which are used for each test (the first and second test of the US procedure), we proceed as follows:

1. The 1,000 figures which make up the first sample are screened to see whether they are lower than or equal to the standard. The percentage of the total sample (1,000) which is lower than or equal to the standard (e.g. 600 figures, i.e. 60 %) represents the probability of passing the certification test with the first test run, which in our case is $P = 60\ \%$.

2. With these probabilities and the multiplication theorem of probability theory, we compute the probability P^* of failing the first test and passing the second as:

   $$P^* = (1 - P_1)\, P_2 \, . \qquad (9.16)$$

   As a consequence

   $$P_1 = P_2 = 0.6 \mathrel{\hat{=}} 60\ \% \ , \qquad (9.17)$$

   $$P^* = (1 - 0.6)\, 0.6 = 0.24 \mathrel{\hat{=}} 24\ \% \ . \qquad (9.18)$$

   The probability of passing the certification test at least after the re-test is:

   $$P_{US} = P_1 + P^* = 60\ \% + 24\ \% = 84\ \% \ . \qquad (9.19)$$

   A comparison with (9.15) shows the same result for $P_1 = P_2 = 0.6$.

3. The probability thus computed applies only to the given mean μ. For the computation of the probability of the other mean values μ the entire process described above has to be repeated. In this way, we finally arrive at the operational characteristics.

Other test procedures which are more complicated than the US method can be dealt with by means of the same computation process (refer to Fig. 9.25).

The degree of precision to be attained by the Monte Carlo method mainly depends on the quality of the equally distributed random numbers. The error involved in the approximation decreases with the square root of the sample population n, which means that it is down to approximately 3 % if the sample population is 1,000.

Results

Figure 9.11 shows the operational characteristics computed for the former European and current US certification test procedures and of $n = 10$ repeated tests

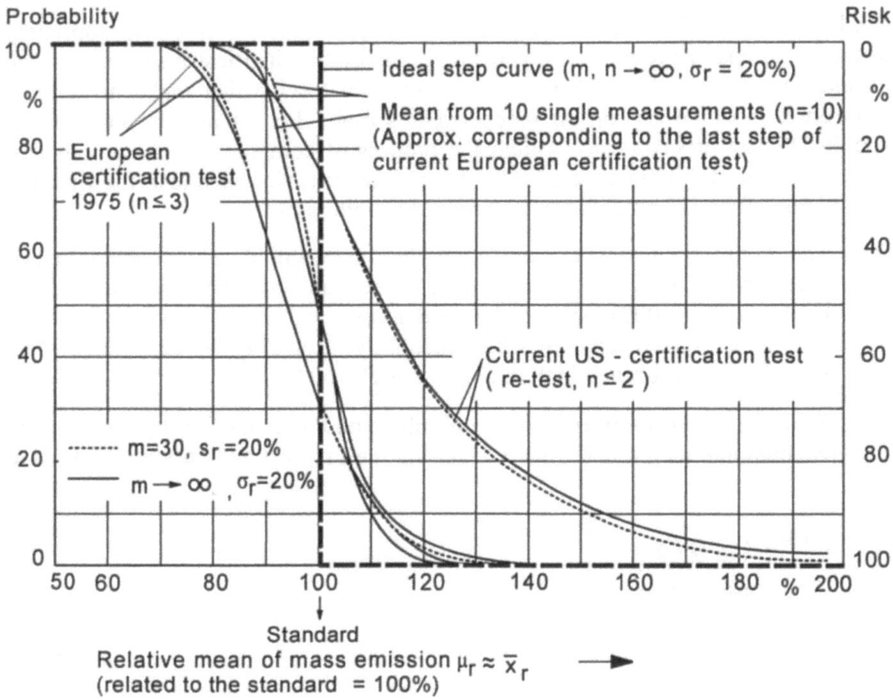

Fig. 9.11: Operational characteristics of various test criteria for ($m \rightarrow \infty$) and (m = 30).

for the last step of the current European test procedure (refer to Fig. 9.25 in Chap. 9.6.1) under the realistic assumption that the number of tests already run by the manufacturer is m = 30.

The curves reiterate the hypothetical case of a large number of tests ($m \rightarrow \infty$) being run by the manufacturer so that the true mean emissions $\bar{x}_r \Rightarrow \mu_r$ of the vehicle are precisely known.

Furthermore, the operational characteristics for the mean from $n = 10$ measurements with $m \rightarrow \infty$ and $m = 30$ are shown. The individual entries represent the probability of passing or failing as a function of the mean $\mu_r \approx \bar{x}_r$, respectively, as a percentage of the standard or the corresponding risk.

The step curve applies if the true mean emissions are known to the manufacturer before the certification testing ($m \rightarrow \infty$). This curve has already been explained.

The differences existing between the certification test procedures concerning the probability of passing or failing them are relatively large. Most suitable would be the mean of $n = 10$ repeated tests, nearly comparable to the last step of the current European certification procedure, refer to Figure 9.25 in Chap. 9.6.1. The US certification test (re-test) has an operational characteristic which is for $\bar{x}_r < 90$ % practically equal to the one for $n = 10$ repeated tests (e.g., last step of the European certification procedure). Due to the necessary safety margins, the curves lie in this range anyway or even lower. The US certification procedure therefore could be used world-wide. The current European certification procedure is much too complicated and was only introduced for political reasons. However, for the automotive manufacturer, the first steps of this procedure are more flexible than the US procedure.

Total probability

The total probability of simultaneous conforming to all exhaust emission standards involved in a certification test (compared to the one-dimensional case) is by definition represented by the integral over a three-dimensional density function $h(x, y, z)$ of the emissions x, y, and z of the three exhaust gas components HC, CO, and NO_X between $-\infty$ and their exhaust emission standards x_G, y_G, z_G.

If there is a correlation between the three exhaust gas components, $h(x, y, z)$ also depends on the correlation coefficients $\rho_{xy}, \rho_{xz}, \rho_{yz}$.

One method of solving it is a process of separation by which the triple integral is reduced to a product of single integrals by transforming the variables. This is always possible when distribution is normal. The objective of the transformation is to arrive at stochastically independent variables, which again are normally distributed. The integrals can then be computed relatively easily in a numerical way.

To illustrate this let us outline the basic procedure to be followed in a two-dimensional case, i.e., that the test involves only two exhaust gas components. The equations for the total probability, on the assumption that there is a linear correlation between these two components, and that only one test run ($n = 1$) has to be performed are as follows:

$$P(x \le x_G;\ y \le y_G) = \int_{-\infty}^{x_G} \int_{-\infty}^{y_G} h(x,y)\, dx\, dy \tag{9.20}$$

with

$$h(x,y) = \frac{1}{2\pi\sigma_x\sigma_y\sqrt{1-\rho_{xy}^2}} \cdot e^{-\frac{1}{2}g(x,y)} \tag{9.21}$$

and

$$g(x,y) = \frac{1}{1-\rho_{xy}^2}\left[\left(\frac{x-\mu_x}{\sigma_x}\right)^2 - 2\rho_{xy}\left(\frac{x-\mu_x}{\sigma_x}\right)\left(\frac{y-\mu_y}{\sigma_y}\right) + \left(\frac{y-\mu_y}{\sigma_y}\right)^2\right]. \tag{9.22}$$

If there is no correlation between the exhaust gas components, i.e. for

$$\rho_{x,y} = 0\ , \tag{9.23}$$

the two-dimensional density function is

$$h(x,y) = h_1(x) \cdot h_2(y) \tag{9.24}$$

and the total probability is

$$P = \iint h(x,y)\, dx\, dy = \int_{-\infty}^{x_G} h_1(x)\, dx \cdot \int_{-\infty}^{y_G} h_2(y)\, dy = P(x) \cdot P(y) \ , \tag{9.25}$$

i.e., the total probability is equal to the product of the individual probabilities.

If there is a correlation, i.e. for

$$\rho_{x,y} \neq 0 \ , \tag{9.26}$$

the transformation

$$\begin{aligned} x &\rightarrow x^* \quad ; \quad x_G \rightarrow x_G{}^* \\ y &\rightarrow y^* \quad ; \quad y_G \rightarrow y_G{}^* \end{aligned} \tag{9.27}$$

is performed, followed by a rotation and subsequent computation of the total probability according to

$$P(x_i \leq x_G \ ; y_i \leq y_G) \ = \int_{-\infty}^{x_G^*} h_1^*(x^*)\, dx^* \cdot \int_{-\infty}^{y_G^*} h_2^*(y^*)\, dy^* \ . \tag{9.28}$$

The computation can be done in a similar fashion if the certification involves several tests (n), the mean results of which constitute the criterion. In this case, $\sigma_x, \sigma_y, \sigma_z$ are replaced by $\frac{\sigma_x}{\sqrt{n}}, \frac{\sigma_y}{\sqrt{n}}, \frac{\sigma_z}{\sqrt{n}}$, x, y by $\bar{x}$, $\bar{y}$ and $h(x, y)$ by the student density function.

9.4 Systematic Errors, Round Robin Tests

As always in measuring techniques, systematic errors constitute another grave problem in all current checks, from certification to field monitoring. Figure 9.12 shows schematically how they occur in different ways.

Systematic errors may occur between laboratories A and B, but also between different test facilities A1, A2, ... or B1, B2 ... and they may occur even between measurement results made at different times on the same test facility, e.g. A1. Often a systematic error can be relatively as high as 100 % and more, i.e., measurement results can differ by more than a factor of two.

In the previous chapter one part of the total error problem, namely the statistical errors were discussed on the assumption that measurement results were free of systematic errors. In exhaust emission measurements, however, systematic errors occur along with the statistical ones. These systematic errors are not easily detected, but often play a deciding role. Errors of this kind frequently occur in practice.

A typical example is shown in Figures 9.13 to 9.15.

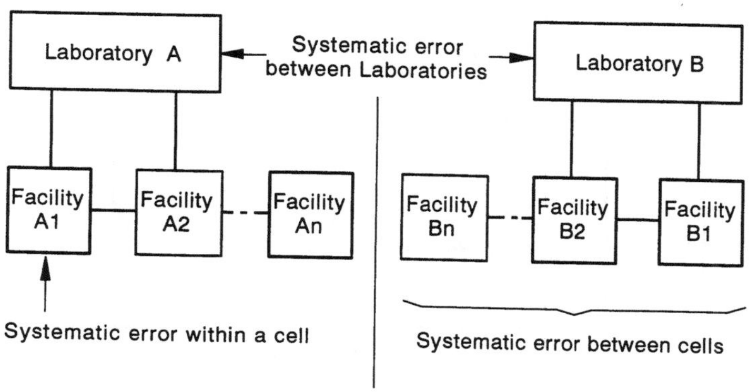

Fig. 9.12: Diagram showing two different test laboratories incorporating various test facilities. Laboratory A might be that of an automobile manufacturer, laboratory B might be a government laboratory

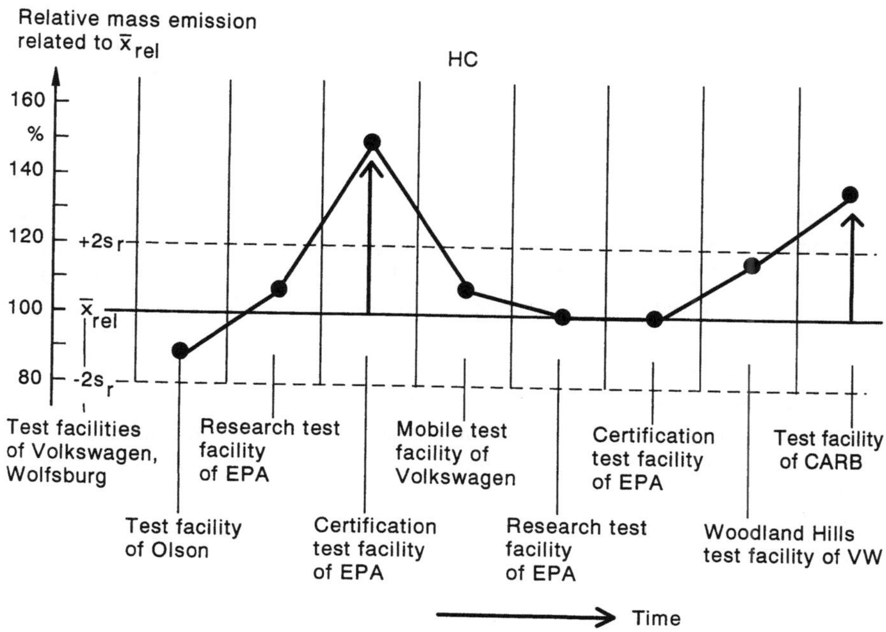

Fig. 9.13: Comparison of exhaust gas emission means for HC obtained from the same vehicle in several different laboratories and test facilities. All mean values given here (drawn from three repeated tests per test facility) are related to the $\bar{x}$ values measured at Wolfsburg (mean drawn from n = 13 repeated tests; $\bar{x}_{rel}$ = 100 %)

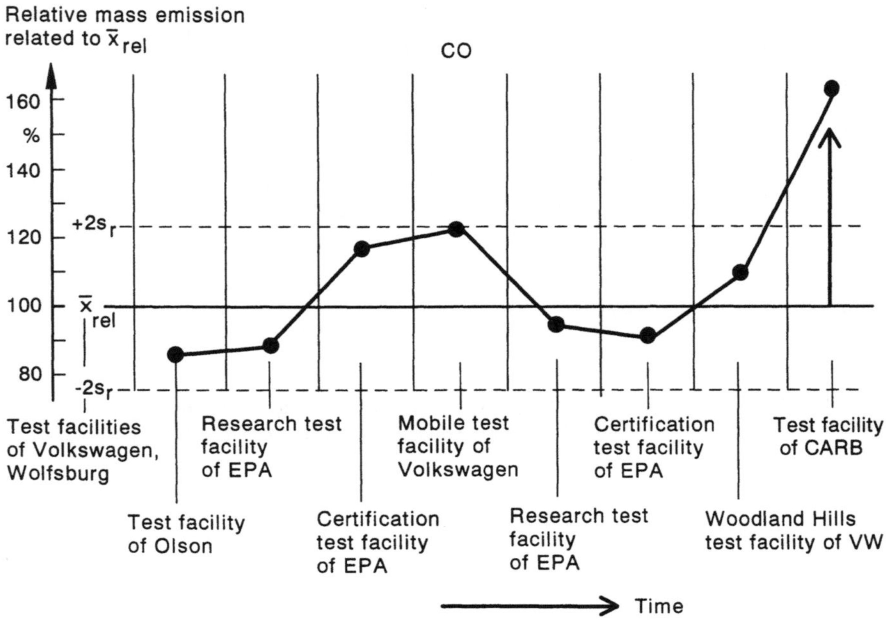

Fig. 9.14: Comparison of exhaust gas emission means for CO obtained from the same vehicle in several different laboratories and test facilities. All mean values given here (drawn from three repeated tests per test facility) are related to the $\bar{x}$ values measured at Wolfsburg (mean drawn from n = 13 repeated tests; $\bar{x}_{rel}$ = 100 %)

In this case, a VW vehicle was tested at the test facilities of several different laboratories, some of them belonging to the official US exhaust emission authorities, the EPA and CARB. The figure shows the mean emissions recorded there in relation to the mean emissions (set to 100 %) recorded at the Wolfsburg test facilities.

Two different EPA test facilities produced HC results (see arrow in Fig. 9.13) which were approximately 40 % apart. Similar deviations occurred between the CO measurements made in Wolfsburg and those made by the Californian exhaust emission authority CARB and the CO deviation (refer to Fig. 9.14) was even more pronounced (see arrow).

Additional measurements taken some weeks later show that the HC and CO values of EPA and Wolfsburg now agree quite well (Fig. 9.13 and Fig. 9.14). All these discrepancies indicate that there are systematic errors inherent in the test facilities themselves. The values mentioned here are not recent (1973); they have merely been used to illustrate the problem. However, only a few years ago, systematic errors of a factor of two to three between values measured at EPA and VW occurred for HC exhaust emissions from diesel vehicles. Therefore systematic errors still play a significant rôle today.

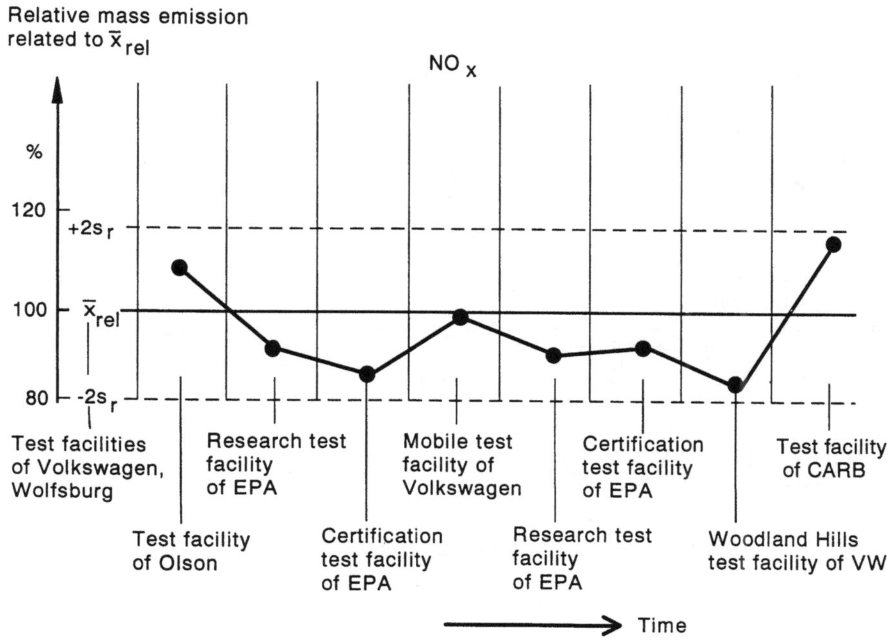

Fig. 9.15: Comparison of exhaust gas emission means for NO_x obtained from the same vehicle in several different laboratories and test facilities. All mean values given here (drawn from three repeated tests per test facility) are related to the $\bar{x}$ values measured at Wolfsburg (mean drawn from n = 13 repeated tests; $\bar{x}_{rel}$ = 100 %)

Current practice is to uncover systematic errors occurring between different test facilities by running what is called a round robin test. For this purpose vehicles specifically selected for their stable emissions are used (Figure 9.16).

The large circles indicate the location of automobile manufacturers' or governmental laboratories. The smaller circles indicate exhaust emission test facilities. Within one larger circle, a continuous series of checks is run, involving one or more vehicles whose emissions are stable. The test facility whose emission readings approximate closest the mean readings of the previous round robin test is defined as the reference test facility (dark circles).

Figures 9.17 to 9.19 illustrate that round robin tests of this kind are capable of pointing out systematic errors between different test facilities.

These figures show the mean mass emissions drawn from between 30 and 70 tests performed on test facilities 4 - 13. If the curves of all three vehicles run parallel, we may assume that any deviations between the various test facilities are not due to the vehicles themselves. Due to the relatively large variability of the test vehicles' emissions, this rough method can only be taken as a first step to develop a reliable procedure for detecting and eliminating systematic errors. The current procedure is far from satisfactory.

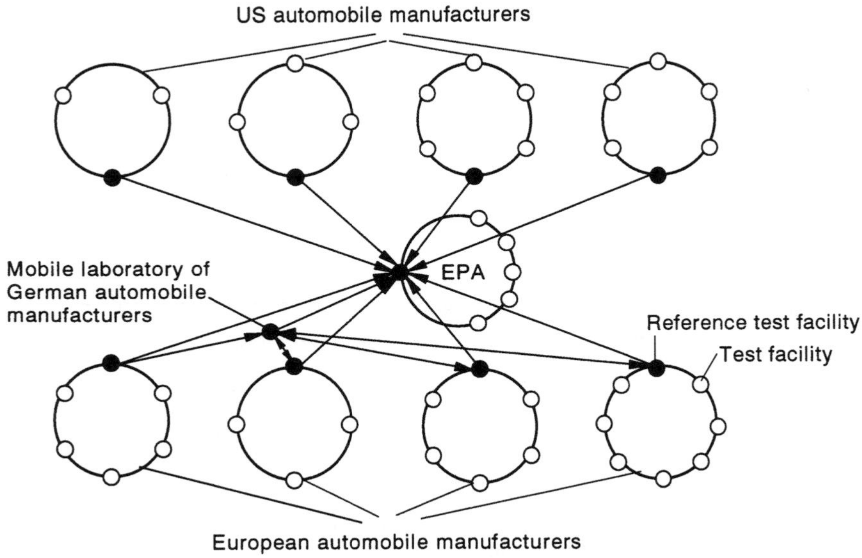

Fig. 9.16: Diagram of a round robin test

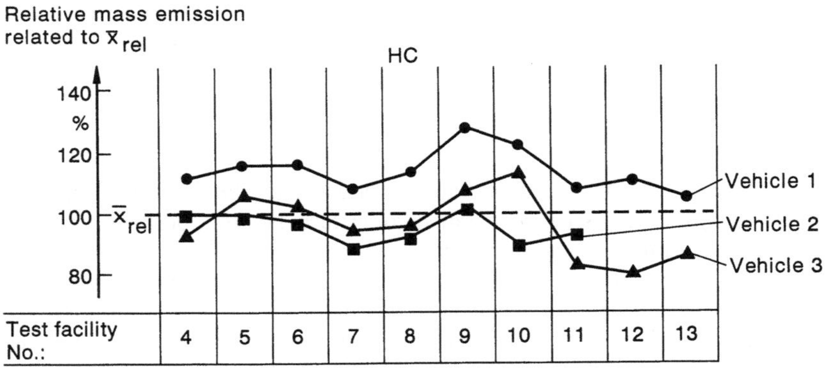

Fig. 9.17: Results of a round robin test for HC involving 10 test facilities and 3 vehicles

In such round robin tests, mobile exhaust emission test facilities are used as well for field monitoring (see Chap. 9.11). Figure 9.20 shows a phantom drawing of a mobile exhaust emission test facility.

Besides the exhaust emission measurements round robin tests require correct calibration of the chassis dynamometer by using a torque measurement unit as described in Chap. 8.2.5.

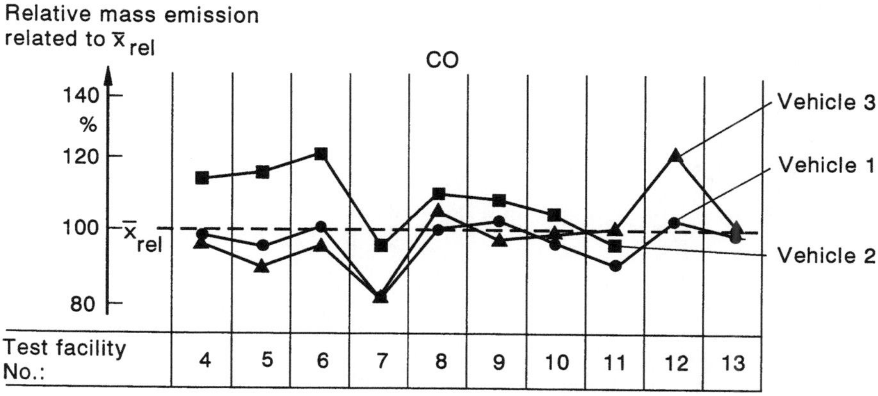

Fig. 9.18: Results of a round robin test for CO, otherwise as in Fig. 9.17

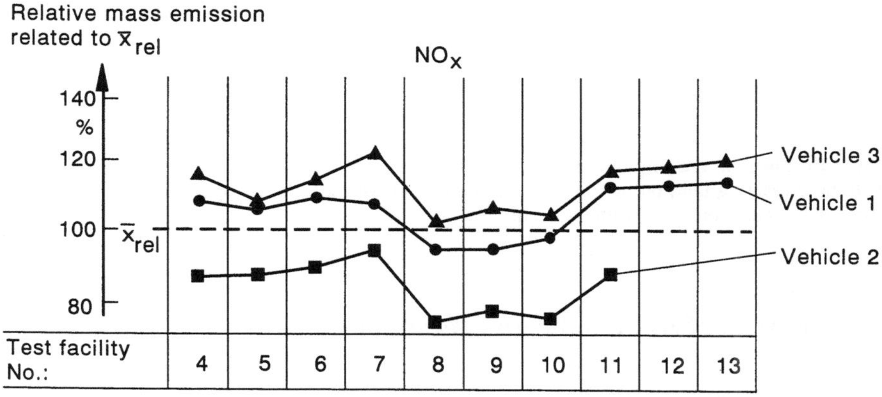

Fig. 9.19: Results of a round robin test for NO_x, otherwise as in Fig. 9.17

The present situation is unsatisfactory because systematic errors are often greater than random errors, the extent of the latter being dependent on the number of reiterative tests used. Unlike random errors, quantitatively unknown systematic errors involved in certain parameters cannot be stated explicitly. Each parameter alone may be the cause of a great systematic error. Therefore, as will be explained in this Chapter, the *safety margin* is governed by systematic errors even more than by random ones. Only the general introduction of calibration vehicles with small random errors of exhaust emissions will improve the situation.

Fig. 9.20: Phantom drawing of a mobile exhaust emission test facility, on the left hand side a bus with the measuring unit, on the right a semi trailer unit with the chassis dynamometer. Measuring and data lines are connected with the setup

9.5 Analysis of Systematic and Random Errors

9.5.1 Both Error Types

Taking into account the measuring uncertainties, the errors of n independent individual measurements on N test facilities are computed according to the following equations:

$$y=\bar{x}\pm u \tag{9.29}$$

$$y=\bar{x}\pm\ t\left(\sqrt{\frac{\sigma^2}{n}+\frac{1}{3N}\sum_{j=1}^{l}\left(\Delta z_j\right)^2}\right) \tag{9.30}$$

or

$$y=\bar{x}\pm\sqrt{\frac{A_1}{n}+\frac{A_2}{N}}\ , \tag{9.31}$$

with

u	measurement uncertainty of the mean $\bar{x}$
Δz_j	systematic portions of parameters such as driver, chassis dynamometer, etc., to be treated in a statistical way,
t	variable of student density function,
$\sigma \approx s$	standard deviation,
l	number of systematic error proportions,
n	number of measurements and
N	number of test facilities or measurement units used.

Practical consequences can be drawn from the formulas shown: If only one measuring instrument, such as one test facility ($N = 1$), is used, the sum total of the systematic errors which can be estimated reaches its maximum. Its contribution to the overall uncertainty will be constant and does not depend on the number of tests actually performed. The contribution of this kind of error can only be reduced by introducing additional measuring instruments (N test facilities, for instance). Certification tests in the US, for instance, should be repeated on another test facility whenever there is any doubt concerning the validity of the first test. As far as drivers are concerned, it is known that experienced drivers cause fewer errors (refer to Figure 8.14 to 8.17 in Chap. 8.2.4). For comparative measurements at least the same driver or an automatic driver (refer to Chap. 8.2.4) should be used throughout. If different drivers are used, then their errors have to be taken into account when comparing results.

Random error, on the other hand, can be minimized by repeated tests ($n > 1$). Because of the expense and time involved in an exhaust emission test, including preparation (cold start after conditioning of 12 h, test time 1 h, refer to Chap. 8), the possible number of repeated exhaust emission tests is limited for practical reasons.

9.5.2 Influence of Parameters of the FTP on Overall Uncertainty

Any computation of statistical errors, which are always present, must start from the equations used for determining mass emissions. According to the Gaussian Law of Error Propagation, the uncertainty involved in measuring mass emissions results from the uncertainties of each individual parameter. However, the problem arises that while the error of the final exhaust emission value can be evaluated, the errors of the individual parameters of the test facility can only be estimated. Schematically, Figure 9.21 shows how parameters contribute to individual measurement uncertainties and how these combine into overall uncertainty.

Figure 9.22 shows the results of computations and estimates averaged over exhaust gas components. The figures given here represent the minimum uncertainties attainable by careful calibration of all measuring instruments and the dynamometer.

The relative errors of 5 % due to the analytical instruments are of little significance as far as the total relative error is concerned, this depending mainly on the influence of engine, driver and chassis dynamometer.

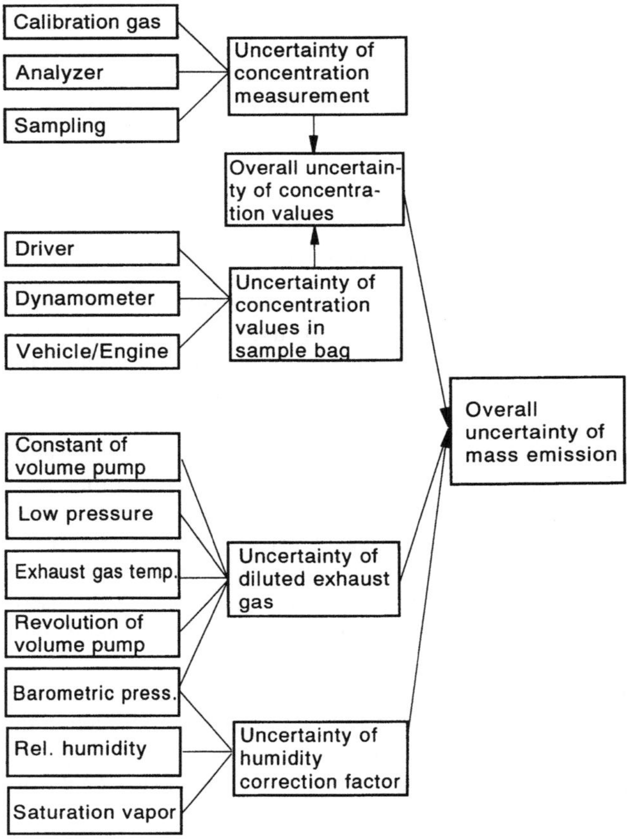

Fig 9.21: Block diagram of the parameters involved in computing the overall uncertainty of mass emission (according to the Gaussian Law of Error Propagation) for the US-75 test procedure. The left column shows the parameters on which the individual uncertainties and, finally, the overall uncertainty depend

Experience has shown that drivers may be responsible for dispersed measured results, even if they carefully try to keep within the tolerance limits of a driving cycle, the critical factor being their way of operating the throttle.

Generally speaking, however, individual figures are of lesser interest compared to the overall measuring uncertainty (engine influence excluded) which, for both test procedures, amounts to approximately 16 % if the environment parameters are taken into account, and to about 12 % if they are not.

Generally, the most significant parameter is the vehicle or engine itself, refer to (9.2). In this context, the term „variability due to engine influence" signifies the dispersion of measured values caused by the instability of the engine itself (variations of the combustion and after-treatment devices, e.g., the catalytic

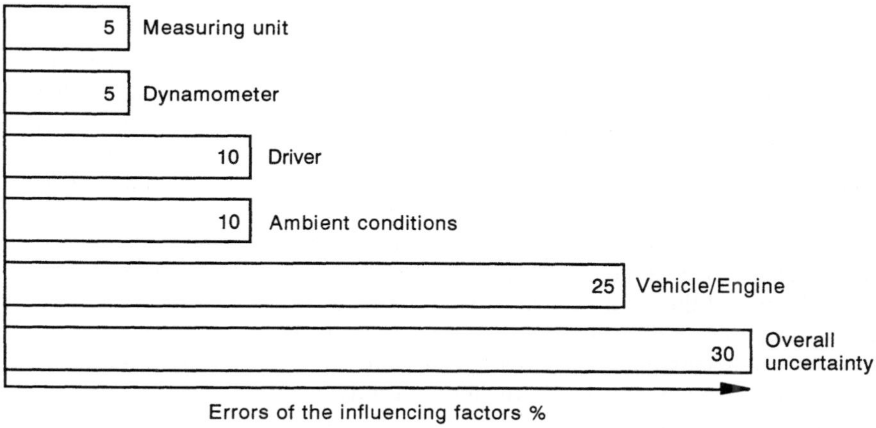

Fig. 9.22: Estimates of the contribution of the different parameters to the overall uncertainty of mass emissions. The individual parameters are to be combined according to the Gaussian Law of Error Propagation

converter), not taking into account the influence of either the driver or the dynamometer. Figure 9.23 shows the influence of the engine on total variability.

If we assume the variability due to the engine to amount to 20 - 40 % ($2s_{rEngine}$, $P = 95$ %) the total variability of measured values is mainly to be attributed to this factor. Only after the engine's proportion has dropped below 15 %, can other parameters become predominant, but see (9.2).

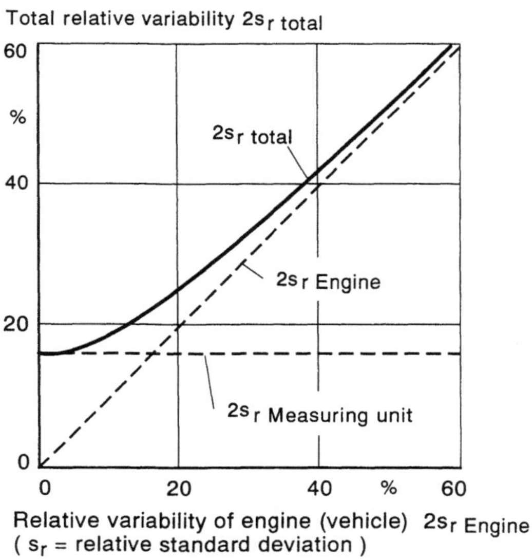

Fig. 9.23: Relative total variability $2s_{r\ total}$ as a function of the relative variability due to the engine $2s_{r\ Engine}$ taking the relative measuring unit variability $2s_{r\ Measuring\ unit}$ into account

9.5.3 Total Safety Margin

The logical consequence of not being able to avoid statistical and systematic errors is that automobile manufacturers are forced to increase the safety margin between mean exhaust emissions and standards even further than shown in Fig. 9.7. The diagram in Figure 9.24 exemplifies the present position. Here again, emission test results are presented in a bar diagram along with a fitted density function (here normal). The systematic error (*F*) is shown as well.

If the test is based on only one or fewer measurements the total safety margin is the sum of the total variability of measured values (2*s*) and of the systematic error (*F*). If the probability of a vehicle passing the statutary government emission test on the basis of one measured result is to amount to P = 95 %, the mean of all emission test results must be below the standard by a safety margin like the one shown in the diagram. This holds true for each emission component.

Therefore the total safety margin can be deduced from statistical and systematical errors. The position of the mean relative to the standard x_G resulting from the total safety margin S_{total}, constitutes the engineering goal *E*.

$$E = x_G - S_{total} = x_G - (F + 2\,s) \tag{9.32}$$

This actually means that the standards indirectly become stricter by a factor of two and more. The public is generally not aware of this restrictive influence. High additional costs arise which are finally paid for by those who buy an automobile. Moreover, this very strictness tends to prolong the periods of development. Further lowering of standard values will aggravate this problem.

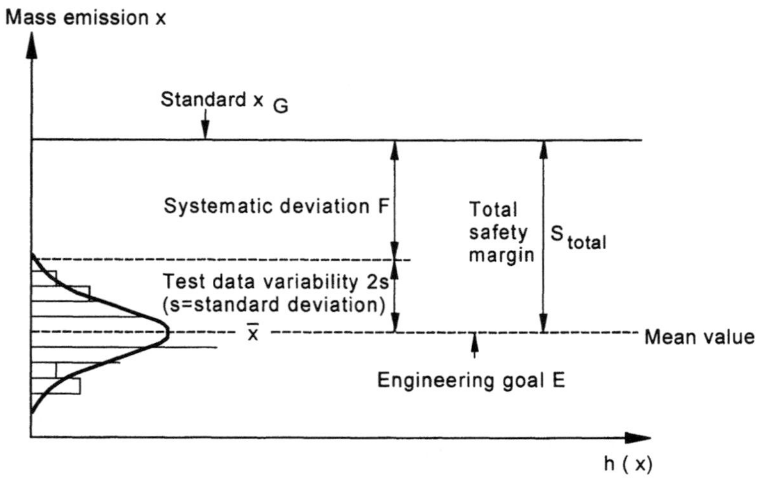

Fig. 9.24: Variability of test data, unrecognized systematic errors, total safety margins, engineering goal, refer to Fig. 9.7

9.6 Certification Testing

9.6.1 European Procedure

In Europe, the official certification of a new type of vehicle depends on the outcome of an exhaust emission test covering HC, CO and NO_X. Figure 9.25 shows a flow chart of the certification test and its relatively complicated evaluation process. As shown in the flow chart, certification is granted after one test if the CO reading and the reading of the sum of HC + NO_X is less or equal to 0.7 of the prescribed standard.

If one of these readings is higher than 0.7 but smaller or equal to 0.85 of the standard, certification is granted after a second test if the readings of this second test are less than the standards and if the sum of the readings of the 1st and 2nd test is less than 1.7 of the standard for CO or HC + NO_X respectively. Further steps are rather complicated.

Finally, the number of tests can go up to ten and certification is passed if the mean of the respective readings is less than the standard. This number of ten repeated tests at a maximum is meant to take the existence of statistical variables into account.

9.6.2 US Procedure

The US procedure or the corresponding flow chart has already been described in Chap. 9, refer to Figure 9.9. To obtain the certification of a vehicle class (engine family) in the United States the manufacturers have to prove that the vehicle will comply with the emission requirements for HC, CO and NO_X over its entire useful life, which is officially defined at 50,000 miles (80,470 km) or 100,000 miles (164,940 km) with different standards for each.

To pass the certification tests, automobile manufacturers have to operate two fleets of vehicles:

A) The durability data fleet, which serves to determine the deterioration factors. All vehicles of this fleet are run for 50,000 or 100,000 miles, with their exhaust emissions measured every 5,000 miles. Following a predetermined program, the vehicles are operated either on the road or on dynamometers. The special driving cycles and the test routes used must be approved beforehand by the supervising authority. The scope of maintenance and inspection is limited, and the maintenance intervals are at present fixed at 12,500 miles. Emission measurements at each maintenance interval are mandatory. All unscheduled maintenance work must be reported. Any unusual measures on the durability test vehicles must be approved in writing by the EPA administrator in charge.

Using all HC, CO and NO_X data obtained, deterioration factors K_i are computed for each engine family. From the actual mileage and the emissions recorded, the emissions at 4,000 and 50,000 miles are computed by means of a regression line.

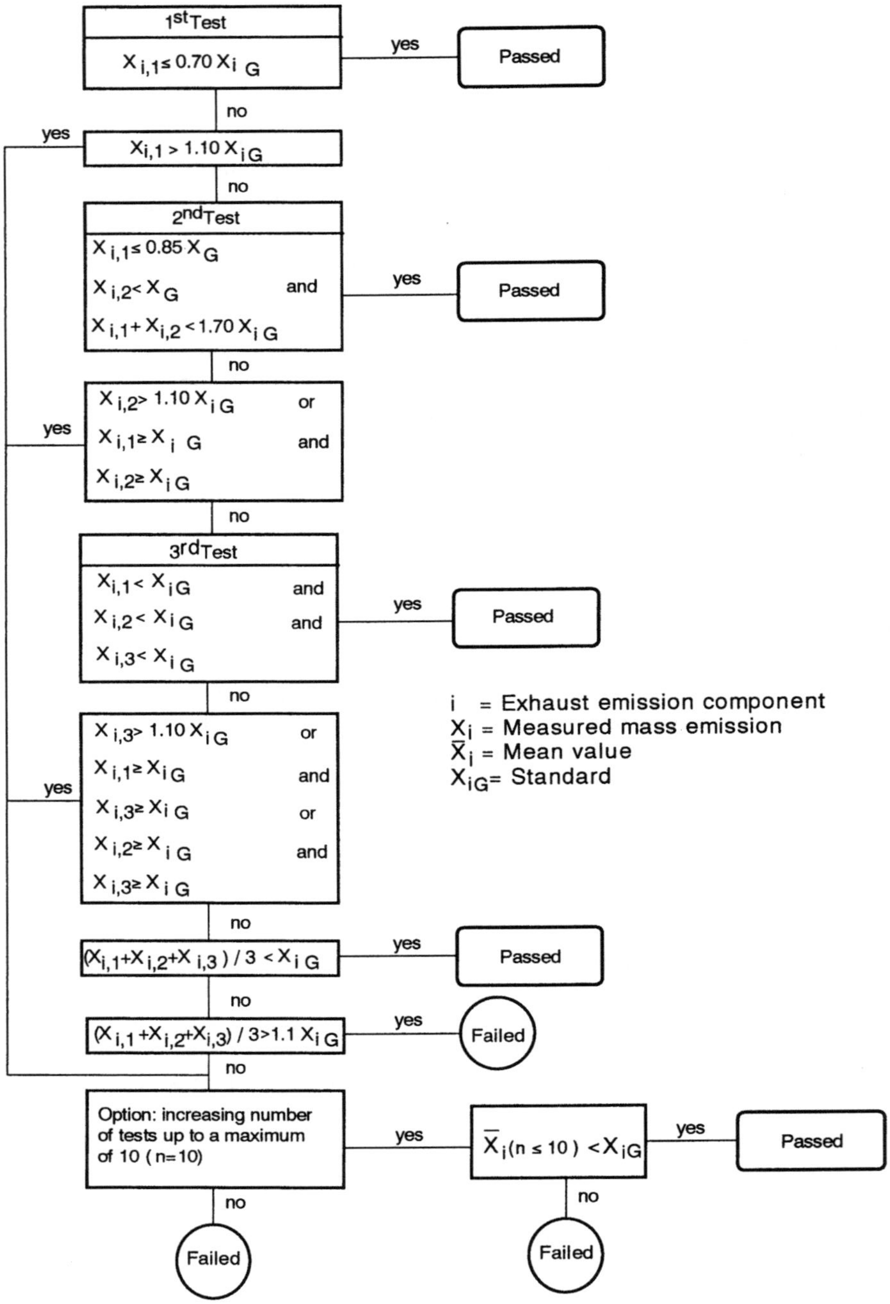

Fig. 9.25: Flow chart of the European certification procedure for limited exhaust gas components, i.e. CO, HC + NO_x (x_{iG} = standard)

Using the values thus obtained (see Figure 9.26), the deterioration factor (K_i) is determined according to the following equation:

$$K_i = \frac{x_i\,(50.000\ \text{mi})}{x_i\,(4.000\ \text{mi})} \tag{9.33}$$

i = exhaust gas component
x_i = mass emission

Figure 9.26 shows the evaluation of a deterioration factor K using CO as an example.

The regression line is defined by the individual readings. The standard (x_{iG}) is divided by the deterioration factor to form a new „standard" (x_{iG} / K_i) for evaluating the exhaust emissions. It is also possible to multiply the emission readings by the deterioration factor, leaving the standard unchanged. Figure 9.26 shows that it is questionable to determine a deterioration factor from a small number of data obtained from a few vehicles only. Regression lines are very uncertain if only a few widely dispersed values are involved, so that a horizontal line ($K_i = 1$) would have the same probability. Consequently this procedure for determining K_i is pointless.

B) The emission data fleet is used to determine the emissions of vehicles made with tools for production. Before being actually tested, all vehicles belonging to this fleet are run for 4,000 miles (6,436 km).

All vehicles belonging to the emission data fleet are checked by the authorities. This certification test consists of up to two test runs (refer to Figure 9.9). If more

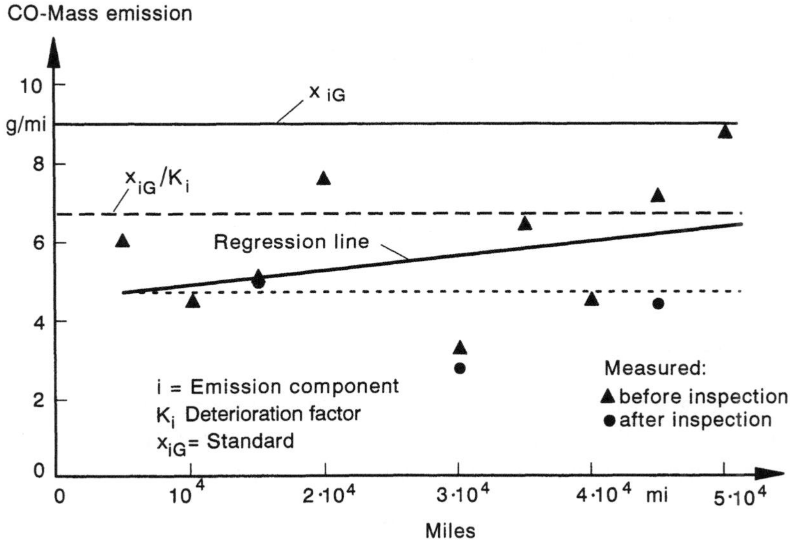

Fig. 9.26: Evaluation of the deterioration factor K_i of the mass emissions measured in grams/mile

than 9,000 vehicles are sold per year, vehicles belonging to the durability fleet will be tested as well. The decisive factor in each case are the emissions as measured by the government authority at 50,000 and 100.000 miles.

In the United States an extremely extensive bureaucratic structure has developed around the process of applying for certification testing.

9.7 Assembly Line Testing

From many tests on US and European vehicles (random samples of at least 100 vehicles per model) it is well known that the histograms of the readings can be best fitted by the log-normal density functions, described in Chap. 9.2. In the following, this density function will be used for vehicles taken from the assembly line.

If a larger random sample of at least 100 vehicles from the assembly line is used for testing, considerable variations occur for the exhaust emissions of the individual vehicles. The log-normal density function is therefore stretched in length (refer to, for instance, Figure 9.34).

9.7.1 European Test, Inspection by Variables

In Europe, the normal procedure for assembly line testing is to begin by testing one production vehicle. The standards to be met are higher than the certification test standards, by 20 % for NO_X, and by 30 % for CO and HC. If the vehicle should fail the test, the manufacturer may insist on a sample from several vehicles being tested; the "failure", however, has to be one of the vehicles in the sample. It is left to the manufacturer to fix the total number of vehicles which make up the sample.

Taking into account the eventual size of the sample, production vehicles are evaluated on the basis of the mean emissions and standard deviations obtained. The emission mean of each limited exhaust gas component plus the uncertainty of the mean must be smaller than or equal to the standard, thus:

$$\bar{x}_i + k\, s_i \leq x_{iG} \tag{9.34}$$

with

i	exhaust gas component (HC, CO, NO_X)
$\bar{x}_i$	mean value
k	statistical factor
s_i	standard deviation
x_{iG}	standard.

The factor k, which depends on the size of the sample, is contained in a legislated table based on a probability rating of $P = 80$ %. The table includes $n \leq 19$; $n \geq 20$ involves a k factor of $0.860/\sqrt{n}$. The term $k \cdot s$ represents the unilateral confidence range of the mean which was identified by $\frac{ts}{\sqrt{n}}$ in the preceding sections. In these sections, however, a probability of $P = 95$ % was assumed.

These equations, however, apply only if all values are approximately normally distributed. Yet even if the distribution is not normal, this procedure still represents a satisfactory approximation, because the means of samples drawn from a parent population are always distributed in a close-to-normal fashion, provided the samples are not too small ($n \geq 10$). This follows from the "central limit theorem." As n increases, the distribution of t (student density function) approaches the normal density function (standard density function) more and more.

The European test procedure is a so called inspection by variables, i.e., the mean is decisive. In practice, however, assembly line testing is not carried out.

9.7.2 Assembly Line Testing in the USA, 49 States, Attributive Method

With the exception of California, a procedure called Selective Enforcement Auditing (SEA) has become mandatory throughout the United States since the beginning of 1977. This procedure was conceived for occasional spot checks by the authorities, and it is also intended to be an incentive to manufacturers to institute quality control programs of their own, the extent of which is not prescribed in detail. These spot checks would normally be run once a year.

The size of an SEA sample depends on the size of the production batch to be checked (engine-transmission combinations), the maximum being 60 vehicles. However, as the number of vehicles involved in a test is staggered in accordance with a sequential sampling plan, SEA will involve considerably less than 60 vehicles in most cases.

What exactly constitutes a production batch is defined by the authorities. In most cases, one test is run on one vehicle, but the manufacturers may insist on up to three tests per vehicle, the performance of the vehicle being represented by the means obtained from the three tests. Unlike the European test procedure, results are evaluated by the so called *attributive* method, i.e., each vehicle is labeled „good" or „defective", depending on whether or not its emissions comply with the standards. The deterioration factor is taken into account. In principle, this attributive method is adequate for tests requiring a „yes" or „no" test result such as for incandescent lamps but it is not adequate for tests where values have variations.

According to the sampling plan which is assumed to reflect the actual emission properties of the production line vehicles, the maximum permissible percentage of „defective" vehicles in a sample is 40 % of all vehicles involved, which means that 40 % of all vehicles may have HC, CO and NO_X emissions exceeding legal standards. This percentage was decreased in the past years to 10 %. The following, however, do not depend on this percentage. Figure 9.27 schematically shows the density function and the permissible percentage of defective vehicles in a sample.

The basis of SEA testing is the Clean Air Act, the de jure official interpretation of which is that *each* individual vehicle must comply with the emission standards. De facto, permitting a percentage of „defective" vehicles constitutes a departure from this maxim.

On the other hand, the Clean Air Act is also used to substantiate the demand that each individual vehicle which was tested and failed to pass should be modified and retested until it passes the test, and that the vehicle thus identified must not be sold

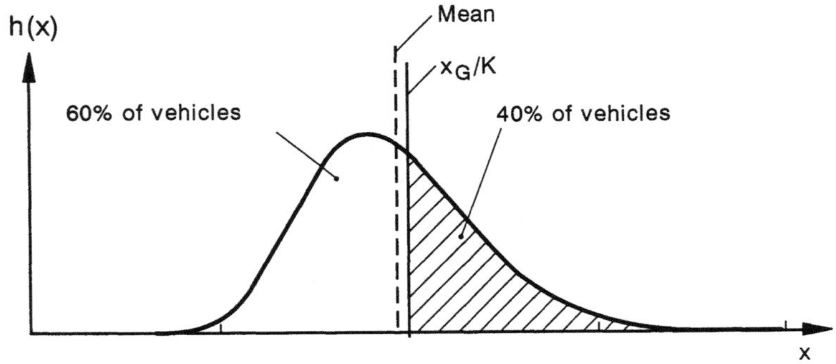

Fig 9.27: Maximum permissible percentage of defective vehicles in SEA for a log-normal density function $h(x)$ (x_G - standard, K - deterioration factor); valid for HC, CO and NO_X

otherwise. As described below in Chap. 9.7.4, this procedure is illogical and not feasible in practice.

9.7.3 California Assembly Line Testing, Inspection by Variables

California has had an assembly line test procedure since 1970. It involves testing the exhaust emission at idle of both hydrocarbons and carbon monoxide. This Steady State Inspection Test is performed by the manufacturers on all produced vehicles to be sold in California and the results must be submitted to the California authorities. None of the readings may exceed the limits which are derived from the first 100 vehicles of an engine family produced. The limit is represented by the mean value derived from the first 100 tests plus twice the standard deviation, provided that the values thus computed do not exceed any legal standards:

$$\bar{x}_i + 2s \leq x_G \ . \tag{9.35}$$

All inspection limits are reviewed quarterly.

Moreover, manufacturers are obliged to subject no less that 2 % of all vehicles produced quarterly to an additional exhaust emission test, which is the US-75 test. The test results must be submitted to the authorities. This procedure, which is applied to the 1977 and all following model years, is based on variables. The criterion is the mean of all values obtained from an engine family within quarter of a year. The condition is that, taking into account deterioration factors, the mean must be below the standard. This is comparable to the European assembly line test procedure.

This procedure permits subtracting the methane component from the hydrocarbon values. If an idle test or a complete test shows that the emissions of an individual vehicle exceed the applicable standards, the vehicle must be modified and retested as in the SEA procedure.

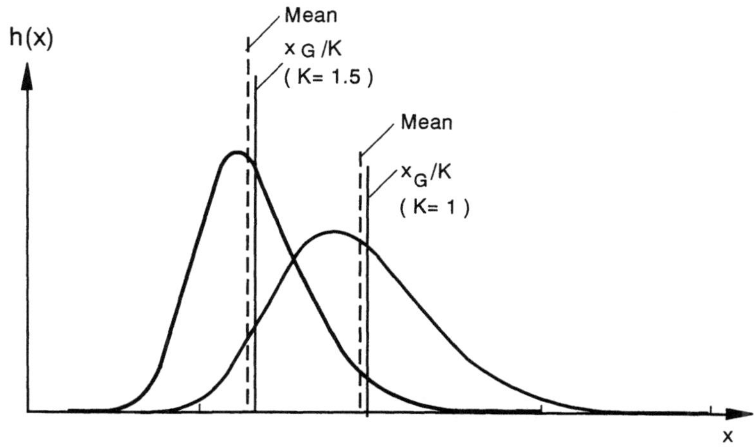

Fig. 9.28: California inspection by variables (x_G - standard, K = deterioration factor), two density functions and two deterioration factors for two different vehicle makes

The evaluation procedure is illustrated qualitatively in Figure 9.28. The mean must be below x_G / K. In addition to this, the authorities occasionally spot check small numbers ($n = 20$, for instance) of brand new vehicles.

9.7.4 Discussion of Existing Assembly Line Test Procedures

The critical quantity on which pollution depends is the total amount of emissions from all the vehicles on the road. The attributive method defining the percentage of vehicles whose emissions may not exceed certain legal standards may lead to wrong conclusions. Figure 9.29 shows two density functions of two vehicle makes with 10 % of the readings in each case being above the standard. Their mean values, however, are clearly different and consequently their pollutant effect. Such density functions are realistic ones.

We may say, therefore, that the SEA attributive procedure is basically not a good one as far as reducing pollution is concerned.

In view of the fact that both the SEA and the California tests permit a certain percentage of vehicles to exceed the legal standards, it is illogical to demand that the emissions of *each individual* vehicle tested should comply with the legal standards. It may even be impossible to meet this demand for the simple reason that this constitutes a sudden transition from judging a multitude of vehicles to judging a single one. This can be illustrated by Figure 9.30.

Assuming that within the permissible limit of 40 % a few vehicles in the sample fail to meet the legal standards, it is, for instance, possible that the emissions of a vehicle Y exceed the legal standard by 40 % (The following discussion is the same for percentages of defective vehicles of 10 %). Now this same vehicle is to be modified so that it will pass the test after no more than three tests. However, as has been explained in Figure 9.24, a large safety margin $S = F + 2s$ is required to

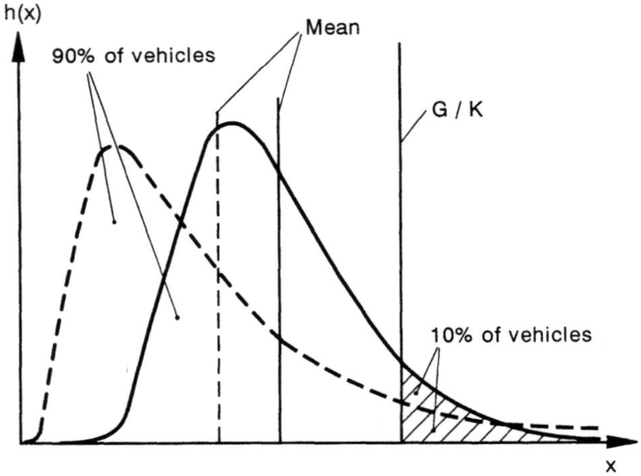

Fig. 9.29: Two density functions (log-normal) of two vehicle makes exceeding the standard by identical percentages of 10 % but with different means. The uninterrupted curve represents a much higher load on the environment, yet both vehicle makes pass the test

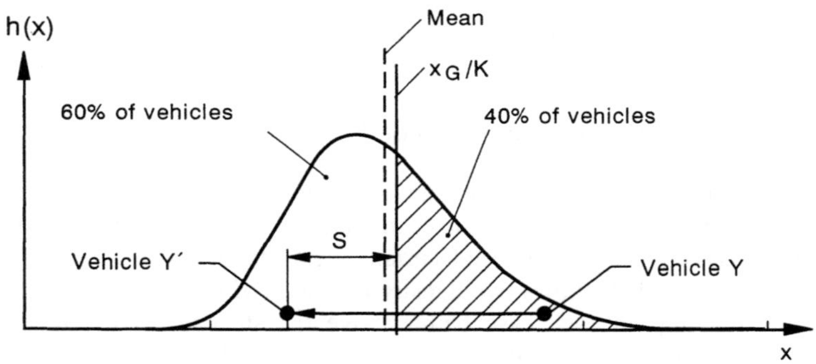

Fig. 9.30: SEA assembly line testing in practice. The emissions of the "failed vehicle" *Y* must be shifted below the emission standard by a certain safety margin (*S*) to ensure a sufficiently high probability of passing the re-test (x_G - standard, K = deterioration factor)

ensure a sufficiently high probability of a single vehicle passing the test. Consequently, the vehicle must be modified to bring its emissions below the standard by the safety margin, for instance, by 50 %.

Now emissions can only be reduced in a very limited way by changing the engine adjustment, and often the only way to lower emissions further is to exchange certain parts. On the other hand, it is not permissible to install parts or components different from serial parts or components, so that the manufacturers run the risk of being left with unsaleable vehicles when performing assembly line tests. Yet all vehicles *not tested in any way*, 40 % (or 10 %) of which may have emissions exceeding the legal standards, may be sold and this is illogical.

There is another consequence to this situation. If each individual vehicle is to be judged separately, manufacturers are bound to feel constrained to avoid additional tests as much as possible. If they do not, additional vehicles may be identified and, once identified, they may not be sold according to the law. This completely perverts the original intention of SEA testing, which was to encourage manufacturers to institute quality control programs of their own.

A sudden transition from testing groups of vehicles, where the mean is relatively certain statistically by virtue of the large number of vehicles tested, to running no more than three tests on an individual vehicle, which involves a high uncertainty factor, must be rejected.

Contradictions of this kind are not involved when sampling by variables (European and Californian assembly line tests) which, unlike the attributive method, incorporates both the mean and the standard deviation.

Just like a certification test procedure, the efficiency of an assembly line test procedure is best assessed by judging its operational characteristic, which is the curve signifying the probability of acceptance. Similar to certification, the position of the mean relative to the standard is decisive. Unlike certification testing, however, the criterion of all attributive evaluation procedures is the percentage of „defective" vehicles in the parent population. The operational characteristics of the variable and the attributive methods can only be compared if the density function is known.

Computing the operational characteristics starts with given parent populations and known parameters. The true mean μ and the relative standard deviation σ_r are given in the case of the variable method, and the percentage of „defective" vehicles and the relative standard deviation in the case of the attributive methods, i.e. the SEA and of the Californian assembly line test. The operational characteristics can be taken from literature. They corroborate the discussion described above.

9.8 Interrelationship between Certification, Assembly Line, and Field Monitoring Tests

As shown by a variety of investigations, the means of results of assembly line tests are nearly equal to their certification test counterparts obtained from vehicles of the same engine family. Without this interrelation a certification test would be meaningless. This interrelation is physically understandable, since according to the legal requirements, all parts of a vehicle influencing emissions must be similar for the pre-production vehicles used for certification and for the produced vehicles. The optimizing development process varying the parts influencing emissions happens *per se* during production because of the tolerances of the parts.

Figure 9.31 shows a comparison between mean HC assembly line test results and mean certification test results

The means of the results of assembly line tests are nearly equal to their certification test counterparts. The same holds good for other exhaust emission components and other model years.

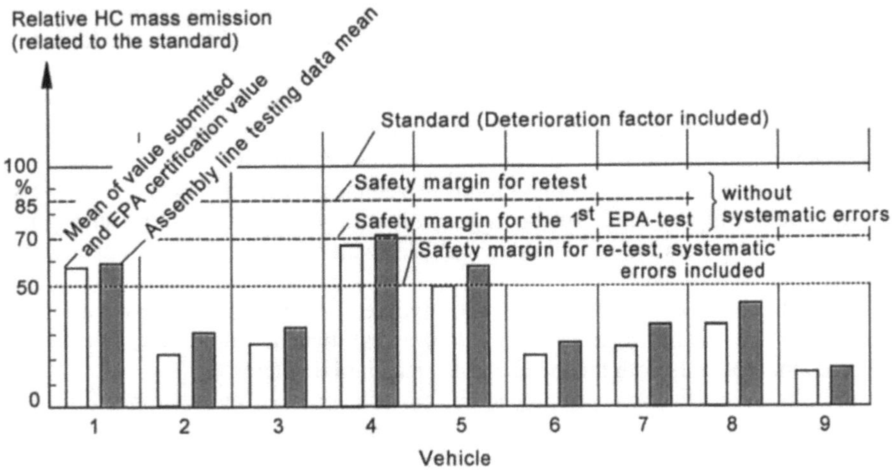

Fig. 9.31: HC certification test results and HC mean values of assembly line tests for California, both means in percent related to the standard for 9 vehicle makes

The knowledge of the interrelation is very important because assembly line test procedures are of higher significance with regard to ambient air quality. Therefore certification test procedures should be selected and optimized under the aspect of the assembly line test procedure. This has not happened since historically the certification test procedure was introduced earlier than the assembly line test procedure.

Field monitoring test procedures are interrelated in a similar way to the certification and assembly line test procedures. Ultimately, the exhaust emission of vehicles on the road are decisive for ambient air quality. The main problem here is the maintenance quality of the vehicles. Investigations have shown that the mean exhaust emissions of customers' vehicles often were considerably higher than the standards permitted, e.g., because of improper adjustments of the engine. After correct adjustment according to the manufacturer's specifications the mean exhaust emissions corresponded to those of their counterpart vehicles leaving the assembly line.

Figure 9.32 and Figure 9.33 show corresponding CO or HC values measured of 50 vehicles in Germany. The means of the vehicles on the road after correct adjustment correspond to the means of new vehicles from assembly line when taking the confidence limits into account.

Figure 9.34 again shows the interrelation between the means of certification, assembly line and field monitoring tests.

Although different density functions may be valid (certification test results → normal density function, assembly line and field monitoring test results → log-normal density function) the means for the density functions are approximately the same. Due to the safety margin S required for the certification test the means of assembly line tests have approximately the same distance S from the standard. This

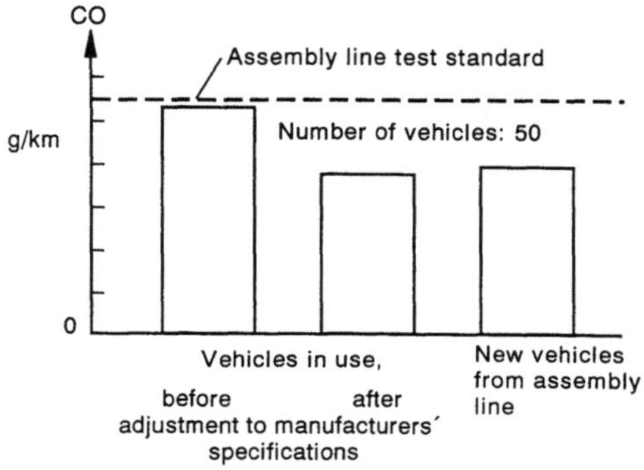

Fig. 9.32: Comparison of CO mean values of 50 new production vehicles with 50 vehicles on the road (TÜV Rhineland). The mean of the vehicles on the road after adjustment is equal to that of new vehicles. Vehicles are of German production

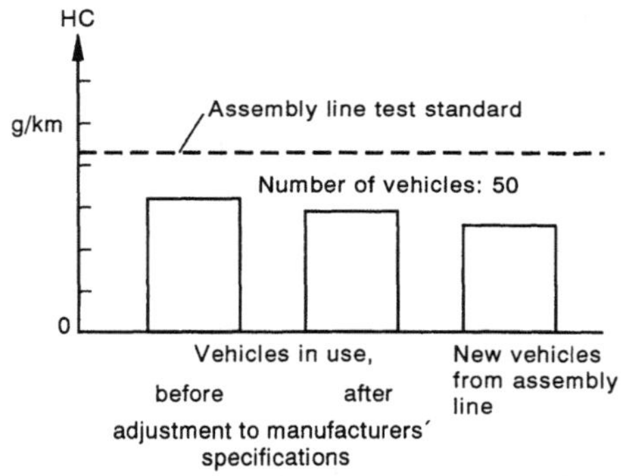

Fig. 9.33: Comparison of HC mean values of 50 new production vehicles with 50 vehicles on the road (TÜV Rhineland). The mean of the vehicles on the road after adjustment is equal to that of new vehicles. Vehicles are of German production

applies to field monitoring tests results only if the inspection and maintenance of vehicles are done properly and the vehicles' parts properly adjusted. In Europe, standards for assembly line testing are higher than the standards for certification and were introduced for reasons which are hard to understand. These higher standards for assembly line testing are not necessary because of the interrelation mentioned above.

Knowing this interrelationship between the three test procedures, it is not difficult to define a unified exhaust emission test procedure, so that the discussed interrelation of the three tests is taken appropriately into account and environmental protection is also kept in mind. As already mentioned several times, the influence of exhaust emissions on air quality, i.e., on the means of the concentrations of pollutants in the ambient air in the defined situations in traffic and ambient conditions, only depends on the means of the exhaust emissions of all

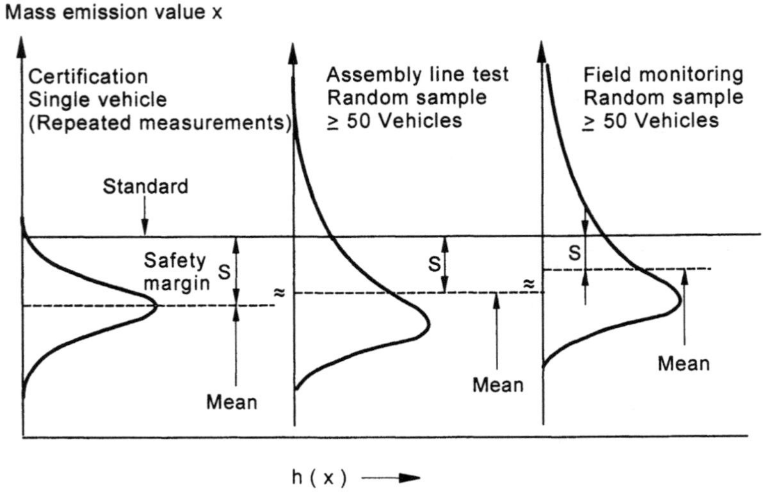

Fig. 9.34: Typical density function of exhaust emission readings of certification, assembly line and field monitoring. The three means S are approx. equal to each other

vehicles on the road. This interrelationship can be for any exhaust emission component can be expressed as follows:

$$\bar{y} = f(x_{total}) \quad (9.36)$$

$$x_{total} = \sum_{j=1}^{n} x_j = n \cdot \bar{x} \quad (9.37)$$

with

- $\bar{y}$ - mean of the ambient air concentration determining the air quality,
- x_{total} - total emission of all vehicles concerned, e.g., in a street canyon over 8 hours,
- x_j - emission of an individual vehicle *j* when driving past,
- n - number of vehicles considered, e. g., over 8 hours,
- $\bar{x}$ - exhaust emission mean of all vehicles considered.

Consequently, the ambient air concentration $\bar{y}$ is merely a function of the total number *n* and the exhaust emission $\bar{x}$ of the vehicles considered:

$$\bar{y} = f(n \cdot \bar{x}) \ . \quad (9.38)$$

It would be sufficient if the total mean of the exhaust emissions of a fleet of vehicles on the road was kept below the standard. In reality, however, this mean lies below the standard by a safety margin S (Figure 9.34). This procedure causes an unnecessary tightening of the standards of which the public and the politicians are unaware and which was caused by the fact that the certification and the assembly line test procedures were not developed in logical interrelationship.

9.9 Monitoring the Vehicles in Use

9.9.1 Introduction

As mentioned earlier several times, the exhaust emissions of the vehicles on the road or in use are decisive for ambient air quality. Let us discuss as an example, the development of monitoring procedures in Germany for vehicles in use.

In Germany, general checks of vehicles in use are regulated and are performed on a regular basis by the TÜV (TÜV = Technischer Überwachungsverein - a commercial agency for monitoring technical installations). Therefore the obvious step was to introduce a more intensive check of the unit influencing exhaust emissions besides the already obligatory CO-measurement in idle state. To achieve this, a testing procedure had to be developed fulfilling several partly contradictory requirements.

The German government already asked in 1976 for an extension of the so called „Hauptuntersuchung der Kraftfahrzeuge nach §29 StVZO“ (the general check of vehicles in use mentioned above) by an extensive check of their exhaust emission characteristics. In 1976 the committee for „Kraftfahrzeugabgase“ (Vehicle exhaust emissions) of the commission called „VDI-Kommission-Reinhaltung der Luft“ (Commission of the Association of Engineers concerning air quality) proposed a functional short test, to check for proper functioning of the vehicle components influencing exhaust emissions. Members of this committee were representatives of the automotive industry, the suppliers, the associations, the ministries, the UBA (Umweltbundesamt - German Federal Environmental Protection Agency) and of different TÜVs.

This proposal was adopted by the VDI commission. First specifications were issued, and in 1977 it was decided to found a working group under the overall control of the „Forschungsvereinigung Verbrennungskraftmaschinen (FVV)“ (Research Association of Engine Manufacturers), including representatives of the UBA and the TÜVs.

This working group initiated two research projects partly given to Porsche, partly to RW-TÜV in Essen and to TÜV Rhineland in Cologne. Both projects were 100 % financed by the UBA. Its findings were published in 1980.

The results of these two research projects formed the basis for the procedure called „Abgassonderuntersuchung (ASU)“ (Exhaust Emission Special Check), called ASU I in the following, which was to be part of the general checks of vehicles in use mentioned above.

In this procedure, the CO concentration in idle, the idle speed, the ignition point and - for ignition systems with contact control - the closing angle are checked to ensure that they are set according to the specifications of the automobile manufacturer. If there are no specifications defined by the manufacturer, the setting is judged according to the technical standards of when the car was built and considered to be correct if the exhaust emissions are a minimum during safe operation of the engine.

The German Federal Government intended to extend this ASU I to the inspection of vehicles with catalytic converters and diesel vehicles by the end of 1987.

Table 9.3: Comparison and evaluation of the proposals discussed for the exhaust emission tests (AU) for gasoline-powered vehicles with catalytic converter

Proposal	Work done	Standards	Measures to be taken	Evaluation
System check (VDA Proposal)	- Visual check - ASU, §47a StVZO (functional check of the exhaust emission recirculation or the secondary air system) - CO measurement ahead of the catalytic converter (if there is a sample device); as an alternative: electrical function check of the λ sensor including the control circuit	According to manufacturers specifications	Replacing faulty parts or adjusting to specifications	- Only a few parts (responsible for approximately 90 % of the exhaust emissions) need to be checked - Detection of malfunction allows for immediate repairs
	- CO measurement at the end of the exhaust pipe at zero load	- Controlled systems: max 0.5 Vol.-% CO - Uncontrolled systems: according to manufacturers specifications		- CO measurement in idle state behind the catalytic converter is not sufficient
UBA Proposal	- Steady state measurement of the gasous components (HC, CO, NO_x) in idle state and using a chassis dynamometer at operational state of 7 kW and at 50 km/h - For controlled systems: measurement at end of exhaust pipe	- Reference values for each make and for HC, CO, NO_x have to be established by the manufacturer before testing, requiring a lot of effort	- Vehicle must be taken to workshop - No indication of possible source of malfunction	- Only steady state measurements are possible - Great disadvantage: determining of standards typical for each make, requiring a lot of effort by the manufacturer (additional measurements with prescribed tests are necessary)
	- For uncontrolled systems: measurements ahead of and behind the catalytic converter	For the evaluation, the conversion rates η are used. There are no standards		Does not function properly since only steady state in the λ interval can be recorded

Continuation of Table 9.3

Table 9.3: Comparison and evaluation of the proposals discussed for the exhaust emission tests (AU) for gasoline-powered vehicles

Proposal	Work done	Standards	Measures to be taken	Evaluation
Component test/Catalytic converter (VW research group) instead of the CO measurement in idle state behind the catalytic converter, otherwise retaining system check (VDA Proposal)	- Exhaust emission measurement ahead of and behind the catalytic converter (load is produced based on the principle of „free acceleration“) - Catalytic converter with λ control : one measurement after adjustment; Determination of the conversion rate η for the gas components HC, CO, NO_x	Conversion rate η: For controlled systems: $\eta_{HC} = 70\,\%$ $\eta_{CO} = 70\,\%$ $\eta_{NOx} = 30\,\%$	If η is smaller than the standard for all three gas components, the catalytic converter must be replaced; if η is smaller than the standards of two components, the engine adjustment needs to be checked.	- Load comparable to the UBA proposal using a dynamometer; however it is dynamic - No standards typical for each make are required. - Outliers easily identifiable - Low investment costs - Because of the dynamic load and measurement in two λ intervals outliers ($\lambda > 1$, $\lambda < 1$) are identified with high probability
	- Catalytic converter without λ control: one measurement of CO, HC each in „lean interval“ ($\lambda > 1$) and of NO_x in „rich interval“ ($\lambda < 1$) is required	For uncontrolled systems: at $\lambda > 1$ $\eta_{HC} = 50\,\%$ $\eta_{CO} = 50\,\%$; at $\lambda < 1$ $\eta_{NOx} = 30\,\%$		

Consequently, the UBA formulated a proposal (called UBA Test in the following) based on CO, HC and NO_X measurements behind the catalytic converter in two operating states (in idle and running on a chassis dynamometer at a constant 50 km/h). This proposal was put to the test by the TÜV in a larger test program with 4000 vehicles. For diesel vehicles the proposal included particle measurement according to a Bosch filter method at full load while running on a chassis dynamometer.

Since this proposal was not acceptable to the automotive industry and to the workshops for economical and technical reasons, another proposal was put forward by the „Verband Deutscher Automobilhersteller (VDA)“ (Association of German Automotive Manufacturers). In this proposal the ASU I was to be extended by a CO measurement behind the catalytic converter in idle state and check of the λ-control. In the meantime this proposal has become mandatory, called AU.

As an alternative to the CO measurement behind the catalytic converter and the VDA proposal, the VW-Research Group developed a proposal based on the principle of free acceleration (load on the engine by inertial forces of the engine's rotating parts during acceleration, starting from idling speed) and based on CO, HC and NO_X measurements ahead of and behind the catalytic converter.

A more profound understanding of the prescribed legislation for vehicles in use can be reached more easily by contemplating the historical evolution of the different proposals.

Table 9.3 shows an overview of the proposals for gasoline vehicles discussed before one was legislated.

9.9.2 Summary

As already described, it is known from the results of investigations that exhaust emissions show large variations in repeated measurements with an individual vehicle as well as in measurements with many vehicles.

The density function of exhaust emission readings of repeated measurements with one and the same vehicle is a normal one in good approximation. The density function of exhaust emission readings, measured for a random sample of many vehicles is a log-normal one in good approximation. Fig 9.35 shows qualitatively this log-normal density function, its mean and the standard for assembly line vehicles. The mean value and density function of an individual vehicle taken from the random sample is also indicated.

The exhaust emissions of individual assembly line vehicles from the random sample can be much higher than the standard. According to the European regulation for assembly line testing, only the mean $\bar{x}_A$ of the exhaust emissions of a random sample defined by the manufacturer must be below the standard, taking the confidence range of the mean into account.

If the exhaust emissions of an individual vehicle in use has to be checked an insoluble problem arises - *a true dilemma*. This is demonstrated in Fig 9.35.

If a specific individual vehicle in the upper part of the log-normal density function is considered (in the figure indicated by a dot), this vehicle shows an exhaust emission value (here for one gas component) which is higher than the

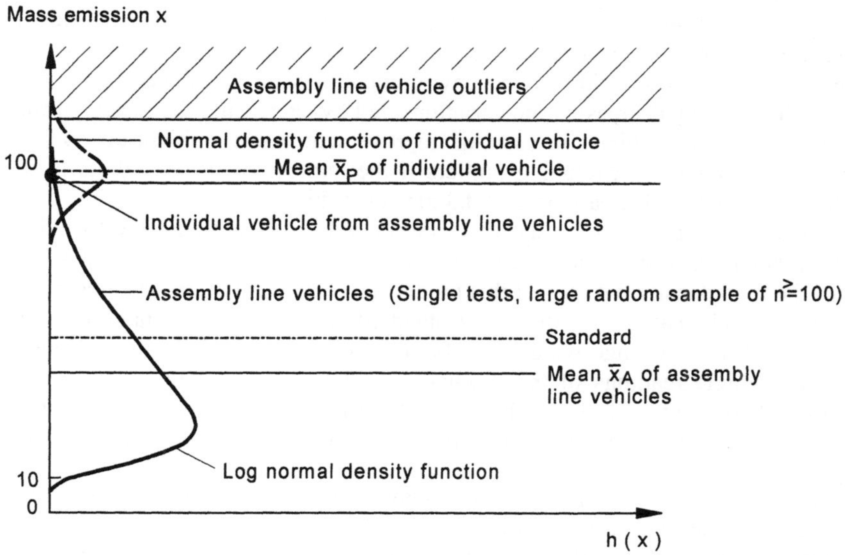

Fig 9.35: Log-normal density function of mass emissions of a large random sample ($n \approx 100$) of assembly line vehicles. The mass emissions are plotted on the axis of coordinates. The normal density function in valid for the individual vehicle from this random sample

standard, e.g., by a factor of two but still within the density function valid for the corresponding model. Thus it is still a „good" vehicle according to the prescribed assembly line test procedure.

The results of repeated measurements for this individual vehicle would also show large variations and would have a normal density function, as shown in the upper part of Figure 9.35 as a dashed curve.

The mean $\bar{x}_p$ of these repeated tests with the individual vehicle (also plotted in Figure 9.35) approximates the "true" exhaust emission value of this vehicle. If only one or two measurements are performed, as is probably the case in practice, the variability of repeated measurements has to be taken into account as the statistical uncertainty, not to mention systematic errors.

The fact that the exhaust emissions of an individual vehicle may lie far above the prescribed standard leads to the dilemma mentioned. Obviously only those vehicles can be defined as "bad", whose true exhaust emissions lie outside the density function derived from the corresponding new vehicles. For the given example, only those vehicles can be disqualified - taking the above-mentioned measurement uncertainty into account - the exhaust emission value of which lie in the shaded area of Figure 9.35

Consequently, only the vehicles whose exhaust emissions are larger than the standard plus four to five times the standard deviation of the exhaust emissions for assembly line vehicles, i.e.,

$$x_j > x_G + 4 \cdots 5\, s \tag{9.39}$$

can be defined as „bad" vehicles (outliers).

These big variations between the assembly line vehicles or for the individual vehicle for repeated measurements, are, however, - as already mentioned - unimportant for the protection of the environment. The total exhaust emissions cannot be lowered by decreasing the variations, but by decreasing the means of the exhaust emissions of vehicles in use. An improvement of the ambient air quality can only be achieved by excluding outliers if the number of these outliers is so large that a significant decrease of the total mean is achieved.

A further dilemma is that, as shown in Figure 9.36, it is not possible to judge from such single measurements whether the number of outliers is so large that the mean of the total density function of this model has changed at all or whether it has changed so much that it lies above the standard (dashed density function 2 in Figure 9.36). Such a change would only be discernible if a statistically sufficient number of vehicle readings of the discussed model were available at a station collecting all relevant information before making a decision. This station would have to evaluate the total number of outliers. In practice such a procedure would hardly be possible.

These conclusions are valid for the regulated ECE or US-75 test, i.e., for following a prescribed driving cycle with idle, acceleration, constant, and deceleration phases after cold start, but also for any other test. This means, that only outliers have to be identified.

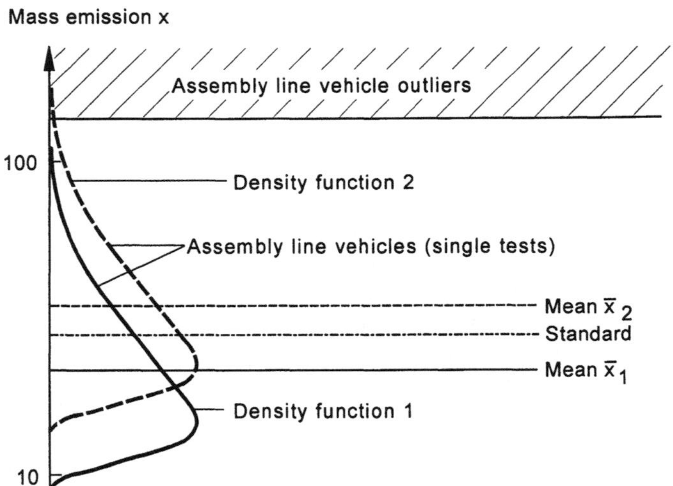

Fig. 9.36: Example of a change of exhaust emissions. The mean $\bar{x}_2$ lies above the standard

9.9.3 Short Tests

After these basic considerations the question of a suitable test procedure to identify outlying vehicles arises. This test procedure would have to fulfill several partly

contradictory prerequisites:
- it should not take much time
- the investment and time required should be reasonable
- it should be dependable
- the operational costs should be low.

Obviously these requirements are not given for the known regulated test procedures (e.g. US, ECE, Japan tests) especially with regard to the effort and ensuing cost. For this reason, a work is being done worldwide to develop procedures which reduce the effort involved drastically. These procedures are therefore called *Short Tests*.

A substantial criterion to evaluate such short tests used to be whether there was a sufficient correlation between the results of the short tests and those of the official tests. This is, however, impossible for a short test - if it is a short test in the strict sense - for physical reasons because of the simple fact that the regulated tests are generally based on the cold start with precisely defined large-scale conditioning. One has to consciously avoid demanding a correlation between the regulated test results and the short test results.

Short tests already exist. If it is a test in idle state or at constant higher speed (zero-load-test) where normally the concentrations of CO or HC measured and used as evaluation criteria for the exhaust emission behavior of the vehicle then these procedures have a grave disadvantage, mainly because of the low load on the engine. For HC the test is of little use, for NO_X of no use to judge the emission behavior of the vehicle under realistic driving conditions.

The main problem of such non-correlating short tests is the requirement that no vehicle should be disqualified by a short test if they pass the regulated test. To achieve this, specific standards have to be introduced for each model, which in most cases have to be such high values that practically no outliers are identified, as shown schematically in Figure 9.37. Furthermore, the automobile manufacturer

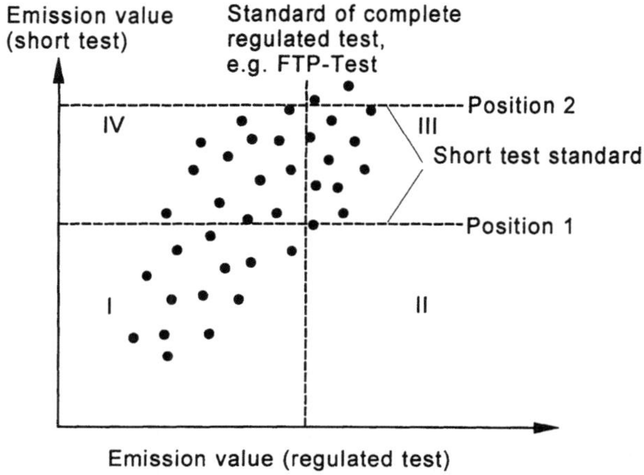

Fig. 9.37: Interrelation between values of a short test and a regulated test

would have to invest a large measurement effort to define these standards typical of the model.

In Figure 9.37 the measured values indicated by dots can be grouped into four areas I-IV. The standard for a short test, typical of a model, must be adapted so that in no case vehicles are disqualified by the short test if they pass the regulated test. As indicated in Figure 9.37 it can happen that most of the values lie below the standard for the specific model (position 2), so that only one or two vehicles are identified as outliers. Furthermore, the comments regarding the criteria for an outlier have to be taken into account (refer to 9.39), which proves that such a test is unrealistic.

CO measurements in idle state

CO measurements in idle state behind the catalytic converter have been performed. Figure 9.38 shows the CO concentrations as function of the readings of the US-75 cold start test values of vehicles according to Section XXIII of the German Automobile Safety Act.

If the short test standard is defined so that the vehicles are below the specific CO idle standards (Position 1) and if the doubled US test standard is taken as a basis, two incorrect judgments are the consequence. One vehicle passes the regulated test, but fails the short test; only one vehicle fails the regulated test, but passes the short test. If one tries to exclude the first incorrect judgment by adapting the specific idle standard, Position 2 has to be chosen. In this case, however, no outliers would be disqualified, as shown when discussing short tests. With this procedure, and as a consequence of these results, a definite disqualification of the vehicle concerned is not possible. The same is also true of the UBA test, which requires a much higher effort.

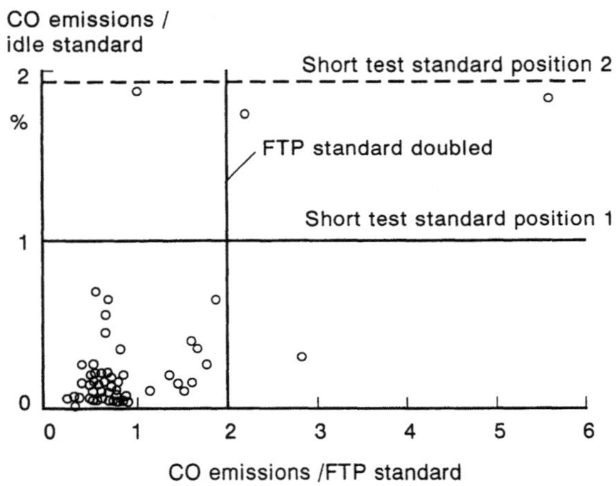

Fig. 9.38: Relative CO results of the US-75 cold start test and of the CO measurements in idle state of vehicles according to Section XXIII, shown corresponding to Figure 9.37 and related to the US standard or to the specific idle standard

9.9.4 System Check (Parts Check)

From experience a practical solution to this problem seems to be functional checks of the parts influencing exhaust emissions, i. e., a system check or parts check.

A large number of measurements of vehicles in use and the experience of the years show that the number of outliers can be kept low by checking the functioning of the parts influencing emissions and setting them according to the manufacturers specifications before testing them, refer to Figures 9.32 and 9.33. Faulty parts, if any, must be replaced.

Experience has led to discussions of possible tests for vehicles in use, the so called system check which was also the basis of the VDA proposal.

The advantage of this monitoring procedure is that not only are the emission outliers identified, but also the cause of the high emissions, namely the faulty part, can be localized for the customer or the workshop. As all the other tests only provide exhaust emission measurement values this cause is not apparent and needs to be identified at the workshop.

The idle CO measurement according to AU is also a parts check in this sense and useful for checking the performance of the catalytic converter.

9.9.5 The AU in Germany

Since December 1993, exhaust emission checks (AU) according to §47a and b of the StVZO have been prescribed in Germany. The AU mainly includes the following four measures:

a) Input of the vehicles' specifications into the test system

b) Visual test of the parts influencing emission:
Catalytic converter, exhaust system, λ sensor should be complete and without damage.

c) Conditioning of the vehicle:
The vehicle is operated until a specified oil temperature has been reached and then operated at higher speed to warm up the catalytic converter and λ sensor

d) Function tests:
- Determine engine speed and CO emissions in idle as a parts check,
- Determine CO emissions and the λ value at higher idle speed ($n \geq 2000$ rpm) as a parts check,
- Check of the λ control circuit by inducing a disturbance (e.g. injection of a certain amount of extra air by taking off the hose or by activating the brake boosters). First this disturbance must show an effect which is then canceled by the control circuit.

The only technical effort necessary for the AU is the injection of extra air into the engine. The inspection station, however, needs a special measuring system which collects the measurement data by a menu controlled computer, compares the values with the specified data, and stores and plots the results.

The AU has to be performed every two years, or for new vehicles after three years of operation, by an authorized inspection station and has to be paid for.

The drawback of the AU is that it provides only limited information on the condition of the exhaust system as only one check of the catalytic converter and of the λ sensor is performed. Only a large interruption of the control circuit can be identified after checking the λ sensor. The catalytic converter is tested at low loads in idle state which corresponds to a conversion rate of approximately 50 % for CO.

9.9.6 On-Board Diagnosis (OBD)

The logical further development of the parts check is on-board diagnosis(OBD), introduced for the first time in California.

The California Air Resources Board first drafted legislation on to OBD in 1984 and this led to the introduction of OBD I from model year 1988 on. In OBD I all the parts directly connected to the engine control circuit have to be controlled.

The requirements for OBD II are much more extensive. Here, all the exhaust emission-relevant parts and systems must be checked. A typical example is the catalytic converter.

In the meantime the Environmental Protection Agency (EPA) has also drafted legislation for an OBD system for the other 49 US states, which can be satisfied by the Californian OBD II until model year 1998.

In the EU there are plans to introduce an OBD II or - as an alternative - OBM (On-Board Measuring) with a suitable HC-, CO- and NO-sensor (yet to be developed).

The goal of the OBD requirement is to allow a function check of all exhaust emission-relevant parts in the vehicle during normal operation. Identified malfunctions should be located as accurately as possible and stored. If the malfunction is confirmed after continued driving, it should be possible to inform the driver via a signal lamp in the instrument panel that he should consult a service station.

Legislation has the following goals:

- *Monitoring* all exhaust emission-relevant parts and systems,
- *Protection* of all endangered parts (catalytic converter),
- *Storing* of information on malfunctions which occur,
- *Indication* of the cases when harmful exhaust emission components exceed prescribed limits,
- *Transferring* the stored information to the workshop's computer system.

These goals require:

- monitoring the catalytic converter,
- monitoring the λ sensor,
- identifying when a cylinder does not ignite,
- monitoring the fuel system,

- monitoring the secondary air system,
- monitoring the exhaust emission recirculation,
- monitoring the fuel tank venting,
- monitoring other systems.

Required for this are

- standardized interfaces,
- storing operationing conditions,
- standardized control circuit for the signal lamp,
- indication that an inspection is necessary,
- tamper-proofing the engine control system.

Besides the parts and systems specifically defined by the legislator, the requirements for monitoring the fuel system and to detect when a cylinder has not ignited lead to the monitoring of all parts of the engine control system.

In Figure 9.39 these parts are shaded for emphasis.

Let us take the monitoring of the catalytic converter as an example, since this is one of the most important requirements of the OBD II. The purpose is to monitor whether the conversion rate of HC is lower than a specified limit

- in dynamic operation (US-75 test) or
- in steady state operation (optional for model year '94 and '95).

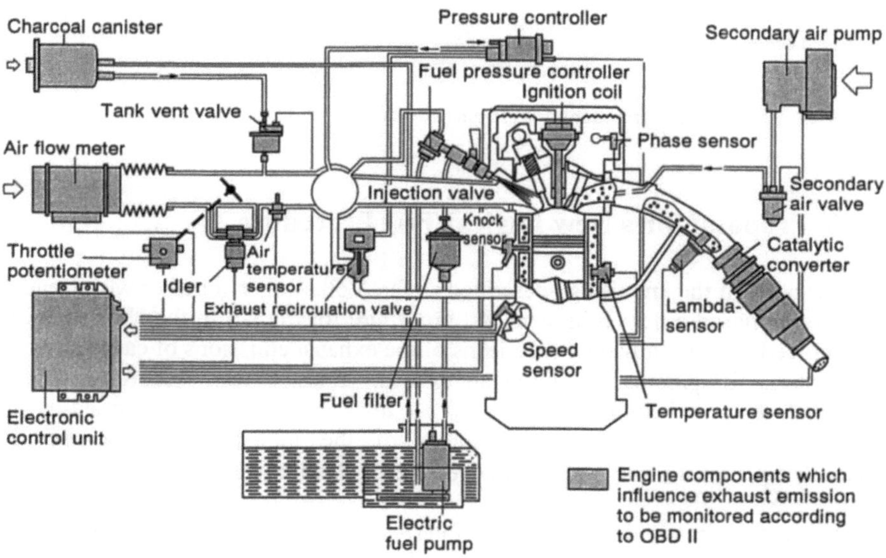

Fig. 9.39: Engine control system MOTRONIC with all exhaust emission-relevant parts shaded for emphasis. (Source: Bosch)

The technical solutions are

- to use the oxygen storage capacity of a catalytic converter,
- the use of one additional λ sensor behind the catalytic converter and
- the comparison of the signal amplitudes of the two λ sensors ahead of and behind the catalytic converter.

Figure 9.40 shows typical signal curves of λ sensor for a new and a very old catalytic converter. Because of the technical principle of the λ control used, there is a relatively constant signal amplitude of the regulating oscillation for the λ sensor ahead of the catalytic converter. In the upper part of the figure the oxygen storage capacity of the new catalytic converter is clearly shown by the signal of the λ sensor behind the catalytic converter. As shown in the lower part of the figure, an old catalytic converter can have a distinctly lower oxygen storage capacity, shown by the signal of the regulating oscillation behind the catalytic converter.

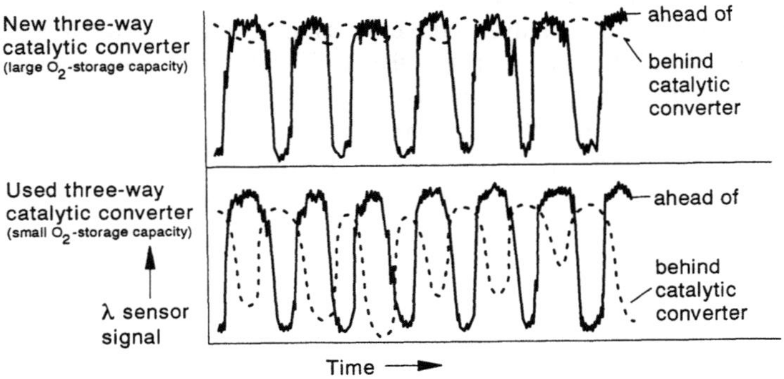

Fig. 9.40: Catalytic converter monitoring, signal curves

9.10 Proposal for a New Logical Test Procedure

On the basis of the knowledge described so far, Volkswagen together with General Motors proposed a new consistent test procedure for the US to the US Senate. In this procedure, basically only the means of the exhaust emissions of cars in use are the criteria for passing or failing an exhaust emission test. According to this proposal, the certification and assembly line tests should not be performed by the authorities. Both are the responsibility of the manufacturer, because the manufacturer must make sure that the vehicles will still meet the standards when in use.

Due to the already described close interrelationship between the three exhaust emission monitoring procedures, the manufacturer can guarantee that the mean of all the vehicles in use actually stay below the standard by keeping to the safety margin for prototypes and produced vehicles.

Of course, the authorities have to check this. To do this, a sufficiently large random sample of vehicles in use could be tested according to an official exhaust emission test. The random sample would have to be a representative cross-section of the manufacturer's whole vehicle fleet. The vehicles should be maintained properly and if necessary adjusted to manufacturer's specifications.

Figure 9.41 clearly demonstrates the basic philosophy of this proposal. The mean of the total vehicle production of the manufacturer is weighed against the number of vehicles sold, compared to the standards and fees levied or a bonus granted. Fees and bonuses depend on the position of the mean in relation to the standards. If the mean is higher than the upper limit, standard plus confidence range of the mean, a fee has to be paid. If the mean is below the lower limit, standard minus confidence range of the mean, the manufacturer gets a bonus.

The extent of the fee should be so high that it would mean a severe competitive disadvantage. A manufacturer should be permitted to sell a bonus to another because ultimately only the total mean of the emissions of all vehicles in traffic determine the ambient air concentrations, refer to (9.36) to (9.38).

This proposal is known in the US as averaging and is frequently discussed. For diesel particles this procedure has been partly used for model 1987 on, but without

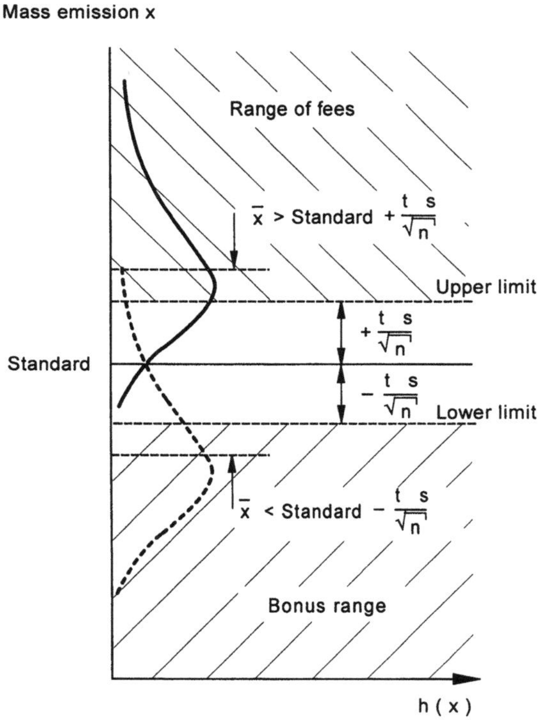

Abb. 9.41: Proposal for a consistent exhaust emission test procedure

any fees. Generally it can be stated that this proposal, though logical and consistent has the problem of acquiring representative random samples of vehicles in use. Of course, this problem can be solved. In the US, however, people are hesitant to solve this problem by regulations.

10 Exhaust Emission Measuring Techniques-Quo Vadis?

10.1 General Aspects

The effort necessary for exhaust emission measuring techniques is oriented to the requirements and regulations of legislation. Research into the potential effects of different components on the environment, including the global ecosystem, will result in an increase of the number of exhaust emission components which will be regulated in the form of emission standards or air quality standards. To express it in a few words:

Unregulated components will become regulated ones.

Furthermore, the legislator is following a preventive policy by making the standards stricter in order to avoid or reduce the impact on the environment.

Prescribing standards is only sensible if the compliance with them can be monitored, i.e., if the emission and air quality concentrations can be measured. Measurement, however, is getting more and more difficult because of the increasing number of regulated components and low concentrations to be measured. The consequence is that measuring the latter is not possible with the currently available measurement techniques or in some cases only with considerably higher effort. In principle the question

Exhaust emission measuring techniques - Quo Vadis?

ought to be preceded by the question

Exhaust emission regulation - Quo Vadis?

That standards are getting tougher and more components are being regulated is demonstrated impressively by exhaust emissions legislation in California.

Technological progress has also influenced exhaust emission measuring techniques. However, one must also consider how to fulfill the extreme requirements expected in future as also the economic aspects.

One step into the future of exhaust emission measuring techniques could be multi-component measuring units for the simultaneous examination of larger numbers of components which affect the environment (refer to Chap. 6). Furthermore, there has been for some time efforts to measure exhaust emission components directly and continuously on board a vehicle, i.e., to use on-line sensors. Such on board measurements are partly already regulated by legislation.

10.2 On-Board Measuring Procedures

On-board measuring procedures would be mainly applied for the direct measurement of automobile exhaust emission gases during the operation of a vehicle. However, this application is only possible if small and inexpensive sensors are available. In this way the installation of relatively complex and expensive measuring equipment would unnecessary, and the exhaust emission could be measured continuously during the operation.

Chemical or electrochemical sensors can be seen as a specific sensor technique. The physical gas sensors on the other hand are generally highly miniaturized and simplified gas analyzers based on known measuring principles such as the IR absorption spectrometry. An example of a possible set up for an infrared gas sensor mounted to the exhaust pipe is shown in Figure 10.1.

The detector is an infrared-sensitive solid-state detector. The selectivity for a certain component in the sample gas is achieved by using an optical interference filter. Both filter and solid-state detector can be restricted to a few millimeters in dimension. The thermal radiation of the hot exhaust pipe can be used as an infrared light source.

Developmental work on miniaturized gas analyzers in connection with optical fibers and microsystems has been going on for a long time. The latter includes, for instance, the realization of a spectrometer on a chip.

The realization of such a sensor would be of economic interest because a complicated measuring and testing unit requires high effort in terms of investment, maintenance and personnel. For example, a European automotive manufacturer operating world-wide invests approximately three hundred million German Marks only for exhaust emission testing laboratories.

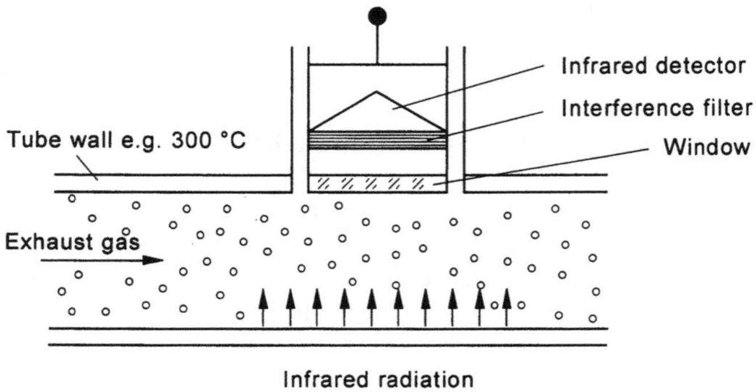

Fig. 10.1: Scheme of an infrared gas sensor mounted to an exhaust pipe

10.3 Outlook

In the service of environmental protection measuring technology has always fulfilled the tasks at hand.

Sometimes only the improvement in measuring techniques, especially in the area of extreme trace analysis, has made the detection of certain substances in the environment possible. This has resulted in new, usually unjustified, demands from environmental protection organizations.

Legislation on the limitation of emissions and ambient air concentrations of pollutants will become more stringent. The demand for continuous control will increase, resulting in more being required from measuring techniques. Assuming that legislation will not impose regulations which cannot be fulfilled, the requirements placed on measuring techniques can probably be satisfied, regardless of the effort, time and expense involved. However, a reasonable cost/benefit analysis should not be made taboo. The lack of standardized regulations should be kept in mind here.

Taking these aspects into account and related to the of ambient air quality, the answer to the above-formulated question: exhaust emission measuring technique-Quo Vadis? is

- *Standardization of test procedures.*
- *Restriction to mean fleet emission.*
- *Advanced development and increasing use of multi-component measuring technique.*
- *More effort in the development of selective exhaust gas-on board sensors.*

Bibliography

Books

1 Baumbach, G.; Baumann, K.; Dröscher, F.; Gross, H.; Steisslinger,B.: Luftreinhaltung. Entstehung, Ausbreitung und Wirkung von Luftverunreinigungen - Meßtechnik, Emissionsminderungen und Vorschriften.- Berlin; Heidelberg: Springer-Verlag, 1990 ISBN 3-540-52677-3

2 Brosthaus, J.; Jost, P.; Sonnborn, K.S.: Abgasemissionsprognose für den Pkw-Verkehr in der Bundesrepublik Deutschland im Zeitraum von 1970 bis 2000.- Köln: Verlag TÜV Rheinland, 1983

3 Budzikiewicz, H.: Massenspektrometrie: Eine Einführung. 3.Aufl.- Weinheim: VCH Verlagsgesellschaft, 1992 ISBN 3-527-26870-7

4 Chantry, G.W.[ed.]: Modern Aspects of Microwave Spectroscopy.- London, New York, Tokyo: Academic Press, 1979

5 Der Rat von Sachverständigen für Umweltfragen: Waldschäden und Luftverunreinigungen (Sondergutachten, März 1983).- Stuttgart; Mainz: W.Kohlhammer Verlag, 1983 ISBN 3-17003265-8

6 Deutscher Bundestag [ed.]: Schutz der Erdatmosphäre: Eine internationale Herausforderung. Zwischenbericht der Enquete-Kommission des 11. Deutschen Bundestages "Vorsorge zum Schutz der Erdatmosphäre" am 9.3.1989.- Bonn: Deutscher Bundestag, Referat Öffentlichkeitsarbeit, 1989 ISBN 3-924521-27-1

7 Deutscher Bundestag [ed.]: Schutz der Erde: eine Bestandsaufnahme mit Vorschlägen zu einer neuen Energiepolitik. 3. Bericht der Enquete-Komission "Vorsorge zum Schutz der Erdatmosphäre" des Deutschen Bundestages. Vol. I and II.- Bonn: Economica Verlag, 1991 ISBN 3-926831-91-X

8 Ehhalt, D.H.; Drummond, J.W.: The Tropospheric Cycle of NO_X.-In: Georgii, H.W.; Jaeschke, W.: Chemistry of the Unpolluted and Polluted Troposphere, p.219-251.- Dordrecht: D.Reidel Publ.Co., 1982

9 Georgii, H.W.; Jaeschke, W. [eds.]: Chemistry of the Unpolluted and Polluted Troposphere.-Proc. of the Nato Advanced Study Inst., Held on the Island of Corfu (Greece), Sept. 28 - Oct. 10, 1981.- Dordrecht: D.Reidel Publ. Co.,1982

10 Hauschulz, G.; Heich, H.-J.; Leisen, P.; Raschke, J.; Waldeyer, H.; Winckler,J.: Emissions- und Immissionsmeßtechnik im Verkehrswesen.- Köln: Verlag TÜV Rheinland, 1983 ISBN 3-88585-58-3

11 Herzberg, G.: Molecular Spectra and Molecular Structure. II. Infrared and Raman Spectra of Polyatomic Molecules.- Malabar, Fla., USA: Krieger Publ. Co., 1991
12 Ishinishi, N.; Koizumi, A.; McClellan, R.O.; Stöber, W.: Carcinogenic and Mutagenic Effects of Diesel Engine Exhaust.-In: Development in Toxicology and Environmental Science, Volume 13 (Congress Report).- Amsterdam; New York; Oxford: Elsevier Science Publishers, 1986
13 Klingenberg, H. et al.: Versuchs- und Meßtechnik. Kapitel 7 in: Bussien: Automobiltechnisches Handbuch. Suppl. 18. Edition.- Berlin; New York: Walter de Gruyter, 1979
14 Lewtas, Y. [Ed.]: Toxicological Effects of Emission from Diesel Engines (Congress report).- Amsterdam: Elsevier Science Publishing Co., 1981
15 Lindinger, W.: Internal and Translational Energy Effects. In: Futrell, J.H. [Ed.]: Gaseous Ion Chemistry.- New York: John Wiley & Sons Inc., 1986
16 Meyer, V.: Praxis der Hochleistungs-Flüssigkeitschromatographie (Laborbücher Chemie). 7.Aufl.- Frankfurt am Main: Otto Salle Verlag, 1992 ISBN 3-7935-5452-X
17 Nießlein, E; Voss, G.: Was wir über das Waldsterben wissen.- Deutscher Institutsverlag, 1985 ISBN 3-602-14158-6
18 N.N.: CO_2-Minderung durch staatliche Maßnahmen.-VDI-Berichte 997.- Düsseldorf: VDI-Verlag, 1992
19 Sachs, L: Angewandte Statistik - Anwendung statistischer Methoden. 7.Aufl.- Berlin, Heidelberg: Springer-Verlag, 1992 ISBN 3-341-00753-9
20 Schröder, E.: Massenspektrometrie: Begriffe und Definitionen.- Berlin; Heidelberg: Springer-Verlag, 1991 ISBN 3-540-53329-X
21 Seiffert, U.; Walzer, P.: Automobiltechnik der Zukunft.- Düsseldorf: VDI-Verlag, 1989 ISBN 3-18-400836-3
22 Smith, R.A.: Air and Rain. The Beginning of a Chemical Climatology.- London: Longmans, Green and Co., 1872
23 Staab, J.: Industrielle Gasanalyse.- München; Wien: R.Oldenbourg, 1994 ISBN 3-486-22808-0
24 Storm, R.: Wahrscheinlichkeitsrechnung, mathematische Statistik und statistische Qualitätskontrolle. 9.Auflage.- Leipzig: Fachbuchverlag, 1988 ISBN 3-343-00181-3
25 Unger, K.K. [Ed.]: Handbuch der HPLC, Vol. 1: Leitfaden für Anfänger.- Darmstadt: GIT Verlag, 1989 ISBN 3-921956-84-6
26 Willeke, K.; Baron, P.A.: Aerosol Measurement. Principles, Techniques, and Applications.- New York: Van Nostrand Reinhold Co., 1993 ISBN 0-442-00486-9
27 Toxic Air Pollutants from Mobile Sources: Emissions and Health Effects. International Specialty Conference, Air & Waste Management Association, Pittsburgh, Pennsylvania (1992)

References

Chap. 2

1 Berg, W.: Aufwand und Probleme für Gesetzgeber und Automobilindustrie bei der Kontrolle der Schadstoffemissionen von Personenkraftwagen mit Otto- und Dieselmotoren (Dissertation).- Braunschweig: Techn. Universität, 1982

2 Brosthaus, J.; Jost, P.; Sonnborn, K.S.: Abgasemissionsprognose für den Pkw-Verkehr in der Bundesrepublik Deutschland im Zeitraum von 1970 bis 2000.- Köln: Verlag TÜV Rheinland, 1983

3 Franke, H.-U.; Witzenhausen, K.; Klingenberg, H.: Variation of the Particulate Size Distribution in the Exhaust Pipe of a Stationary Diesel Engine.- Paper on 4thInt. Aerosol Conf., Session 7E5, Los Angeles, 1994

4 Hassel, D.; Dursbeck, F.; Jost, P.; Hofmann, K.; Waldeyer, H.: Das Abgasemissionsverhalten von Pkw in der Bundesrepublik Deutschland im Bezugsjahr 1985 (UBA-Bericht 07/87).- Berlin: Erich Schmidt Verlag, 1987

5 Klingenberg, H.; Kuhler, M.; Schürmann, D.: Stand und Entwicklungstendenzen der Abgasgesetzgebung in den USA. - Technische Akademie Esslingen: „Meßverfahren und Meßtechnik für Kraftfahrzeugabgase“, 18.10.82

6 Klingenberg, H.; Schürmann, D.: Abgasemissions- und Kraftstoffverbrauchsprognosen für den Pkw-Verkehr in der Bundesrepublik Deutschland im Zeitraum von 1970 bis 2000 auf der Basis der verschiedenen Grenzwertsituationen.- Schriftenreihe der Forschungsvereinigung Automobiltechnik e.V. (FAT) Nr. 45, 1985

7 Klingenberg, H.; Schürmann, D.; Staab, J.: Einfluß verschiedener Katalysatorkonzepte auf die Abgasemissionen bei realen Straßenfahrten.- MTZ 49 (1988) 2

8 Klingenberg, H.; Schürmann, D.; Lies, K.-H.: Dieselmotorabgas - Entstehung und Messung.- VDI-Berichte Nr. 888 (1991) p. 119-131

9 Klingenberg, H.; Wagner, G.: Umweltschutzstrategien der Automobilindustrie am Beispiel der Volkswagen AG. In: Steger, U.: Handbuch des Umweltmanagements.- München: C.H.Beck'sche Verlagsbuchhandlung, 1992

10 Klingenberg, H.; Franke, H.-U.; Witzenhausen, K.: Größenverteilung der Dieselpartikeln längs des Abgasstranges.- UMTEC '94, Magdeburg, 1994

11 Lach, G.; Winckler, J.: Kraftstoffdampf-Emissionen von Personenwagen mit Ottomotor.- ATZ Automobiltechnische Zeitschrift (Stuttgart) 92 (1990), 7/8

12 Meier, E.; Plaßmann, E.; Wolff, C. et al.: Großversuch zur Untersuchung der Auswirkung einer Geschwindigkeitsbegrenzung auf das Abgas-Emissionsverhalten von Personenkraftwagen auf Autobahnen (Abgas-Großversuch).- Essen: Vereinigung der Technischen Überwachungs-Vereine e.V., 1985

13 Metz, N.: Entwicklung der Abgasemissionen des Personenwagen-Verkehrs in der BRD von 1970 bis 2010.- ATZ (Stuttgart) 92 (1990), 4, p. 176-183

14 Obländer, K.; Kräft, D.: Abgasgesetzgebung, Grenzwerte und Meßverfahren im Kraftfahrzeugwesen.- MTZ 34 (1973) 3, p.84-91

15 Schürmann, D.; Wagner, G.; Klingenberg, H.: Social Impact of the Car. - 93rd General Assembly of the Alliance Internationale de Tourisme, Sydney, Australia (1991)
16 von Goßlar, V.: Grenzen der Motorisierung in Sicht.- Aktuelle Wirtschafts-analyse (Deutsche Shell), Hamburg, (1989) 20
17 Witzenhausen, K.; Franke, H.-U.; Klingenberg, H.: Morphology Changes of Diesel Particulate in the Exhaust Pipe at Different Static Loads.- Paper on 4thInt.Aerosol Conf., Session 7E7, Los Angeles, 1994

Chap. 3

1 Bolin, B.; Arrhenius, E.: Nitrogen - An Essential Life Factor and a Growing Environmental Hazard (Report from Nobel Symposium No. 38).- AMBIO VI (1977), 2-3
2 Böttger, A.; Ehhalt, D.H.; Gravenhorst, G.: Atmosphärische Kreisläufe von Stickoxiden und Ammoniak.- Berichte der KFA Jülich, Nr. 1558 (1978)
3 Crutzen, P.J.: The Role of NO and NO_2 in the Chemistry of the Troposphere and Stratosphere.- Ann.Rev.Earth Planet.Sci. (1979), 7, 443-472
4 Crutzen, P.J.: Atmospheric Interactions - Homogeneous Gas Reactions of C, N, and S Containing Compounds.-In: Bolin, B.; Cook, R.B. (eds.): The Major Biogeochemical Cycles and Their Interactions.- SCOPE (1983)
5 EPA: National Air Quality and Emission Trends Report 1981.- US-EPA (1983)
6 Fishman, J.; Seiler, W.: Correlative Nature of Ozone and Carbon Monoxide in the Troposphere: Implication for the Tropospheric Ozone Budget.- J.Geophys.res. 88 (1983), C6, p.3662-3670
7 Fricke, W.: Großräumige Verteilung und Transport von Ozon und Vorläufern.- VDI-Berichte Nr.500 (1983) p.55-62
8 Hahn, J.: The Cycle of Atmospheric Nitrous Oxide.- Phil.Trans.R.Soc.London A 290 (1979), 495-504
9 Kuhler, M.; Kraft, J.; Klingenberg, H.; Schürmann, D.: Natürliche und anthropogene Emissionen.- Automobil-Industrie 30 (1985) 2, 165-176
10 Logan, J.A.: Nitrogen Oxides in the Troposphere: Global and Regional Budgets.- J.Geophys.Res. 88 (1983), C15, 10785-10807
11 Maxwell, C.; Martinez, J.R.: Performance Evaluation of three Chemical Models Used with the Empirical Kinetic Modeling Approach (EKMA) APCA 82-20.2 (1982)
12 Mosier, A.R.; Hutchinson, G.L.: Nitous Oxide Emissions from Cropped Fields.- J.Environ.Qual. 10 (1981), 2
13 Nolting, F.; Zetzsch, C.: Smogkammeruntersuchungen zur Luftchemie biogener Kohlenwasserstoffe in Gegenwart von Ozon, NO_x und SO_2 (Abschlußbericht). KfK-PEF 58, Kernforschungszentrum Karlsruhe (1990)

14 Paul, E.A.: Nitrogen Cycling in Terrestrial Ecosystems.-In: Environmental Biogeochemistry, Vol.1 (ed. by J.O. Nriagu).- Ann Arbor: Ann Arbor Science Publishers Inc., 1976
15 Rotty, R.M.: Uncertainties Associated with Global Effects of Atmospheric Carbon Dioxide.- DOE-Contract EY 76C050033, Oak Ridge Associated University, Institute for Energy Analysis, (1979)
16 Schönwiese, C.-D.: Das Problem menschlicher Eingriffe in das Globalklima ("Treibhauseffekt") in aktueller Übersicht.- Frankfurter Geowiss. Arbeiten, Serie B, Metereologie und Geophysik Bd.3.- 1991 ISBN 3-922540-34-1
17 Schwela, D.: Vergleich der nassen Deposition von Luftverunreinigungen in den Jahren um 1970 mit heutigen Belastungswerten.- Staub - Reinhaltung der Luft 43 (1983) 4, 135-139
18 Seiler, W.: The Cycle of Atmospheric CO.- Tellus XXVI (1974), 1-2, 116-135
19 Seiler, W.; Zankl, H.: Man's Impact on the Atmospheric Carbon Monoxide Cycle.-In: Nriagu, J.O. [ed.]: Environmental Biogeochemistry, Vol.1.- Ann Arbor: Ann Arbor Science Publishers Inc., 1976
20 Seiler, W.; Crutzen, P.J.: Estimate of Gross and Net Fluxes of Carbon Between the Biosphere and the Atmosphere from Biomass Burning.- Climatic Change (D.Reidel Publ.Co.), Dordrecht (1980), 2, 207-247
21 Seiler, W.; Fishman, J.: The Distribution of Carbon Monoxide and Ozone in the Free Troposphere.- J.Geophys.Research 86 (1981), C 8, p.7255-7265
22 Seiler, W.; Conrad, R.; Scharffe, D.: Field Studies of Methane Emission from Termite Nests into the Atmosphere and Measurements of Methane Uptake by Tropical Soils.- J.Atmosph.Chemistry (1984), 1, 171-186
23 Seiler, W.: Contribution of Biological Processes to the Global Budget of CH_4 in the Atmosphere.-In: Klug, M.J.; Raddy, C.A.[eds.]:Current Perspectives in Microbiological Ecology, S.468-477.- Washington D.C.: American Society for Microbiology, 1984
24 Seiler, W.; Holzapfel-Pschorn, A.; Conrad, R.; Scharffe, D.: Methane Emissions from Rice Paddies.- J.Atmosph.Chem. (1984), 1, 241-268
25 Stein, N.: The Role of the Terrestrial Vegetation in the Global Carbon Cycle.- In: Georgii, H.W.; Jaeschke, W. [eds.]: Chemistry of the Unpolluted and Polluted Troposphere, S.185-202.- Dordrecht: D.Reidel Publ.Co., 1982
26 Wagner, H.G.; Zellner,R.: Abbau von Kohlenwasserstoffen in der Atmosphäre.- Erdöl und Kohle - Erdgas - Petrochemie. Brennstoffchemie 37 (1984), 212
27 Wagner, H.G.; Zellner, R.: Die Geschwindigkeit des reaktiven Abbaus anthropogener Emissionen in der Atmosphäre.- Angew.Chem. 91 (1979), 707
28 Wang, W.Ch.; Sze, N.D.: Coupled Effects of Atmospheric N_2O and O_3 on the Earth's Climate.- Nature 286 (1980), 589-590

Chap. 4

1 Chock, D.P.: A Simple Line Source Model for Dispersion Near Roadways.- Atmospheric Environ. 12 (1978), 823
2 Johnson, W.B.; Ludwig, F.L.; Dabberdt, W.F.; Alle, R.J.: An Urban Diffusion Simulation Model for Carbon Mooxide.- JAPCA 23 (1973), 490
3 Kuhler, M.; Kraft, J.; Koch, W.; Windt, H.. Dispersion of Car Emissions in the Vicinity of a Highway. In: Grefen,K.; Löbel,J.: Environmental Metereology.- Dordrecht: Kluwer Academic Publishers, 1988
4 Verein Deutscher Ingenieure, Kommission Reinhaltung der Luft: VDI-Richtlinie 3210: Maximale Immissions-Werte.- Berlin: Beuth-Verlag, 1974

Chap. 5

1 Aufderheide, M.; Mohr, U.; Thiedemann, K.U.; Heinrich, U.: Quantification of hyperplastic areas in hamster lungs after chronic inhalation of different cadmium compounds.- Toxicol. Environ. Chem. 27 (1990), 173-180
2 Bellmann, B.; Muhle, H.; Creutzenberg, O.; Mermelstein, R.: Irreversible pulmonary changes after dust overloading of lungs in rats.- J. Aerosol Med. 3 (1990) 1, 68-69
3 Bellmann, B.; Muhle, H.; Creutzenberg, O.; Mermelstein, R.: Recovery behaviour after dust overloading of lungs in rats.- J. Aerosol Sci., 21 (1990) p.377-380
4 Bellmann, B.; Muhle, H.; Creutzenberg, O. and Mermelstein, R.: Irreversible Pulmonary changes induced in rat lung by dust overload.- Environ. Health Persp. 97 (1992), 189-191
5 Creutzenberg, O.; Bellmann, B.; Klingebiel, R.; Heinrich, U.; Muhle, H.: Toxic and carcinogenic effects of solid particles in the respiratory tract. - International Life Sciences Institut-ILSI (Washington/D.C), p. 549-552, 1994
6 Creutzenberg, O.; Bellmann, B.; Heinrich, U.; Fuhst, R.; Koch, W.; Muhle, H.: Clearance and retention of inhaled diesel exhaust particles, carbon black, and titanium dioxide in rats at lung overload conditions.- J. Aerosol Sci. 21 (1990) 1, p. 455-458
7 Gerlich, G.: Die physikalischen Grundlagen des Treibhauseffektes und fiktiver Treibhauseffekte.- Proc. Int. Symposium „Treibhaus-Kontroverse und Ozon-Problem“, Leipzig 9.-10. Nov. 1995.-Tübingen: Europäische Akademie für Umweltfragen, 1996
8 Ishinishi, N.; Koizumi, A.; McClellan, R.O.; Stöber, W.: Carcinogenic and Mutagenic Effects of Diesel Engine Exhaust.-In: Development in Toxicology and Environmental Science, Volume 13 (Congress Report).- Amsterdam; New York; Oxford: Elsevier Science Publishers, 1986
9 Deutsche Forschungsgemeinschaft: Maximale Arbeitsplatzkonzentrationen und Biologische Arbeitsstofftoleranzwerte 1991 (Mitteilung XXVII der Senatskomission zur Prüfung gesundheitlicher Arbeitsstoffe).- Berlin; Heidelberg: VCH Verlagsgesellschaft, 1991 ISBN 3-527-27398-0

10 Heinrich; U.; et al.: Chronic Effects on the Respiratory Tract of Hamsters, Mice and Rats after Long-Term Inhalation of High Concentrations of Filtered and Unfiltered Diesel Engine Emissions.- Journal of Applied Toxicology, Vol. 6 (6), 383-395, 1986

11 Heinrich, U.: Gesundheitliche Wirkung der Dieselabgasemission; Stand der Forschung. In: VDI-Berichte 885: Abgas- und Geräuschemissionen von Nutzfahrzeugen.- Düsseldorf: VDI Verlag, p.1-100 (1991)

12 Heinrich, U.; Pott, F.; Roller, M.: Polycyclische aromatische Kohlenwasserstoffe - Tierexperimentelle Ergebnisse und epidemiologische Befunde zur Risiko-Abschätzung. In: VDI-Berichte 888.- Düsseldorf: VDI Verlag, p. 71-92 (1991)

13 Heinrich, U.; Fuhst, R.; Mohr, U.: Tierexperimentelle Inhalationsstudien zur Frage der tumorinduzierenden Wirkung von Dieselmotorabgasen und zwei Teststäuben. In: Ökologische Forschung, Auswirkungen von Dieselmotorabgasen auf die Gesundheit (BMFT-Broschüre), p. 21-30, (1992)

14 Heinrich, U.: Toxic and carcinogenic effects of solid particles in the respiratory tract.- Intern. Life Sciences Institut-ILSI (Washington/D.C) 1994, p. 57-73

15 Henschler, D.: Verhältnismäßigkeit im Umweltschutz.- Nachr. Chem. Techn. Lab. 36 (1988) 9

16 Klingenberg, H.; Stöber, W.; Heinrich, U.: Stand der Arbeiten zur biologischen Wirkung von Abgasemissionen aus Verbrennungsmotoren - ein Statusbericht.- FISITA-congress, Hamburg, 1980

17 Klingenberg, H.: Automotive Emissions and Health Effects Research Need: The European View.- SAE-Paper 840906 (1984)

18 Klingenberg, H.: Wirkungsforschung - Aufgabe der Automobilindustrie?.- Automobil-Industrie 32 (1987) 2, 105-114

19 Klingenberg, H.: Studies on Health Effects of Automobile Exhaust Emissions.- Proc. of the Third International Highway-Pollution Symposium, Munich, 1989

20 Klingenberg, H.: Wie gefährlich sind Dieselmotorabgase?. Österreichische Ingenieur- und Architekten-Zeitschrift (ÖIAZ), 134 (1989) 11

21 Klingenberg, H.; Winneke, H.: Studies on Health Effects of Automotive Exhaust Emissions.-How Dangerous are Diesel Emissions?.- The Science of the Total Environment (Elsevier Sci. Publ.) 93 (1990) p.95-105

22 Levsen, K.; Behnert, S.; Prieß, B.; Winkeler, H.D.; Zietlow, J.: Der Eintrag organischer Verbindungen in den Boden durch Niederschläge. -VDI Bericht 837 (1990) p.401-435

23 Mohr, U.; Bader, R.; Ernst, H.; Ettlin, R.; Gembardt, C.; Harleman, J.H.; Hartig, F.; Jahn, W.; Kaliner, G.; Karbe, E.; Kaufmann, W.; Krieg, K.; Krinke, G.; Küttler, K.; Landes, C.; Mettler, F.; Morswitz, G.; Notman, J.; Püschner, H.; Qureshi, S.; Reznik, G.; Rittinghausen, S.; Tuch, K.; Urwyler, H.; Weisse, G.; Weisse, I.; Zehnder, J.: Tumor REGISTRY Data Base, Suggestions for a systematized nomenclature for pre-neoplastic and neoplastic lesions in rats.- Exp. Pathol., 38; 1-18 (1990)

24 Mohr, U.; Rittinghausen, S.; Takenaka, S.; Ernst, H.; Dungworth, D.L.; Pylev, L.N.: Tumours of the lower respiratory tract and pleura. In: Turusov, V.S.; Mohr, U. (eds.): Pathology of tumours in laboratory animals.- Vol I. Tumours of the rat. 2nd edition. -IARC Scientific Publications. 99: 275-299 (1990)
25 Morrow, P.E.; Muhle, H.; Mermelstein, R.: Chronic inhalation study findings as a basis for proposing a new occupational dust exposure limit.- J. Am. Coll. Toxicol. 10: 279-290 (1991)
26 Muhle, H.; Creutzenberg,-O.; Bellmann,-B.; Heinrich,-U.; Mermelstein,-R.: Dust overloading of lungs; investigations of various materials species differences and irreversibility of effects. -J. Aerosol Med. 3: 111-128 (1990)
27 Muhle, H.; Bellmann, B.; Creutzenberg, O.: Toxic and carcinogenic effects of solid particles in the respiratory tract.- International Life Sciences Institut-ILSI (Washington/D.C) 1994, p. 29-41
28 N.N.: Verbundforschungsprogramm "Auswirkungen von Automobilabgasen auf die Gesundheit und Umwelt".- Bundesanzeiger Bonn, 38 (1986) 1121
29 Stöber, W.: On the Health Hazards of Particulate Diesel Engine Exhaust Emissions.- SAE Technical Paper Series 871988 (1987)
30 Stöber, W.; Rosner, G.: Health aspects of indoor air pollution by organic matter and combustion products. In: Kasuga, H. (ed.): Indoor air quality.- Berlin, Heidelberg: Springer Verlag, p.403-414 (1990)
31 Thüne, W.: Zum natürlichen Treibhauseffekt - sind die 33°C ein faktischer oder fiktiver Wert?- Elektrizitätswirtschaft 94 (1995), 26, p.1800-1807

Chap. 6

1 Heise, H.M.: Infrarotspektrometrische Gasanalytik - Verfahren und Anwendungen. In:Analytiker-Taschenbuch Band 9.- Berlin, Heidelberg: Springer-Verlag, 1990
2 Heller, B.; Klingenberg, H.; Lach, G.; Winckler, J.: Dynamic Exhaust Emission Measurement System (SESAM) According to Legal Demands and for Development Purposes.- ISATA-Proceedings, Vol. III, Stream c, p. 17-24 ISATA Vienna, 1990
3 Heller, B.; Klingenberg, H.; Lach, G.; Winckler, J.: Performance of a New System for Emission Sampling and Measurement.- SAE-Paper 900275, 1990
4 Klingenberg, H.: Meß- und Prüfverfahren für Automobilabgase - Übersicht und Kritik.- atm 44 (1977) 1-4, p. 3-14, 53-60, 95-102, 133-138
5 Klingenberg, H.; Winckler, J.: Multicomponent Automobile Exhaust Measurements.- p.108-115.- In: Proc. Intern. Symp. "Monitoring of Gaseous Pollutants by Tunable Diode Lasers".- Freiburg: Fraunhofer Institut, 1986
6 Klingenberg, H.; Staab, J.: Zeitaufgelöste Messung von Automobilabgasen - Grenzen der Technologie und Notwendigkeit der Anwendung.- ATZ 90 (1988) 10

7 Klingenberg, H.; Jecht, U.: Mehrkomponentenmessung an Kfz-Abgasen mittels eines On-Line-FTIR-Spektrometers.- VDI-Berichte 838 (1990) p.111-146
8 Martin, K.: Spurenanalytik anorganischer Verbindungen mittels Gaschromatographie/Codestillation/Matrixisolation/Fouriertransform-Infrarotspektroskopie.- Dissertation, Universität Hannover, (1993)
9 Riedel, W.J. et.al.: Infrared Diode Laser Exhaust Gas Analyser for Trace Components.- ISATA paper 910309, Florence (Italy), 1991
10 Riedel, W.J.: Optics for Tunable Diode Laser Spectrometers.- Proc. SPIE Vol. 1433 (1991) p.179-189
11 Staab, J.; Fabinski, W.; Zöchbauer, M.: Ein Betriebsfotometer nach dem Resonanzabsorptionsverfahren zur Messung von Stickstoffoxid.- Technisches Messen (1978) 1, p.11-75
12 Staab, J.; Klingenberg, H.; Schürmann, D.: Strategy for the Development of a New Multicomponent Exhaust Emissions Measurement Technique.- SAE-Paper 830437, 1983
13 Staab, J.; Klingenberg, H.; Herget, W.F.; Riedel, W.I.: Progress in the Prototype Development of a New Multicomponent Exhaust Gas Sampling and Analysing System.- SAE-Paper 840470, 1984
14 Staab, J.; Klingenberg, H.; Pflüger, H.; Herget, W.F.; Tromp, M.I.: First Experiences in Testing a New Multicomponent Exhaust Gas Sampling and Analysing System.- SAE-Paper 851659, 1985
15 Staab, J.; Klingenberg, H.: Ein neues Vielkomponenten-Meßsystem zur gleichzeitigen Messung der limitierten und von nicht limitierten Automobilabgasemissionen.- Automobil-Industrie 31 (1986), 359-369

Chap. 7

1 Cadle, S.H.; Mulawa, P.A.: Sulfide Emissions from Catalyst-Equipped Cars.- SAE-Paper 780200, 1978
2 Grimmer, G.; Hildebrandt, A.; Böhnke, H.: Probennahme und Analytik von PAK in Kfz-Abgasen.- Erdöl und Kohle - Erdgas - Petrochemie 25 (1972), 442
3 Grimmer, G.; Glaser, A.; Böhnke, H.: Probenahme und Analytik polyzyklischer aromatischer Kohlenwasserstoffe im Kraftfahrzeugabgas.- Proc.Third Intern. Clean Air Congress Düsseldorf, p.C 50 - C 53, 1973
4 Hartung, A.; Kraft, J.; Lies, K.-H., Schulze, J.: Messen polycyclischer aromatischer Kohlenwasserstoffe im Abgas von Dieselmotoren.- MTZ 43 (1982), 263
5 Hartung, A.; Kraft, J.; Schulze, J.; Kieß, H.; Lies, K.-H.: The Identification of Nitrated PAH in Diesel Particulate Extracts.- Chromatograph. 19 (1984), 269
6 Hartung, A.; Kraft, J.; Schulze, J.; Lies, K.-H.: Identifizierung und quantitative Bestimmung von Nitroderivaten der polycyclischen aromatischen Kohlenwasserstoffe in Diesel-Partikel-Extrakten.- Z.Anal.Chem. 320 (1985), 702

7 Hartung, A.; Schulze, J.; Kieß, H.; Lies, K.-H.: Nitroderivate der PAK als Artefakte bei der Probennahme aus dem Dieslabgas.- Staub - Reinh. Luft 46 (1986), 132

8 Hertel, R.F.; König, H.P.; Inacker, R.; Malessa, R.: Nachweis der Freisetzung und Indentifizierung von Edelmetallen im Abgasstrom von Katalysator-fahrzeugen. In: Ökologische Forschung.- GSF/Neuherberg and KFZ/Jülich. p.16-21 (1990)

9 Klingenberg, H.: Dieselpartikelemission: Entstehung, Schädlichkeit.- Schweizer Automobil-Revue 47 (1987), 27-33

10 Klingenberg, H.; Schürmann, D; Lies, K.H.: Air Pollutants from European Passenger Cars - Emissions and Ambient Air Concentrations.- Air & Waste Management Association, EPA, Environmental Criteria and Assessment Office, Motor Vehicle Manufacturers Association, American Petroleum Institute, Engine Manufacturers Association, Detroit, 1991

11 Klingenberg, H.; Franke, H.-U.; Witzenhausen, K.: Untersuchungen der Partikelemission von Dieselmotoren.- GIV-Kolloquium, Frankfurt/Main, 1994

12 Kraft, J.; Hartung, A.; Lies, K.-H.: Quantitative Determination of Polycyclic Aromatic Hydrocarbons by Means of Glass Capillary Chromatography in Exhaust Gases of Automobiles.-In: Fourth International Symposium on Capillary Chromatography, Hindelang 1981.- Heidelberg: Hüthig, 1981

13 Kraft, J.; Lies, K.-H.: Polycyclic Aromatic Hydrocarbons in the Exhaust of Gasoline and Diesel Vehicles.- SAE paper 810082, 1981

14 Kraft, J.; Hartung, A.; Schulze, J.; Lies, K.-H.: Determination of PAH in Diluted and Undiluted Exhaust Gas of Diesel Engines.- SAE paper 821219, 1982

15 Kraft, J.; Kuhler, M.: Aldehydes from Motor Vehicles: Toxicity and Air Quality.- SAE-Paper 851661, 1985

16 Levsen, K.; Schilhabel, J.: Identification of nitrated polycyclic hydrocarbons in diesel particulate extracts by tandem mass spectrometry.- Proceeding of the 38th ASMS Conference on Mass Spectrometry and Allied Topics.- Tucson (USA), p. 637-638 (1990)

17 Lies, K.-H.; Postulka, A.; Gring, H.: Particulate Emissions from Diesel Engines - Evaluation of Measurement and Results.- SAE paper 830455, 1983

18 Lies, K.-H.; Postulka, A.;Gring, H.: Characterization of Exhaust Emissions from Diesel-Powered Passenger Cars with Particular Reference to Unregulated Components.- SAE paper 840361, 1984

19 Lies, K.-H.; Klingenberg, H.; Schürmann, D.: Meßmethoden und Meßergebnisse nicht limitierter Abgaskomponenten.- VDI-Fortschritt-Berichte Reihe 12, Nr.86 (1987) 57-77

20 MacDonald, J.S.; Plee, S.L.; D'Arcy, J.B.; Schreck, R.M.: Experimental Measurements of the Independent Effects of Dilution Ratio and Filter Temperature on Diesel Exhaust Particulate Samples.- SAE paper 800185, 1980

21 Schuetzle, D.: Sampling of Vehicle Emissions for Chemical Analysis and Biological Testing.- Environ.Health Perspect.J. 47 (1983), 65
22 Schulze, J.; Hartung, A.; Kieß, H.; Kraft, J.; Lies, K.-H.: Identification of Oxy-PAH in Diesel Particulate Matter by GC and GC/MS.- Chromatographia 19 (1984), 391
23 Schulze, J.; Hartung, A.; Kieß, H.; Lies, K.-H.: Oxy-Derivate der PAK in Dieselabgas und Artefakt-Bildung während der Probennahme.- Staub-Reinh. Luft 47 (1987), 135
24 Schürmann, D; Lies, K.-H.; Schulze, J. et al.: Nicht limitierte Automobil-Abgaskomponenten.- Wolfsburg: Volkswagen AG, R&D, 1988
25 Schürmann, D.; Klingenberg, H.; Lies, K.-H.; Schulze, J.: Nicht limitierte Automobilabgaskomponenten.- Automobil-Industrie 34 (1989) 5
26 Ten Noever De Brauw, M.C.: Combined Gas Chromatography - Mass Spectrometry: A Powerful Tool in Analytical Chemistry.- J.Chromatogr. 165 (1979), 207
27 Truex, .J.; Windawi, .; Ellgen, .C.: The Chemistry and Control of H_2S Emissions in Three-Way-Catalysts.- SAE-Paper 872162, 1987
28 Wiedemann, B.; Neumann, K.-H.: Vehicular Experience with Additives for Regeneration of Ceramic Diesel Filters.- SAE-Paper 850017, 1985

Chap. 8

1 Juneja, W.K.; Horchler, D.D.; Hasken, H.M.: Exhaust Emission Test Variability.- Proc. APCA-Meeting, New Orleans, 1975
2 Klingenberg, H.; Fock, M.; Lies, K.-H.,Pazsitka, L.: A Critical Study of the United States Exhaust Emission Certification,Test Error Analysis for the Test Procedure.- Paper No. 74-242.- 67th Annual Meeting of the Air Pollution Control Association (APCA), Denver, Col., 1974
3 Klingenberg, H.; Lies, K.-H.: Critical Survey of the Exhaust Emission Test Procedures for USA and Europe.- Proceedings, Vol.1.- ISATA 75: Fourth International Symposium on Engine Testing Automation Performance, Emission and Diagnostics.- Naples (Italy), 1975
4 Klingenberg, H.; Kinne, D.; Schürmann, D.; Will, R.: A Method for Correlating Different Exhaust Emission Test Cells.- Proceedings, Vol.1, ISATA 1976, Rome (Italy), 1976
5 Klingenberg, H.; Kinne, D.; Schürmann, D.: Torque Measurements and Mechanized Driver for Correlating Exhaust Emission Test Facilities.- SAE-Paper 770139, 1977
6 Klingenberg, H.: Driving Cycles. Session of UNO-Commission: GRPA Genf, 13.02.79
7 Kuhler, M.; Karstens, D.; Klingenberg, H.; Schürmann, D.; Krause, N.; Kinne, D.: Should Europe adopt a Modified FTP? Automot. Engng, June 1978
8 N.N.: VDI-Richtlinie 3490: Messen von Gasen: Prüfgase.- Düsseldorf: VDI-Verlag, 1980

Chap. 9

1 Karstens, D.; Klingenberg, H.; Kuhler, M.: Entwicklung eines verbesserten Fahrprogramms für Abgasmessungen.- ATZ Automobiltechnische Zeitschrift 80 (1978) 6, p.251-257

2 Klingenberg, H.; Legro, S.: Meeting of EPA (Stan Legro, Assistant Administrator for Enforcement) with VW AG (H. Klingenberg et al.) to discuss the Selective Enforcement Auditing Program, Washington D.C., February 18, 1976

3 Klingenberg, H.; Kuhler, M.; Lies, K.-H.; Pazsitka, L.; Schürmann, D.: Comparison and Optimization of Exhaust Emission Test Procedures.- SAE-Paper 770137, 1977

4 Klingenberg, H.; Schürmann, D. et al.: Analyse der in Europa und in den USA gesetzlich vorgeschriebenen Prüfmethoden und Meßverfahren für Automobilabgase.- Wolfsburg: Volkswagenwerk AG, R&D, 1977

5 Klingenberg, H.: Harmonization of Testing Procedures for Automotive Exhaust Gas.- SAE-Paper 780647, 1978

6 Klingenberg, H.: Effects of Exhaust Emissions Test Data Variability and Interrelation between the Certification-, the Assembly-Line- and the Field testing.- Proc., APCA-Meeting: Quality Assurance in Air Pollution, 1979

7 Klingenberg, H.; Kuhler, M.: Influence of Variability and Distribution of Exhaust Emission Values on Assembly Line Tests.- APCA 1979, New Orleans 1979

8 Klingenberg, H.; Schürmann. D.: Proposal of New Exhaust Emission Compliance and Testing Procedure Based on Averaging.- SAE-Paper 821192, 1982

9 Klingenberg, H.; Neumann, K.-H.: Überprüfung der Abgasemissionen des Einzelfahrzeuges in Kundenhand - ein unlösbares Problem?.- ATZ 86 (1984) 1, 13-16

10 Klingenberg, H.; Schürmann, D.: Eine Idee zur Überprüfung der Effizienz von Abgaskatalysatoren im Personenwagen.- MTZ Motortechnische Zeitschrift 46 (1985) 7/8, 261-262

11 Klingenberg, H.; Neumann, K.-H.: Exhaust Gas Emission Control at the Car-Owner Level - An Insoluble Problem?.- SAE-Paper 851658, 1985

12 Klingenberg, H.: Probleme der Abgasmessungen an neuen und gebrauchten Automobilen.- Automobil-Industrie 31 (1986) 1, 23-33

13 Klingenberg, H.; Müller, R.-H.: Vorschlag zur Überprüfung der Effizienz von Abgas-Katalysatoren in Personenwagen ohne Rollenprüfstand.- MTZ 47 (1986) 5, 181-184

14 Klingenberg, H.; Kluczynski, G.: Abgasfelduntersuchungen an Katalysatorfahrzeugen mit dem neuen Testverfahren "Bauteiletest / Katalysator".- MTZ 48 (1987) 2, 55-58

15 Klingenberg, H.: Inspection and Maintenance of Cars in Use.- SAE-Paper 871101, 1987

16 Klingenberg, H.: ASU-Methode für Fahrzeuge mit Katalysator.- Automobil-Industrie 33 (1988) 1, 29-36

Chap. 10

1 Klingenberg, H.: Meßtechnik im Dienste des Umweltschutzes - Quo vadis?.- VDI-Tagung Umweltmeßtechnik, Leipzig, 1992
2 Staab, J.; Klingenberg, H.; Schürmann, D.: Strategy for the Development of a New Multicomponent Exhaust Emissions Measurement Technology. SAE paper 830437, 1983

Index

Springer-Verlag and the Environment

We at Springer-Verlag firmly believe that an international science publisher has a special obligation to the environment, and our corporate policies consistently reflect this conviction.

We also expect our business partners – paper mills, printers, packaging manufacturers, etc. – to commit themselves to using environmentally friendly materials and production processes.

The paper in this book is made from low- or no-chlorine pulp and is acid free, in conformance with international standards for paper permanency.

FSC
www.fsc.org
MIX
Papier aus verantwortungsvollen Quellen
Paper from responsible sources
FSC® C105338

If you have any concerns about our products,
you can contact us on
ProductSafety@springernature.com

In case Publisher is established outside the EU,
the EU authorized representative is:
Springer Nature Customer Service Center GmbH
Europaplatz 3, 69115 Heidelberg, Germany

Printed by Libri Plureos GmbH
in Hamburg, Germany